Food Science Text Series

The *Food Science Text Series* provides faculty with the leading teaching tools. The Editorial Board has outlined the most appropriate and complete content for each food science course in a typical food science program and has identified textbooks of the highest quality, written by the leading food science educators.

More information about this series at http://www.springer.com/series/5999

Marc C. Sanchez

Food Law and Regulation for Non-Lawyers

A US Perspective

 Springer

Marc C. Sanchez
Contract In-House Counsel &
Consultants, LLC
Washington, DC
USA

ISSN 1572-0330
Food Science Text Series
ISBN 978-3-319-12471-1   ISBN 978-3-319-12472-8 (eBook)
DOI 10.1007/978-3-319-12472-8

Library of Congress Control Number: 2014956385

Springer Cham Heidelberg New York Dordrecht London

Printed on acid-free paper

Springer is part of Springer Science+Business Media (www.springer.com)

First and foremost to Blake, my constant source of inspiration and encouragement. And to all my family, Rosanne, Dora, and others, for their loving support. Then to the countless clients who opened their facilities and operations to my counsel. Thank you for your confidence. Finally, to the FDA and USDA agents with whom I've all too often found a kindred spirit to share my passion for the regulations and commitment to ensuring the health and safety of the public. This text would not be possible without this trinity of support, comity, and learning.

Contents

Abbreviations

ALJ	Administrative Law Judge
APA	Administrative Procedures Act
AMA	Agricultural Marketing Act
AMS	Agricultural Marketing Services
APHIS	Animal and Plant Health Inspection Service
APHS	Animal and Plant Health Service (forerunner to APHIS)
ARS	Agricultural Research Service
BAI	Bureau of Animal Industries (forerunner to FSIS)
CFSAN	Center for Food Safety and Applied Nutrition
CBER	Center for Biologics Evaluation and Research
CDRH	Center for Device and Radiological Health
CDC	Center for Disease Control and Prevention
CDER	Center for Drug Evaluation and Research
CTB	Center for Tobacco Products
CVM	Center for Veterinary Medicine
CPSI	Center for Science in the Public Interest (non-governmental organization)
CFR	Code of Federal Regulations
COOL	Country of Origin Labeling
DHS	Department of Homeland Security
DOJ	Department of Justice
DSHEA	Dietary Supplement Health and Education Act
DM	District Manager (FSIS)
DWPE	Detention Without Physical Examination
ECJ	Evaporated Cane Juice
EPI	Egg Products Inspection Act
EIR	Establishment Inspection Report (FDA)
EPA	Environmental Protection Agency
FALCPA	Food Allergen Labeling and Consumer Protection Act
FBI	Federal Bureau of Investigation
FCN	Food Contact Notification
FCS	Food Contact Substance
FD&C	Food Drug and Cosmetic Act
FDA	Food and Drug Administration
FDAMA	Food and Drug Administration Modernization Act
FIFRA	Federal Insecticide, Fungicide, and Rodenticide Act
FMI	Federal Meat Inspection Act

FTC	Federal Trade Commission
FQPA	Food Quality Protection Act
FR	Final Rule
FSIS	Food Safety Inspection Service
FSMA	Food Safety Modernization Act
FS&QS	Food Safety and Quality Service (forerunner to FSIS)
FSVP	Foreign Supplier Verification Program (FSMA rule)
GAO	Government Accountability Office
GMO	Genetically Modified Organism
GMP or cGMP	Current Good Manufacturing Practices
GRAS	Generally Recognized As Safe
HACCP	Hazard Analysis and Critical Control Point
HFCS	High Fructose Corn Syrup
HMSA	Humane Methods of Slaughter Act
IOM	Inspection Operations Manual (FDA)
NELA	Nutrition Labeling and Education Act
NIH	National Institutes of Health
NMFS	National Marine Fisheries Services
NPIS	New Poultry Inspection System
NOAA	National Oceanic and Atmospheric Administration
NR	Noncompliance Record (FSIS)
OCI	Office of Criminal Investigations (FDA)
OFO	Office of Field Operations (FSIS)
OIEA	Office of Investigation, Enforcement and Audit (FSIS)
ORA	Office of Regulatory Affairs (FDA)
PDP	Principal Display Panel
PPI	Poultry Products Inspection Act
PREDICT	Predictive Risk-based Evaluation for Dynamic Import Compliance Targeting
PHSA	Public Health Service Act
RAC	Raw Agricultural Commodity
RACC	Reference Amount Customarily Consumed per Eating Occasion
RDA	Recommended Daily Allowance
RPM	Regulatory Procedures Manual (FDA)
PRP	Rules of Practice Regulation (FSIS)
RTE	Read-to-Eat (FSIS)
SOP	Standard Operating Procedure (component of GMPs)
SSA	Significant Scientific Agreement
SOI	Standard of Identity
USCA or USC	United States Code or United States Code Annotated
USDA	United States Department of Agriculture

Introduction to Statutory Framework and Case Law

1

Abstract

This chapter begins a seven-part analysis introducing food law and regulation in the USA. It looks at the basic structure and function of the US government, including key concepts of federalism and the structure of judicial opinions. The chapter distinguishes the role of the USDA and FDA through statutory definitions, responsibilities, and the legislative happenstance that resulted in a two-agency-model. Special emphasis is placed on defining "food" and the judicially created concept of "intended use." The chapter ends with a comparative look at food regulatory bodies around the world.

1.1 The Need for Food Law

1.1.1 Our Own Experience with Food

Readers may arrive to this textbook from a broad range of experiences and backgrounds, but we all share the common experience of food and drinks. To some extent, that makes us all experts in food regulations. Whether that experience is legal or regulatory or simply our daily meals and snacks, we have all encountered a food label or fought a bout of mild food poisoning. We have all developed expectations about what our food should be—organic, local, or free range, for example—and maintain a keen interest in news of outbreaks or instances of new or hidden ingredients such as "pink slime." This textbook will take that experience and peel back the curtain for the reader to gain a deeper insight into the basics of the US system of food safety laws and enforcement mechanisms. This text book will explore both the legislative history of this system and the contours of current law and the case precedent that shapes its interpretation.

Understanding the food enforcement agencies is vital given the size of the industry. The US Food and Drug Administration (FDA) estimates that a person spends 75 cents of every dollar on an FDA-regulated product (CFSAN 2014).[1] For an agency that regulates food, beverage, dietary supplements, cosmetics, drugs, medical devices, and animal feed and drugs, it is not hard to see how this is possible. The vast majority of that spending, however, goes towards food. The Center for Food Safety and Applied Nutrition (CFSAN), the branch within the FDA responsi-

---

[1] FDA and CFSAN (2014).

M. C. Sanchez, *Food Law and Regulation for Non-Lawyers,* Food Science Text Series,
DOI 10.1007/978-3-319-12472-8_1, © Springer International Publishing Switzerland 2015

ble for food, beverage, and dietary supplements, estimates that 75 % of the spending is on the products it regulates (CFSAN).[2] This means that consumers spend roughly 57 cents of every dollar on products they consume. Still unaccounted for are their purchases on United States Department of Agriculture (USDA) regulated meat products. As can be seen, the consumer relies on the food safety system to protect a bulk of the purchases one may make. This places food law and regulation as an area of paramount importance.

1.1.2 In Food We Trust

In its broadest terms food law is about protecting the public. Nearly everyone in modern society relies on someone else growing or making the vast majority of the food we eat. We trust this food will not make us sick and will be exactly as declared on the label. In food we trust. This protection against harmful products in the enforcement or postmarket surveillance context is known as adulteration. The safeguard against fraudulent products, those whose labels do not accurately describe what the product contains, are known as misbranded, in the enforcement context. Prior to an enforcement action, both are considered under a broader umbrella known as the premarket approval process. The overarching aim remains the same—protect the consumer. The primary mechanism to achieve this goal is to ensure that the agencies tasked with enforcing food safety laws can inspect facilities and use enforcement tools to remove harmful or fraudulent products from the market. The enforcement tools also act as a deterrent. Deterrence and inspection together build trust and allow consumers to shop with confidence. The two also protect a brand's reputation.

Food law also plays an important function for society as a whole. The USA lacked a robust system of product liability laws and enforcement for much of its history. Prior to 1906, the main power consumers wielded lay in organizing boycotts. Consumers thus could indirectly influence players in food production through the power of bad publicity, protests, and weakened sales. It is widely noted that meat sales fell by nearly half following Upton Sinclair's *The Jungle* (FDA 2006).[3] Still, for those seriously hurt or killed by a bad food product, there remained little accountability. Beginning in roughly 1906, the USA codified a system of food laws, which could be found in some form in the common law as far back as 1860, and matched the statutes with the machinery of enforcement. The machinery came in the form of federal agencies imbued with powers to regulate pre- and postmarket activities. Prior to the creation of these agencies, citizens were left to initiate private lawsuits that were often impractical and expensive. This lead one commentator to ask, "…who would or could go to court over a single can of peas?" (Friedman 1985).[4] The new system of laws and bureaucracy added public confidence to the market by introducing an element of accountability to bad actors.

Accountability not only means responsibility to those who were directly harmed, but to those indirectly harmed or burdened by the action. Economist call such indirect costs "externalities" that is the ripples created from an action (see Fig. 1.1). The FDA and Center for Disease Control and Prevention (CDC) conducted a number of studies on the number of foodborne illnesses and the economic burden of those illnesses. The CDC estimates that annually one in six Americans become ill with a foodborne illness (CDC Findings).[5] Annually, foodborne illness costs the US economy upwards of $ 100 billion (FSMA Economic Assessment).[6] Widely publicized recalls offer another excellent example of externalities.

---

[3] FDA (2006).

[4] Friedman (1985).

[5] CDC 2011 Estimates: Findings, Centers for Disease Control (2011) available at http://www.cdc.gov/foodborneburden/2011-foodborne-estimates.html.

[6] See e.g., Economic Assessment Report for the Proposed Rule on Hazard Analysis and Preventative Controls for Human Food under the Food Safety Modernization Act (FSMA) (identifying five sources of illness under the proposed rule as imposing an economic burden of over $ 2 billion).

---

[2] *Id.*

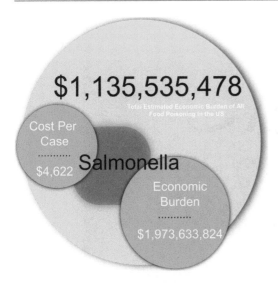

Fig. 1.1 Externalities associated with foodborne illness

During the cantaloupe recall initiated by Jensen Farms, all melons—watermelon, honey dews, cantaloupes—measured some impact from reduced prices to cutting the season short (Kroger 2014).[7] The Jensen recall impacted not only the producers of those unaffected melons, but also their workers, and the economy as a whole that depended on their confidence and ability to spend (Kroger 2014).[8] These ripples far from Jensen Farms are what the current system of laws and enforcement tools works to control.

1.1.3   History and the Courts' View of Food Drug and Cosmetic Act (FD&C)

The FDA's authority over food derives from the Food Drug and Cosmetic Act (FD&C). The Act at the center of the FDA's authority stands at the end of a long history beginning in 1906. Understanding this history is important to gain insight into the drafter's intent as to what provisions

and definitions mean and how the Act should be construed. This is an exercise routinely undertaken by courts.

A Brief Overview of Food Law History

There is a long history of ineffective state regulation in the USA prior to 1906. The common law made it a crime to sell diseased meat, for example, as early as 1860. The trouble with the common law approach was twofold. First, it relied on individuals to bring a suit in law or equity against a seller. There was no State enforcement mechanism to assist in the litigation, thus leaving the full cost and burden of enforcement, including investigations on private individuals. This proved impractical for the average consumer. Second, the States were powerless when the poison of tainted food poured across State lines. The USA was a collection of sovereign States. The private litigant struggled to take their cause of action across State lines. Furthermore, the States relied on the federal government under its authority to regulate interstate commerce. In the Absence of a federal statute or federal agency, the interstate sale of food went unregulated.

The end of the Civil War and the start of the Industrial Revolution are for many the roots of federal food safety legislation. Following the Civil War, the family farm was largely replaced by the impersonal corporation. Accountability faded and corruption flourished. In the period between 1879 and 1905, over 100 food and drug bills were debated in Congress (FDA Milestones).[9] Yet, none of them passed. As it so often is, crisis drives change in politics. In this case, it was not until the "embalmed beef scandal" of the Spanish–American War that Congress spurred to action. The scandal is so named because the commander of the Army, General Nelson A. Miles, in testimony to Congress described the beef as having an "odor like an embalmed dead body." The incident gained widespread media attention

---

[7] Kroger (2014).

[8] *See Kroger Id.* quoting one producer who equated a 3 week reduction in the melon season to a 20% pay cut for seasonal employees.

[9] FDA, Milestones in US Food and Drug Law History (1972) ((Pub. No. (FDA) 73-1018) available at: http://www.fda.gov/AboutFDA/WhatWeDo/History/Milestones/ucm128305.htm.

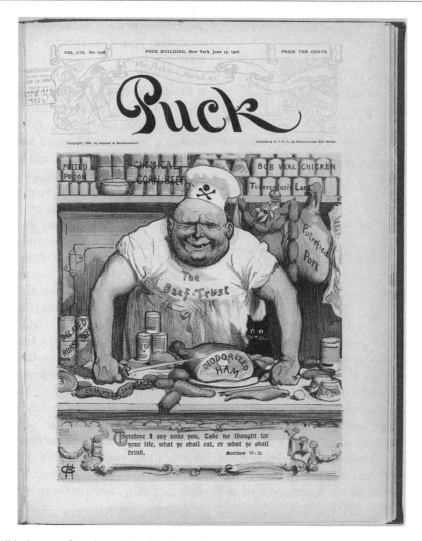

Fig. 1.2 Political cartoon from the embalmed beef scandal

with many claiming the tainted meat killed more soldiers than Spanish bullets (see Fig. 1.2).

In 1906, not long after the embalmed beef scandal, Congress passed the Pure Food and Drug Act. The initial act addressed two issues. It set out to proscribe dangerous foods and drugs, and to curtail deceptive marketing and labeling practices. The initial Act was fraught with issues. Chief among the issues was the failure to enact any premarket testing or review procedures for regulated products.

The 1906 Act persisted for 22 years before another scandal brought a change. In 1937, a new cure-all was relased on the market called, "Elixir

of Sulfanilamide." Dissatisfied with the taste, smell, and appearance of the elixir, the manufacturer added a combination of chemicals used in paint, varnish, and antifreeze (FDA Consumer Magazine).[10] The toxic effects were never considered and with no FDA premarket approval process, the product simply entered the market. By the time the FDA could take action to remove

---

[10] FDA Consumer magazine, June 1981 Issue, Taste of Raspberries, Taste of Death The 1937 Elixir Sulfanilamide Incident available at: http://www.fda.gov/aboutfda/whatwedo/history/productregulation/sulfanilamidedisaster/default.htm.

the product from the market, nearly 100 people died. The victims were predominately children. Congress clamoured to act, and in 1938 passed the Food Drug and Cosmetic Act. This is still the statute in effect today. The statute that served as the basis for amendments and now the Food Safety Modernization Act (FSMA) builds on.

There are three lessons from the passage of the 1938 Act which serve as the tripritie lense through which we must view all agency actions. These lessons are necessary to also effectively interpret the Act and understand how courts will do the same. These lessons are: (1) the drafter's indicated courts should broadly contrue the new Act to protect the public, (2) Congress rejected the 1906 Act's assumption that consumers were capable of protecting themselves, and (3) the objective of the 1938 Act not only continued to carry the mantale of protecting the public health, but also added emphasis to defending consumers by preventing fraud (see Fig. 1.3). We will see the broad construction and deference, courts give the agency and the Act, as we discuss nearly every topic. The ignorant consumer standard will be particularly noticeable as we disucuss labeling and restrictions on speech.

1.1.4   Organization of this Text

It is from this perspective—the personal, the societal, and the historical—we will build our knowledge and understanding of the US food safety laws and enforcement system. We will begin by looking at the postmarket surveillance aspects of food law starting with the two largest preliminary issues to any enforcement action. The first issue is determining whether an agency can exercise jurisdiction based on the classification of a regulated product. This issue asks the question, "What is food?" The second is assessing whether an agency conducted a valid inspection within the confines of the US Constitution and the agency's enabling acts. In the case where the agency cannot clear these hurdles, the other questions about whether the product was harmful or fraudulent are irrelevant.

Chapters 3 and 4 that follow expound on the two principal prohibited acts. Chapter 3 looks at

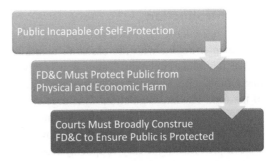

Fig. 1.3 Three goals of the 1938 Act

adulteration or the contamination of food, while Chapter 4 looks at misbranding. Misbranding will outline common labeling violations and defenses under the First Amendment. This text will then shift to the premarket context to explore issues of new ingredients and food packaging under various Amendments. Dietary supplements and other areas of specialized regulation will focus on both premarket and postmarket aspects unique to those products. The text ends with a discussion on private actions for labeling violations and foodborne illnesses and the pinnacle of enforcement—criminal sanctions. Food law and regulation involves nearly all facets of food production and this text will touch on each topic, some in greater detail than others.

As an introduction to food law and regulation, it begins to consider your own relationship with food. How much do you trust the food you buy? Begin by identifying one or two laws or regulations you think may be involved with your favorite food. There are a number of questions you may have about how it is made, marketed, or where it comes from. Some elements you can easily research and find discussions about.

For example, one of my favorite foods is dark chocolate. Did you know the FDA sets a threshold on the amount of insect parts that can be present in chocolate? If a chocolate maker keeps the insects inadvertently collected during harvesting below the threshold, the product is deemed safe for consumers. But if the amount passes the threshold, it is considered adulterated. This is known as Food Defect Action Levels, and it applies to all foods and all types of contaminants.

1.2 Introduction

This chapter begins with the basics of administrative law and regulation. There is a compelling temptation to jump directly into the details of the food law and regulation. This urge stems largely from an intuitive sense of what one may expect as a consumer. The regulations, however, are complex making it easy to quickly lose vital context. Rather than beginning with a close-up discussion of specific topics, this chapter will take a birds-eye view of administrative law and regulation. From this height, the reader can see how all the dots connect and become part of a larger pattern of law and policy.

This chapter starts with an overview of the US legal system. A synopsis of the US legal system offers key insights into the unique separation between the States and federal government. Understanding this relationship will be important for later topics such as labeling litigation. It also sheds light on why federal agencies are the primary regulators of food identity and safety standards. Other aspects of the US legal system will provide tips on understanding judicial opinions and the role of the Constitution in protecting individual liberties.

The chapter then shifts away from the US legal system to narrow its focus on the two primary food product regulators. Beginning with a broad overview of the regulatory landscape, the relationship of agencies charged with food regulation and sources of law will come into focus. From this vantage point, the reader will have a better sense of where the ensuing chapters fit into the entire food regulatory scheme. For example, jurisdictional boundaries will emerge through key definitions of "food" and "meat." This introductory material provides the springboard for the remaining chapters on substantive food law.

Food law enjoys a rich complex history. Given the common experience of food, some of the concepts will feel intuitive and familiar. Other topics will place new language and boundaries on our expectations as connoisseurs of food. The material in this section aims to plant firm roots that can always be followed when the concept becomes unfamiliar or complicated. Retracing the steps provided in this chapter from identifying sources of law to understanding key definitions, one can act as a guide to grasping the later concepts of this text.

1.3 Overview of the US Legal System

1.3.1 Federalism and the Structure of Government

For readers outside the USA, it is necessary to take a brief moment to introduce several key concepts of US law and the structure of government. The USA as its name suggests is a union of several sovereign States. Sovereign that is to the extent allowed by the US Constitution. The Constitution is the foundational document not only both for the creation of a federal government, but also for stipulating how in certain circumstances States govern their citizens. The US Constitution plays three pivotal roles. In the Absence of the Constitution, there would be neither federal agencies nor any federal laws like the Food Drug and Cosmetic Act or the FDA. Also without the Constitution, there would be no legal grounds for any entity other than a state to regulate food production or safety within its borders. And if never ratified, many individual rights and freedoms, like the protection from unreasonable search and seizure, would be missing. The Constitution, therefore, provides a framework for federal regulation that plays a role in interstate commerce and individual and corporate freedoms and liberties.

The US Constitution establishes a federal government. Meaning it provides specific powers to the federal government and the remaining powers to the States. Like a corporate merger, the States forfeited rights and reserved others to a new superior organization. This concept is often referred to as federalism, but can simply be thought of as a vertical division between the federal government and the states (see Fig. 1.4).

Federalism places limits on those laws States can enact. Generally speaking, the States, reserved "police powers," those powers necessary to protect its citizens' health and welfare. Still federal law can enter this zone of State au-

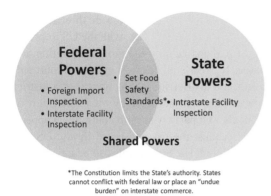

Fig. 1.4 Federalism applied to food law

thority using a secondary basis such as the power to regulate commerce. This is a one-way street. Only the federal government can enact laws where States normally exercise authority. Within certain narrow exceptions, States cannot enact laws where federal law already exists. This is a concept known as preemption. States can enact limited legislation where federal law exists so long as it does not "unduly burden" interstate commerce. Preemption provides a narrow space for States to regulate. In some instances, States like California may pass more stringent laws for activities, but the laws will only impact intrastate food production. The more stringent law may be a model for future federal legislation, but cannot otherwise exercise any effect outside its borders.

The US Constitution does more than play a role in the relationship between the federal government and the States. It also provides the primary framework for the laws and structure of the federal government. From this document, the USA derives its three branches of government— the legislative, the executive, and the judicial. A term often used around this structure is the "separation of powers" or "checks and balances." The essential purpose of the design was to disperse power among several institutions and leaders.

In order to understand the concepts in the chapters to follow, it is important to have a basic idea of the role of each branch of government. The legislative branch consists of the bicameral Congress. Congress is comprised of a House of Representatives and a Senate. The US Congress is solely responsible for passing new legislation,

such as it did in 1906 with the Pure Food and Drug Act. The executive branch is comprised of the President, Vice President, and their administration. In terms of domestic policy, the role of the President is to ensure a duly passed law is implemented and executed. The President organizes the executive branch through a series of federal agencies that will be responsible for an area of regulation. The two primary food agencies we have today are the US Food and Drug Administration and the US Department of Agriculture's Food Safety Inspection Service (FSIS). Once a law is passed, typically, the agency implementing the law will promulgate a series of rules to provide structure and clarity to vague or broad statutory provisions. The Pure Food and Drug Act, for example, when passed was a mere five pages long! The federal court system makes up the judicial branch. Federal courts are organized into three layers—district, circuit, and Supreme Court. Nearly every case begins at the district court level and in a series of appeals may be heard by the Supreme Court. Some districts operate with magistrate judges providing an initial review or decision before the district court. It is the role of the judiciary, the federal judges across the USA, to hear challenges of both the statutory provisions and the administrative rules.

1.3.2   The Role of the Constitution in Food Law

The US Constitution not only plays a role in the structure of the federal government, but also constrains the limits of federal laws. The Constitution enumerates both the limits of federal power with the States and with the individuals. The first ten amendments to the Constitution lists the familiar individual freedoms, commonly known as the Bill of Rights. The Bill of Rights includes the freedom of speech, freedom from unreasonable searches, and the protection against self-incrimination among others. If a law infringes on any constitutional provision, it can be challenged, and if successful, the law can be invalidated. Although constitutional rights are typically thought of as pertaining solely to individuals, they also apply to corporations. In later chapters, we will see the First Amendment challenges to advertising and labeling, Fourth Amendment challenges to facility inspections, and Fifth Amendment Due Process challenges when agencies take enforcement actions.

1.3.3   The Role of Judicial Opinions in Food Law

Introduction to Judicial Opinions

Although regulations play the largest role in this area of law, a complete understanding of the food law and policy of the USA requires an ability to read and interpret judicial opinions. Judicial opinions are the rulings from a judge in a dispute between two parties. The judicial opinion, also known as the rulings, explains the facts of the case, the legal principles involved, and a decision on how the principles apply to the facts at hand. Every judicial opinion follows a formula, which once understood allows the reader to quickly scan and understand the parties involved, the court deciding the case and the basic legal principle in question.

The Structure of Opinions

There are five components to every judicial opinion. All cases begin with the caption. The caption informs the reader who is suing who. If the government brought the suit then the caption will be *United States v. Food Processor A,* and vice versa if industry challenges the government in court. Some cases will name the agency, for example *U.S. Dept. of Health and Human Services, Food and Drug Admin. v. Food Importer X,* or the Secretary of the Agency, such as *ABC Sugar-Pops Inc. v. Sibelius.* Below the case caption there are series of letters and numbers called the legal citation. This citation is necessary for any research and retrieval of a particular case. The citation provides the "name of the course that decided the case, the law book in which the opinion was published, and the year in which the court decided

the case" (Orin 2007).[11] There are three levels of judicial review in the USA, with the US Supreme Court perhaps gaining the most recognition and notoriety. Nearly all lawsuits in federal court will begin in a district court. Any appeal of a district court ruling will be heard by a Circuit Court of Appeals. If a party wishes to continue its appeal, it can request the case be heard by the US Supreme Court. The three levels can be quickly distinguished in the citation. The US Supreme Court cases use the following citation 712 U.S. 111, while the circuit court opinions always state the circuit with the year such as 363 F.2d 465 (6th Cir. 1929). After the legal citation, is the judge's name. If these two are missing from a citation then the opinion is from a district court.

Are All Opinions Treated Equal?

Not all opinions carry the same authority in deciding the future cases. Each level of judicial review carries increasing authority. District court opinions are the lowest rung of authority. It will set a new standard only for that district court often acting only as persuasive authority. Circuit courts are binding precedent both on itself and on district courts, but only for those in the same Circuit. There are eleven circuits with three or more States in each Circuit. The Supreme Court as the highest level of review acts as binding authority on all courts. Read opinions with caution. District court opinions should be cited sparingly and Circuit courts cited persuasively outside the Circuit. State court opinions are subject to a similar structure, but names and the nature of appeals will vary by State.

The final two components of a judicial opinion center on the facts and law of the case. The facts of the case are fairly straightforward. The judge will provide a chronological summary of the events that lead to the lawsuit. The only new

---

[11]   Orin (2007).

feature non-lawyers may encounter is the use of "procedural history." The procedural history explains the path the case took either through the courts, the administrative agency, or both to reach the judge currently deciding the case. For our purposes, it will be important to note the judge's evaluation of whether the administrative agency finished reviewing the case. In this analysis, the judge will conclude whether the case is "ripe for adjudication," a topic explored in the sections to follow, beginning with Chapter 2. Following the facts the judge will state what legal principles are involved in the case, including any statutory sections, administrative rules, or prior case law that is known as judicial precedent. The law of the case typically follows two stages. The first stage provides the reader with a background of jurisprudence for the topic at hand, while the second stage applies the legal principles to the facts of the case (Orin 2007).[12] For non-lawyers, the most important sections will be the summary of facts and the legal principles. Similar facts will assist in determining whether the principles of the case apply and potentially the outcome as well.

The Impact of Judicial Opinions in Food Law

Judicial opinions play limited roles when interacting with agencies. It will be the exceptionally rare case that a judicial opinion will play a decisive part in resolving a compliance matter. In court yes, but in the agency, simply no. Instead, judicial opinions play a role in the background. For example, providing insight into statutory definitions or understanding concepts like Constitutional limits on agency activities. Judicial opinions play an important role, but only when applied correctly.

1.4 Food Law and Regulation

There are numerous federal agencies touching food and food safety in some facets. The Government Accountability Office (GAO) identifies 15 federal agencies administering no less than 30 laws related to food safety. Some agencies and laws serve only an administrative role, while others stand at the heart of food safety. This section will briefly identify those agencies before concentrating its efforts on the two primary agencies.

1.4.1 Sources of Food Law

There are numerous sources of laws. To make the regulatory landscape more manageable, this text will narrow the scope of laws covered. As mentioned above the GAO estimates 30 laws directly involving food topics. From this selection only around five statutes will be discussed. This set of core statutes will cover the most common areas of enforcement and agency approval. A strong understanding of the core statutes serves many valuable purposes. Chief amongst those is understanding the classification of a product. Classification informs many other decisions, such as the controlling agency and the regulatory burden of the product.

Statutes are not the only source of law to consider. The Code of Federal Regulations (CFR) contains the administrative rules issued by various federal agencies. The rules are organized by Titles, which groups rules by agency or statute. In most cases, the CFR title corresponds to the enabling statute. For example, Title 21 covers the rule promulgated to implement and interpret Title 21 of the United States Code, the FD&C Act. Other Titles related to food are Title 7 (agriculture) and Title 9 (animal and animal products). Title 9 contains the bulk of FSIS regulations, whereas Title 7 relates to other USDA activities. When an action is taken by an agency or division of an agency, it is important to identify the appropriate statute and regulations associated with that action. Each agency is only responsive to the particular statutes outlining its authority and the regulations providing meaning to that authority.

---

[12] *Id.* Kerr.

Sources of Food Law		
Food Drug and Cosmetic Act	21 U.S.C. 301 et. seq.	21 C.F.R.
Meat Inspection Act	21 U.S.C. 601 et. seq.	9 C.F.R.
Poultry Products Inspection Act	21 U.S.C. 451 et. seq.	9 C.F.R.
Egg Products Inspection Act	21 U.S.C. 1031 et. seq.	9 C.F.R

1.4.2   Primary and Secondary Agencies

In the array of agencies involved in food safety, two stand as the megaliths in the center. The FDA and FSIS, the inspection arm of the USDA compose the bulk of federal funding and staffing of the federal government's food regulatory system. This makes the FDA and FSIS the primary agencies regulating food safety. As will be discussed in detail below, the size of these two agencies matches the massive mandate for each entity.

The Role of Other Agencies and Divisions
Other federal agencies or divisions of agencies take some minor responsibility for food safety (see Fig. 1.5). These agencies are tertiary to the efforts of the FDA and FSIS. The Department of Homeland Security, for example, is responsible for customs activities, which can include coordinating inspections of food imports with the FDA. The USDA's Animal and Plant Health Inspection Service (APHIS) also indirectly serves as a food safety function. Its programs are aimed to protect plant and animal resources from pests and diseases like bovine spongiform encephalopathy (BSE or "mad cow" disease). The vast array of federal agencies can be dizzying. The central focus of this text will be on the primary two food agencies tasked with implementing and enforcing food law and regulation.

An important lesson to understand about federal agencies is the relationship of divisions within an agency. As outlined above, within the FDA and USDA there are divisions or branches within each of the agencies assigned certain product categories or enforcement responsibilities. FSIS and CFSAN were introduced as the primary food

divisions with the USDA and FDA respectively. These divisions may be referred to by name or simply by their parent agency, but no matter how they are called they are never autonomous. They exist within a framework of bureaucracy and regulation, which dictates the flow of decision making and the hierarchy for decisions, such as appeals.

Introduction to the Food and Drug Administration
The FDA monitors domestic and imported food products. The section below outlines in detail the FDA's authority and how "food" is defined by its enabling acts. Generally speaking, the FDA regulates all food products except for meat and poultry under the authority of FSIS. This necessarily means, any meat product not regulated by FSIS, such as game meats mentioned below, are under the FDA's purview. The boundary between FSIS and FDA can be blurry. Eggs are an excellent example. The FDA is responsible for facilities that sell, serve, or use eggs as an ingredient. It is also responsible for animal feed but not the laying facilities themselves. FSIS, however, is responsible for liquid, frozen and dried eggs, and grading eggs. The primary statutes governing FDA's activities are the Federal Food, Drug, and Cosmetic Act (21 U.S.C. 301 et. seq.); the Public Health Service Act (42 U.S.C. 201 et. seq.) and the Egg Products Inspection Act (21 U.S.C. 1031 et. seq.).

Only two branches within the FDA take part in food safety and regulation. The Center for Food Safety and Applied Nutrition (CFSAN) functions as the central player in food safety for the FDA. It takes on the full range of food safety functions, including: (1) food safety research (2) overseeing enforcement (3) evaluating surveillance and compliance programs (4) coordinating with state's food safety activities, and (5) developing and implementing regulations, guidance documents and consumer safety information (Congressional Research Service).[13] The

---

[13] Congressional Research Service Report To Congress: The Federal Food Safety System: Primer (2007).

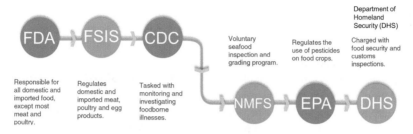

Fig. 1.5 Food safety roles of the primary and some secondary agencies

Center for Veterinary Medicine (CVM) straddles the animal-human food connection for the FDA. For food-producing animals, the CVM is responsible for ensuring animal feeds and drugs do not produce hazards to humans, in particular with animal drug residues (see Chapter 6 Food Additives). This is in addition to its other functions overseeing pet foods, drugs, and devices.

Introduction to the Food Safety Inspection Service for the US Department of Agriculture

FSIS regulates meat and poultry sold for human consumption. Under the Federal Meat Inspection Act (1906; 21 U.S.C. 601 et seq.) and Poultry Products Inspection Act (1957; 21 U.S.C. 451 et seq.), FSIS inspects the slaughter and processing of animals identified in the statutes. Those animals are defined in detail below. FSIS operates over domestic and foreign meat facilities. FSIS alone is responsible for certifying foreign meat and poultry plants produced products for safety as domestic plants before exporting to the USA FSIS also coordinates with State-operated meat and poultry inspection programs. The FMI and the PPI require FSIS to cooperate with the State agencies in developing and administering State meat and poultry inspection programs. In its 2013 fiscal year, FSIS reported coordinating with 27 States, which oversee 1600 small and very small establishments (FSIS Review).[14] FSIS does not inspect the establishment itself. Instead,

FSIS evaluated each State's inspection program to ensure it operates on a level "at least equal to" the FSIS inspection programs.

Three additional agencies though not as big or deeply involved as the FSIS and the FDA offer ancillary services that play a significant role in food safety. Those agencies are the National Marine Fisheries Services (NFMS), which is part of the US Department of Commerce, the Environmental Protection Agency (EPA), and the Center for Disease Control and Prevention (CDC), which is also a part of the Department of Health and Human Services. Seafood, with the exception of catfish, falls under the jurisdiction of the FDA. The NFMS, however, conducts a fee-for-service voluntary seafood inspection and grading program for US fish and shellfish. The NFMS's authority to conduct the voluntary inspections lies in the Agricultural Marketing Act of 1946 (7 U.S.C. 1621 et. seq.). The EPA well known for clean air and water regulations also regulates chemicals used on food crops. The EPA's Office of Pesticide Programs task is to ensure chemicals used on food crops do not pose a risk to public health. The CDC takes on an investigational role in foodborne disease outbreaks and research. It works in concert with the FDA, FSIS, NFMS and state and local health departments to monitor, identify, investigation, and research foodborne illnesses. It operates under the authority of the Public Health Service Act (42 U.S.C. 201 et. seq.).

---

[14] FSIS Review of State Meat and Poultry Inspection Programs: Fiscal Year 2013, Summary Report (December 2013).

1.5   What is Food? FDA Jurisdiction and Authority

To achieve its twin aims of protecting the public's health and its purse, the FDA utilizes a quadripartite set of regulations. As the regulations relate to humans, a product is either a drug, device, food/dietary supplement, or a cosmetic. Technically there is a fifth category—none of the above or unregulated by the FDA. This classification decision lies at the heart of every enforcement matter. It also remains the first question before launching a new product or adding a new ingredient. The classification decision is made using the definitions provided in the Act. Failing to start with the definitions, is a fundamental mistake.

1.5.1   The Role of Definitions

The definition of food is not what a lay person would think it to be. Asked on the street, "What is food?" One could point to a number of examples—fruit, a bag of chips, or maybe even a sports drink. The definition in the Act, however, is more nuanced. It is a term of art. There are in fact three definitions of food in the Act. Those are provided in the table below.

2 Section 201 (f) "food"	"food" means (1) articles used for food or drink for man or other animals, (2) chewing gum, and (3) articles used for components of any such article
Section 201(s) "food additive"	"food additive" means any substance the intended use of which results or may reasonably be expected to result, directly or indirectly, in its becoming a component or otherwise affecting the characteristics of any food (including any substance intended for use in producing, manufacturing, packing, processing, preparing, treating, packaging, transporting, or holding food; and including any source of radiation intended for any such use), if such substance is not generally recognized, among experts qualified by scientific training and experience to evaluate its safety, as having been adequately shown through scientific procedures (or, in the case of a substance used in food prior to January 1, 1958, through either scientific procedures or experience based on common use in food) to be safe under the conditions of its intended use; except that such term does not include— (1) a pesticide chemical residue in or on a raw agricultural commodity or processed food; or (2) a pesticide chemical; or (3) a color additive; or (4) any substance used in accordance with a sanction or approval granted prior to the enactment of this paragraph 4 pursuant to this Act [enacted Sept. 6, 1958], the Poultry Products Inspection Act (21 U.S.C. 451 and the following) or the Meat Inspection Act of March 4, 1907 (34 Stat. 1260), as amended and extended (21 U.S.C. 71 and the following); (5) a new animal drug; or (6) an ingredient described in paragraph (ff) in, or intended for use in, a dietary supplement

2 Section 201 (f) "food"	"food" means (1) articles used for food or drink for man or other animals, (2) chewing gum, and (3) articles used for components of any such article
Section 201 (ff) "Dietary Supplement"	"dietary supplement"—

Section 201 (ff) "Dietary Supplement"

"dietary supplement"—

(1) means a product (other than tobacco) intended to supplement the diet that bears or contains one or more of the following dietary ingredients ingredients :
  (A) a vitamin;
  (B) a mineral;
  (C) an herb or other botanical;
  (D) an amino acid;
  (E) a dietary substance for use by man to supplement the diet by increasing the total dietary intake; or
  (F) a concentrate, metabolite, constituent, extract, or combination of any ingredient described in clause (A), (B), (C), (D), or (E);
(2) means a product that—
  (A) (i) is intended for ingestion in a form described in section 411(c)(1)(B)(i); or
      (ii) complies with section 411(c)(1)(B)(ii);
  (B) is not represented for use as a conventional food or as a sole item of a meal or the diet; and
  (C) is labeled as a dietary supplement; and
(3) does—
  (A) include an article that is approved as a new drug under section 505 or licensed as a biologic under section 351 of the Public Health Service Act (42 U.S.C. 262) and was, prior to such approval, certification, or license, marketed as a dietary supplement or as a food unless the Secretary has issued a regulation, after notice and comment, finding that the article, when used as or in a dietary supplement under the conditions of use and dosages set forth in the labeling for such dietary supplement, is unlawful under section 402(f); and
  (B) not include—
      (i) an article that is approved as a new drug under section 505, certified as an antibiotic under section 507 7, or licensed as a biologic under section 351 of the Public Health Service Act (42 U.S.C. 262), or
      (ii) an article authorized for investigation as a new drug, antibiotic, or biological for which substantial clinical investigations have been instituted and for which the existence of such investigations has been made public, which was not before such approval, certification, licensing, or authorization marketed as a dietary supplement or as a food unless the Secretary, in the Secretary's discretion, has issued a regulation, after notice and comment, finding that the article would be lawful under this Act

Except for purposes of section 201(g), a dietary supplement shall be deemed to be a food within the meaning of this Act

There is another definition that plays a role in classifying a food product. That is section 201(g)(1)(c), which contains part of the definition of a drug. It reads: "the term 'drug' means ... (B) articles intended for use in the diagnosis, cure, mitigation, treatment, or prevention of disease in man or other animals...(C) articles (other than food) intended to affect the structure or any function of the body of man or other animals." Also called the "food exception," it is the oft cited defense to a drug classification of a food-based product claiming some drug-like effect.

Setting aside the concept of dietary supplements as food products, it becomes clear that the term in the Act is unworkable. Nearly any food-based product could be classified as a food, drug, or both. Take honey as an example. Classification as a food seems appropriate as it is often eaten plain or added as a sweetener to other foods like oatmeal, for example. Now imagine a scenario where the label states therapeutic claims. Adding a claim about calming an upset stomach, for instance, and one could also reasonably conclude it states a use to treat indigestion (250 Jars of US

Fancy Pure Honey).[15] This type of claim, often called a "disease claim" falls under the drug definition of section 201(g)(1)(c)(B). Now imagine a more subtle claim. Adding a claim about how honey slows the absorption of trans fats. Section 201(g)(1)(c)(C) addresses this type of claim, which is often referred to as a "structure-function claim." Perhaps, it is both a food and a drug? Such a possibility is not out of the realm of reasonable conclusions.

The federal courts tackled these issues in two cases. Together, the cases provide a more workable definition of "food." In the two-case excerpts below, we see a common theme in statutory interpretation. As is often the case with legislation in the USA, Congress writes statutory provisions intended to capture a wide swath of industry or issues. It then falls on the agency to issue a rule interpreting and narrowing the scope of the language to something manageable for both the agency and industry. If the agency does not take on this role, then courts will. This was the case for the definition of food in the Act. As will be explained below, the courts added a new concept to interpret the definitions—intended use.

All three cases are provided first with a discussion to follow.

Nutrilab, Inc. v. Schweiker, 547 F. Supp. 880 (N.D.Ill.1982))
BUA, District Judge.

STATEMENT OF THE CASE
The instant litigation concerns those products which have become known generically as "starch blockers." The plaintiffs, manufacturers and distributors of the products, initiated the lawsuit seeking a declaratory judgment pursuant to 28 U.S.C. §§ 2201 and 2202 and requesting this Court to declare that starch blockers are "foods" under 21 U.S.C. § 321(f) and not "drugs" as defined by 21 U.S.C. § 321(g).

The lawsuit was initiated in response to the classification by the Food and Drug Administration (FDA) of the products as "drugs" and to the agency's request that all such products be removed from the market until FDA approval was received. Absent substantial scientific evidence demonstrating that the product was generally recognized as safe and effective, the FDA regarded the product as a "new drug" under 21 U.S.C. 321(g) and considered further interstate distribution of the product a violation of 21 U.S.C. § 355(a).

The defendants counterclaimed seeking a temporary restraining order enjoining the plaintiffs from further distributing starch blockers in interstate commerce in violation of the Federal Food, Drug, and Cosmetic Act, ("the Act"), 21 U.S.C. §§ 301–392. The motion for the temporary restraining order was denied and a hearing was held on defendants' motion for a preliminary injunction. At the close of the hearing, the parties stipulated to advancing the hearing as a trial on the merits.

FINDINGS OF FACT AND CONCLUSIONS OF LAW
The Court, having heard the testimony of the witnesses and having examined the exhibits introduced in evidence does hereby make the following findings of fact and conclusions of law pursuant to Rule 52(a) of the Federal Rules of Civil Procedure.

"Starch blocker" is the generic name for the group of products manufactured from the protein contained in a certain type of raw kidney bean. The product is sold in both tablet and capsule form.

It is claimed that the protein which makes up the product acts to prevent the digestion of starch. Specifically, it is claimed, the protein acts as an alpha-amylase inhibitor. Alpha-amylase is an enzyme secreted by the pancreas which is necessary to the

---

[15] See e.g., 250 *Jars of US Fancy Pure Honey,* 218 F. Supp. 208 (E.D. Mich. 1963).

digestion of starch. When one or more starch blocker pills are ingested during a meal, the protein acts to prevent the alpha-amylase enzyme from acting, this allowing the undigested starch to pass from the system. As digestion of starch, a complex sugar, is cited as a cause of weight gain, the passage of starch through the system in an undigested form allows the individual taking the starch blocker to consume foods containing starch and high in carbohydrates without the risk of putting on weight.

The safety and effectiveness of the product has yet to be tested by the FDA. As the plaintiffs consider their products to be foods, no testing as required to obtain FDA approval as a new drug has taken place. No new drug application has been filed for these products nor has any investigational new drug exemption been issued pursuant to 21 U.S.C. 355(i).

The central issue in this case involves a determination of whether starch blockers are a drug under 21 U.S.C. § 321(g) or a food under 21 U.S.C. § 321(f). If a drug, the manufacturers of starch blockers would be required to file a new drug application pursuant to 21 U.S.C. 355 and to be regulated as such. The immediate consequence of such a determination would be the issuance of a permanent injunction requiring plaintiffs to remove the product from the marketplace until approved as a drug by the FDA.

Under 21 U.S.C. 321(g)(1)(C), drugs are defined as "… articles (other than food) intended to affect the structure or any function of the body of man…" Foods, on the other hand, are defined as "articles used for food or drink for man…" 21 U.S.C. § 321(f).

Because of the breadth and necessary vagueness of these statutory definitions, it is incumbent upon this Court to formulate Usable working definitions for these terms which can be applied to the case at bar. In undertaking such a task, the Court is mindful of the policy requiring liberal construction of the terms consistent with the overriding purpose of the Act—the protection of public health …

The plaintiffs have urged the Court to make its determination of whether starch blockers are foods or drugs based upon the source from which the product has been derived and, apparently, upon the common perception of the category into which the main component of the product falls. Thus, it is argued that, as the product is manufactured from beans, indisputably a natural food, and is made up of mere protein, a substance often regarded as a food, the product must be considered a food. This argument must, however, be rejected.

That a product is naturally occurring or derived from a natural food does not preclude its regulation as a drug…Nor does the fact that an item might, in one instance, be regarded as a food prevent it from being regulated as a drug in another… Therefore, that the product is derived from a natural food and is comprised of vegetable protein does not necessitate a finding by the Court that starch blockers are foods.

Resolution of the issue before the Court must come down to a question of intended use. If a product is intended by the user and the manufacturer or distributor to be used as a drug, it will be regulated as such. Conversely, if it is intended that the product be used as a food, it will be so considered…

By its language, the Act contemplates that "food" refers only to those items actually and solely "Used for food." 21 U.S.C. § 321(f)(1). Expert testimony received in the instant case leads the Court to conclude that substances used for food are those consumed either for taste, aroma, or nutritional value. It is clear that starch blockers are used for none of these purposes. The U.S.er of starch blockers U.S.es them as a drug, not as a food.

It cannot be said that the manufacturers or sellers of starch blockers intend the product to be used as food. The intent of the vendor in the sale of the product to the public is a key element in determining which statutory definition a product falls into... The FDA may consider the manufacturers subjective intent as well as actual therapeutic intent based upon objective evidence in this determination... Additionally, regardless of the actual physical effect of the product, for purposes of the Act it will be deemed a drug where the labeling and promotional claims show intended uses bringing the product within the drug definition....

The Court finds that the intent in the marketing of the product is that starch blockers be used as a drug. Starch blockers are marketed for treatment of an overweight condition. Although it has been recognized that products used for overweight could be in the category of foods for "special dietary use," 21 C.F.R. 105.3, starch blockers are not marketed for taste or aroma or for their nutritional value. Instead, they are to be used for treatment of a certain condition. Clearly they are intended to be used as a drug.

That starch blockers are to be considered a drug may further be gleaned from an examination of the promotional materials and labeling associated with the products. The various materials claim the product to be "totally natural and safe," "absolutely safe and exceptionally effective... no side effects," and "tested; approved." Additionally, the U.S.er is warned to "keep out of reach of children." Finally, it is claimed that use of starch blockers can aid in prevention of "degenerative diseases, including arteriosclerosis, arthritis, and diabetes mellitus..." Claims such as these are clearly claims normally associated with drugs and

not food products. Hence, the conclusion that the products were intended to act as a drug and be considered as such by the public is unavoidable...

ORDER

The Court therefore concludes that starch blockers are drugs under 21 U.S.C. § 321(g)(1)(C)...

Nutrilab, Inc. v. Schweiker, 713 F.2d 335 (7th Cir.1983)

CUMMINGS, Chief Judge.

The only issue on appeal is whether starch blockers are foods or drugs under the Federal Food, Drug, and Cosmetic Act, 21 U.S.C. § 301 et seq.

...

In order to decide if starch blockers are drugs under section 321(g)(1)(C), therefore, we must decide if they are foods within the meaning of the part C "other than food" parenthetical exception to section 321(g)(1)(C). And in order to decide the meaning of "food" in that parenthetical exception, we must first decide the meaning of "food" in section 321(f).

Congress defined "food" in section 321(f) as "articles used as food." This definition is not too helpful, but it does emphasize that "food" is to be defined in terms of its function as food, rather than in terms of its source, biochemical composition or ingestibility.

Plaintiffs' argument that starch blockers are food because they are derived from food—kidney beans—is not convincing; if Congress intended food to mean articles derived from food it would have so specified. Indeed some articles that are derived from food are indisputably not food, such as caffeine and penicillin. In addition, all articles that are classed biochemically as proteins cannot be food either, because for example insulin, botulism toxin, human hair and influenza virus are proteins that are clearly not food.

Plaintiffs argue that 21 U.S.C. § 343(j) specifying labeling requirements for food for special dietary U.S.es indicates that Congress intended products offered for weight conditions to come within the statutory definition of "food." Plaintiffs misinterpret that statutory Section. It does not define food but merely requires that if a product is a food and purports to be for special dietary uses, its label must contain certain information to avoid being misbranded... If all products intended to affect underweight or overweight conditions were per se foods, no diet product could be regulated as a drug under section 321(g)(1)(C), a result clearly contrary to the intent of Congress that "anti-fat remedies" and "slenderizers" qualify as drugs under that Section.

If defining food in terms of its source or defining it in terms of its biochemical composition is clearly wrong, defining food as articles intended by the manufacturer to be used as food is problematic. When Congress meant to define a drug in terms of its intended use, it explicitly incorporated that element into its statutory definition. For example, section 321(g)(1)(B) defines drugs as articles "intended for use" in, among other things, the treatment of disease; section 321(g)(1)(C) defines drugs as "articles (other than food) intended to affect the structure or any function of the body of man or other animals." The definition of food in section 321(f) omits any reference to intent...

Further, a manufacturer cannot avoid the reach of the FDA by claiming that a product which looks like food and smells like food is not food because it was not intended for consumption. In *United States v. Technical Egg Prods., Inc.*, 171 F. Supp. 326 (N.D.Ga.1959), the defendant argued that the eggs at issue were not adulterated food under the Act because they were not intended to be eaten. The court held that there was a danger of their being diverted to food use and rejected defendant's argument....

Although it is easy to reject the proffered food definitions, it is difficult to arrive at a satisfactory one. In the absence of clearcut Congressional guidance, it is best to rely on statutory language and common sense. The statute evidently U.S.es the word "food" in two different ways.

The statutory definition of "food" in section 321(f) is a term of art and is clearly intended to be broader than the common-sense definition of food, because the statutory definition of "food" also includes chewing gum and food additives. Food additives can be any substance the intended use of which results or may reasonably result in its becoming a component or otherwise affecting the characteristics of any food.... Paper food-packaging when containing polychlorinated biphenyls (PCB's), for example, is an adulterated food because the PCB's may migrate from the package to the food and thereby become a component of it...

Yet the statutory definition of "food" also includes in section 321(f)(1) the common-sense definition of food. When the statute defines "food" as "articles used for food," it means that the statutory definition of "food" includes articles used by people in the ordinary way most people use food—primarily for taste, aroma, or nutritive value. To hold as did the district court that articles used as food are articles used solely for taste, aroma or nutritive value is unduly restrictive since some products such as coffee or prune juice are undoubtedly food but may be consumed on occasion for reasons other than taste, aroma, or nutritive value. 547 F.Supp. at 883...

American Health ProductsCo., Inc. v. Hayes,
574   F. Supp. 1498 (S.D.N.Y. 1983)

SOFAER, District Judge:
On July 1, 1982, the Food and Drug Administration ("FDA") announced its decision to classify as drugs under section 201(g)(1)(C) of the Federal Food, Drug, and Cosmetic Act (hereinafter the "Act"), a group of products generally known as "starchblockers." 21 U.S.C. § 321(g)(1)(C). A short time later plaintiffs, who manufacture starchblockers, brought this action for a declaratory judgment that their products are a "food" under the Act and therefore exempt from its premarketing approval requirements....

The starchblockers under consideration here are derived from White Northern beans. The FDA maintains that the manufacturers make starchblocker tablets and powder by isolating and extracting the inhibitory protein from its natural source, as the labels of several of their products state. The manufacturers contend that a large proportion of the starch is simply removed from the beans in order to produce a flour with a high concentration of protein; starchblocker pills are then made by adding various binders and excipients...

The beans contain a protein that inhibits the normal functioning of alpha-amylase, an enzyme produced by the pancreas. Alpha-amylase aids in the digestion of starch by breaking it down into glucose, which the body then absorbs and utilizes for energy. Plaintiffs market their product as an aid to weight reduction, claiming that, since the protein prevents the alpha-amylase from acting, starchblockers allow some starch to pass through the body undigested, enabling dieters to avoid calories. The protein thus functions as an antinutrient by interfering with the normal digestion, absorption, and utilization of starch.

Section 201(g)(1)(C) of the Act defines the term "drug" in part to mean "articles (other than food) intended to affect the structure or any function of the body of man or other animals." 21 U.S.C. § 321(g)(1)(C). The immediately preceding subsection defines "food" as "(1) articles used for food or drink for man or other animals, (2) chewing gum, and (3) articles used for components of any such article." Id. § 321(f).

The starchblocker manufacturers concede that their products are intended to affect a human bodily function, but contend that the products are "food" and thus fall within the parenthetical exclusion of subsection (g)(1)(C). The government argues that the statutory definition of food does not encompass starchblockers, and seeks to enjoin their sale until the FDA approves them as a "drug." Unless a manufacturer can demonstrate that its product is "generally recognized as safe and effective," see id. § 321(p), classification as a drug requires the manufacturer to cease marketing its product until the FDA approves its new drug application. See id. § 355(a)2.

In deciding whether the FDA's determination that starchblockers are drugs is "in accordance with law," 5 U.S.C. § 706(2)(A), the FDA's interpretation merits substantial deference... A court must only ensure that the agency's action was "governed by an intelligible statutory principle." Here the government contends that subsection (g)(1)(C) contemplates dual classification; in addition to classifying as drugs all those products that affect bodily structure or functions and are not common foods, the Act is said also to classify as drugs within the part (C) definition even a "food" product, if it is sold with specific representations as to its physiological effects. The manufacturers argue, on the other hand, that the definition of "food" in subsection (f), which refers to common usage, governs the reach of the parenthetical exclusion in part (C). Therefore, they urge, if an article is a "food" under subsection (f), it cannot be regulated as a drug under subsection (g)(1)(C) regardless of any representations as to its structural or functional effects made in connection with its sale.

The government's contention is untenable. Though most sections of the Act countenance dual classification, no other contains a parenthetical like that Congress inserted in part (C). Ignoring that parenthetical would render meaningless the distinctions Congress has attempted to delineate. Nevertheless, the government is correct in claiming that starchblocker pills are a "drug" under the Act, because the pills are not a "food" in any sense cognizable under the statute…

I. PROPRIETY OF DUAL CLASSIFICATION.

…

B. LANGUAGE, STRUCTURE, AND LEGISLATIVE HISTORY OF SECTION 321(G)(1)(C).

The most natural way to read the term "food" in the exclusion under part (C), as the manufacturers urge, is as a reference to the definition of that term in the immediately preceding subsection [section 201(f)], which was added by Congress at the same time. However, the term would then refer most plausibly to that entire definition, not solely its first element, which concerns common usage. That the drafters chose to repeat the defined term in the first part of its own definition—"articles used for food or drink for man or other animals"—suggests only that they intended to include the everyday meaning of food as one component of the statutory meaning. Had they intended the parenthetical exclusion of part (C) to encompass only this everyday meaning, they could easily have repeated

the words "used for food" appearing in subsection (f)(1), instead of employing the defined term, "food."

The second element of the subsection (f) definition of food is chewing gum. Consistent with its interpretive stance generally, the government argues by example that a chewing gum marketed as a laxative would be regulable as a drug under section 321(g)(1)(C). The manufacturers would presumably argue that the parenthetical excludes such a product unless some ingredient independently was classifiable as a drug.

The third element of the subsection (f) definition, however,—"articles used for components of any such articles" as chewing gum and ordinary food—poses a severe problem for the manufacturers' construction. Reading the parenthetical to exclude components of ordinary foods and chewing gum from the coverage of part (C) raises the anomalous possibility that a manufacturer might escape the premarketing approval requirements by distributing an otherwise regulable article not in pure form but as an ingredient of a food or chewing gum. Such a result would convert a provision designed to bring a species of article not otherwise covered by any portion of the Act within the purview of the food provisions… into a boundless escape hatch from drug coverage. That reading would also conflict with the commonly understood import of the term "article" as used in the drug definitions, which the manufacturers recognize reaches any determinate substance, including one distributed as an ingredient of a nondrug product.

Thus, contrary to the manufacturers' contention, the meaning of the term "food" for purposes of the parenthetical exclusion and its technical statutory meaning for purposes of the coverage of the food provisions cannot be identical…

The government's construction, however, is equally problematic. The government in effect asks this court to impute the element of intention into the meaning of "food" in the parenthetical and thereby to confer a third meaning on the term distinct from those it carries in sections 321(f) and 321(f)(1). The language and structure of the relevant provisions seem inconsistent with this view. Neither the definition of food in subsection (f) nor the parenthetical exclusion of food in subsection (g)(1)(C) makes any reference to intention... even though the concept is present in the definition in which the exclusion appears... The definition in section 321(g)(1)(B) instead refers to "articles used for food."

By repeating the defined term in its own definition, and by making use or function the definitional criterion, Congress appears to have intended that this component of the statutory definition of "food" refer to common usage. The ordinary way in which an article is used, therefore, not any marketing claim on the part of the manufacturer or distributor as to a specific physiological purpose of that use, should determine whether it is a food for purposes of the parenthetical exclusion of section 321(g)(1)(C).

The government's suggested construction also renders the parenthetical in section 321(g)(1)(C) meaningless. If the subsection deems articles represented to affect the structure and functions of the human body drugs notwithstanding the common usage of those articles for food, then the coverage of an article by the food definition is irrelevant and the parenthetical excludes nothing that the nonparenthetical language would include. The government urges that, since all foods in some way affect bodily structure and function, the drafters must have inserted the parenthetical to avoid subjecting all foods to regulation as drugs by virtue of their universal effect on bodily structure and function. But the courts have always read the claU.S.es

in the statutory definitions employing the term "intended" to refer to specific marketing representations...so this universal effect would not bring common foods sold without recourse to physiological claims within the statute in any case. As a logical matter, either the parenthetical limits the coverage of the subsection to nonfoods, or it expresses no limit beyond the nonparenthetical definition.

Construing the parenthetical to exclude all articles commonly used for food regardless of physiological claims made in connection with their sale does not defeat the central purpose of the provision as revealed in its legislative history. The original Food and Drug Act, enacted in 1906, classified as drugs only articles intended for therapeutic use. The legislative history of the 1938 Act makes clear that Congress added what became section 321(g)(1)(C) in order to expand the reach of the drug definition to cover products marketed for their physiological effects, which at the time escaped regulation.

Much of the legislative history upon which the government relies, however, suggests that Congress acted not in order specifically to regulate as drugs all substances intended to affect bodily structure or function, but rather to reach those products... which evaded regulation altogether because they were neither foods nor therapeutic agents.

Thus, if an article affects bodily structure or function by way of its consumption as a food, the parenthetical precludes its regulation as a drug notwithstanding a manufacturer's representations as to physiological effect. The Act evidences throughout an objective to guarantee accurate information to consumers of foods, drugs, and cosmetics. The presence of the parenthetical in part (C) suggests that Congress did not want to inhibit the dissemination of useful information concerning a food's physiological properties by subjecting foods to drug regulation on the basis of representations in this regard.

Whether or not the parenthetical of section 321(g)(1)(C) incorporates the entire technical definition of food in section 321(f), however, and whether or not section321(g)(1)(C) contemplates dual classification, the manufacturers cannot prevail here. Notwithstanding the government's contention that the FDA could regulate their products as a food if it chose to do so, the manufacturers cannot demonstrate that starchblockers are a food in any sense cognizable under the statute.

II. APPLYING SECTION 321(G)(1)(C) TO STARCHBLOCKERS.

The Seventh Circuit recently considered the identical question of the status. of starchblockers under part (C) and concluded that they are drugs. ...*Nutrilab*... defined food as articles used "primarily for taste, aroma, or nutritive value,"... properly rejecting also any suggestion that the source of a product makes it a food...

Here the manufacturers contend that starchblockers must be deemed a food because their biochemical composition varies from that of the bean flour used for making bread—a paradigmatic food—only by the percentage of each component and the addition of excipients and binders. This argument fails for the same reasons articulated in Nutrilab. The concentration of certain components during processing effects a significant physical change...Most fundamentally, the argument fails to address the Act's focus on usage.

The government approves the Seventh Circuit's result in *Nutrilab*, but takes issue with the standard of usage by which it was reached. The government argues that the meaning of "food" for purposes of the parenthetical exclusion of section 321(g)(1)(C) diverges from the definition of that term contained in section 321(f). If a type of article is used at all in a manner associated with the ordinary meaning of food—for taste, aroma, or nutritive value—the government States it may be regulated as a food under section 321(f)(1). On the other hand, if the same food were represented to have a particular structural or functional effect on the body, the government claims it may also be regulated as a drug under section 321(g)(1)(C). thus., the government prefers the formulation of the *Nutrilab* district court, which held that only if a food is used solely for taste, aroma, or nutrition and no claims are made that it affects bodily function may it escape regulation under section 321(g)(1)(C).

The manufacturers concur in the Seventh Circuit's criticism of the *Nutrilab* district court's definition, but contend that the appellate court's formulation is itself "unduly restrictive" because it fails to draw appropriate conclusions from the fact it recognized—that some foods, such as coffee and prune juice, are frequently consumed specifically for their physiological effect, but are nonetheless foods.

But this criticism, even if it were valid, has no relevance to the present case. The manufacturers adduced no evidence at trial that starchblockers are ever consumed by anyone for taste, aroma, nutrition, or sustenance. They manufacture their products by a process which concentrates the antinutrient to the exclusion of components which contribute food value. One could adjust the definitional threshold for "food" considerably lower than the Seventh Circuit's "primarily used" standard and still exclude starchblockers, because they have no taste or aroma, their nutritional value is negligible, they provide no sustenance, and they are not consumed for any of these purposes.

An example offered by the manufacturers illustrates well the consequences of the dual classification dispute, but demonstrates at the same time its irrelevance to the case of starchblocker pills. Coffee is often consumed as a stimulant, but is also commonly drunk for its taste and aroma, if not its nutritional value. Were the parenthetical read to preclude dual classification, the common usage of coffee as a food would place it beyond the reach of the part (C) drug definition even though a manufacturer might promote its coffee exclusively for its use in staying awake…This result is unavailable to the manufacturers of starchblockers, however, because they cannot demonstrate that their products are commonly used for food. The manufacturers simply cannot bring their products within the parenthetical of part (C).

The manufacturers also attempt to draw an analogy between starchblockers and saccharin, suggesting that each, though without food value itself, makes other foods more palatable to the dieter. See 21 C.F.R. § 105.3(a)(2) (artificial sweeteners as food for special dietary use). The analogy fails because saccharin is commonly consumed for its taste effect. Though the taste of saccharin may have food value only when used to enhance the flavor of other foods, it is a food characteristic nonetheless. Garlic is no less enticing because people generally do not eat it by the clove.

For the foregoing reasons, starchblocker pills are declared drugs under the Act. Their seizure is therefore permissible. The Clerk will enter judgment accordingly, without costs, in this case and in all presently pending, related cases. Any party that manufacturers not starchblockers but a bean flour actually used for foods may move for appropriate relief upon a proper showing.

1.5.2   Intended Use

Two lessons emerge from *Nutrilab* and *American Health Products*. First, we see the courts quickly focus on the central issue of intent. Prior to *Nutrilab* and *American Health Products* there were a number of cases that looked at the question of whether the food and drug definitions of the Act were mutually exclusive. In a series of cases federal courts reached a near universal conclusion holding the definitions overlapped and intent would be the touchstone to determine classification. This meant the agency could use a sliding scale. If a product, such as honey, began making claims on labeling attributing drug-like properties, mitigating indigestion for example, then the FDA could slide the classification from food to drug. Intent could be subjective or objective. Some courts when evaluating an FDA classification determination may look to subjective intent—did the manufacturer subjectively intend a consumer to purchase a food product for a therapeutic purpose? As can be seen in *Nutrilab* the court raises the ability to use subjective intent only to quickly move on to objective intent (Nutrilab District Court).[16] Objective intent looks to claims in labeling and promotions to determine how the seller intended the consumer to use the product. Second, the concept of intended use emerges. Intended use is a short hand for intent when making a classification decision. It is the favored term of the agency.

Through the lens of intended use *Nutrilab* and *American Health Products* provide a clearer picture of how to define food in the Act. In *Nutrilab* the FDA argued the starch "slenderizer" was a drug because it intended to affect the structure-function of the body. Nutrilab on the other hand argued their product was a food since it was derived solely from kidney beans. The court found both arguments wrong. At the trial court the definition of food is measured on a "sole use" standard. When a manufacturer states an intended use for a product to be consumed "actually and solely" for "taste, aroma, or nu-

---

[16] *Nutrilab* District Court at 883.

tritional value" then it is a food. On appeal the circuit court reasoned the trial court's definition would allow the agency to regulate as drug products food products like coffee or prune juice, which are consumed on occasion for purposes other than taste, aroma, or nutritive value. Thus, the circuit court held that the appropriate test of intended use is a "primary purpose" standard. If a manufacturer intends for consumers to use their product "primarily for taste, aroma, or nutritive value" then it is a food.

The district court in *American Health Products* takes the definition one step further. It begins by reasoning the "food exception" to the drug definition explicitly excludes food products from the classification as drugs. It then states agreement with the *Nutrilab* "primary purpose" standard, but notes section 201 of the Act does not reference intent. Intent is a construction used by courts not the Act itself. Instead the drug definition excludes "articles used for food." The court concludes due to the lack of an intent standard in the definitions of the Act a product a consumer would ordinarily call or use as food could only be classified as a food under the Act. This means no matter the claims, which are a measure of intent or intended use, if a product, like coffee is normally consumed as a food then it can never be classified as a drug. *American Health Products* was appealed to the second circuit appeals court. The second circuit in a *per ciruam* one page opinion affirmed the district court's ruling.

Therefore, we can say that food under the Act is measured by "primary purpose" and no product normally used as a food can be classified as a drug. Classification remains a critical issue to ensure regulated products are subject to the appropriate regulatory burden. The regulatory requirements for some categories, such as drug and medical device, is significantly more burdensome than for other classes of products (see Fig. 1.6).

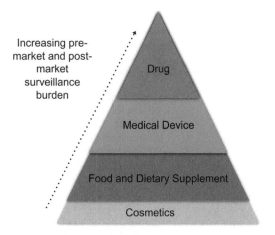

Fig. 1.6 Increasing regulatory burden based on classification of regulated product

1.6   What Are Meat, Poultry, and Eggs? USDA/FSIS Jurisdiction and Authority

The classification decision also plays a role with the USDA. Part of the analysis of determining whether a product is a "food" under the FD&C Act is to determine which agency exercises jurisdiction over the "food." As will be discussed below there are some categories where one agency exercises jurisdiction over the other and some categories, such as color additives, where the agencies reached an agreement on who would exercise jurisdiction. In many cases the boundary of authority is clear, but there are many instances where the line is arbitrary.

Generally speaking, the USDA through FSIS regulates meat, poultry, and egg products. These food products, however, are exempt from the FD&C Act provisions only to the extent the Federal Meat Inspection Act or other Acts apply. Again, this means the definition of meat, poultry, and eggs, is paramount to understanding the outlines of USDA authority to regulate foods. The following Acts provide the definitions needed to define the USDA's authority:

FMI Act § 601(j)	The term "meat food product" means any product capable of use as human food which is made wholly or in part from any meat or other portion of the carcass of any ***cattle, sheep, swine, or goats***, excepting products which contain meat or other portions of such carcasses only in a relatively small proportion or historically have not been considered by consumers as products of the meat food industry, and which are exempted from definition as a meat food product by the Secretary under such conditions as he may prescribe to assure that the meat or other portions of such carcasses contained in such product are not adulterated and that such products are not represented as meat food products. This term as applied to food products of equines shall have a meaning comparable to that provided in this paragraph with respect to cattle, sheep, swine, and goats. (Mandatory inspection for Ratites and Squab under a 2001 Order)
The Poultry Products Inspection Act § 453(e) and (f)	(e) The term "poultry" means any domesticated bird, whether live or dead (f) The term "poultry product" means any poultry carcass, or part thereof; or any product which is made wholly or in part from any poultry carcass or part thereof, excepting products which contain poultry ingredients only in a relatively small proportion or historically have not been considered by consumers as products of the poultry food industry, and which are exempted by the Secretary from definition as a poultry product under such conditions as the Secretary may prescribe to assure that the poultry ingredients in such products are not adulterated and that such products are not represented as poultry products
The Egg Products Inspection Act § 1033(g)	Defines an egg to mean the shell egg of domesticated chicken, turkey, duck, goose, or guinea

The term "domesticated bird" can sweep in a wide range of species. Statutes aim broadly while agency rules, codified in the code of federal regulations, focus narrowly. Thus, as is often the case, the USDA needed to issue a rule to make the statute manageable both for enforcement and compliance. It issued a rule interpreting the PPI Act definition of "domesticated bird" to mean domestic chickens, turkeys, ducks, geese, and guineas.

Other definitions of meat and poultry require further regulations to clarify the USDA's jurisdiction. Note in the definition for poultry for example, the exemption from the definition of poultry products with a "relatively small portion" or historically viewed as consisting of so little meat as not to be "poultry products." The same criterion is set for meat products. How small is small enough to fall outside the definition of meat or poultry products? The USDA promulgated rules establishing a threshold for meat and poultry products (see Investigations Operations Manual (IOM) Exhibit 3-1 in Table 1.1). Any product below the threshold is regulated by the FDA.

1.6.1   Jurisdictional Overlap

This creates areas of overlap and fosters confusion about jurisdiction. For example, a sausage product would be regulated by both the agencies. Assuming it was a meat listed in the FMI, the USDA would govern the meat filling, and the FDA the casing which contains meat of no nutritional value. Eggs are another area where the two agencies intersect. The FDA oversees shelled eggs, chicken feed, and egg labeling. The USDA regulates USDA egg products (liquid, dehydrated, frozen etc.), the laying facilities, and the grading of eggs. This is one example where a single food safety agency would provide a more cohesive approach.

The confusion and double standards are especially challenging for a facility regulated by both the agencies. A facility that makes chicken and tomato soup for example, would be inspected by both facilities. The chicken soup would be regulated differently than the tomato. Understanding the boundaries of each agency is important during inspections, enforcement actions, and proactively building compliance programs.

The table from the IOM provides the best tool in assessing the jurisdictional boundaries.

This table summarizes information concerning jurisdiction overlap for commercial products regulated by either or both FDA and USDA. It does not cover products made for on-site consumption such as pizza parlors, delicatessens, fast food sites, etc.

Table 1.1 Investigations operations manual 2013 Exhibit 1.1

FDA JURISDICTION	USDA JURISDICTION		
21 U.S.C 392(b) Meats and meat food products shall be exempt from the provisions of this Act to the extent of the application or the extension thereto of the Meat Inspection Act. FDA responsible for all non-specified red meats (bison, rabbits, game animals, zoo animals and all members of the deer family including elk (wapiti) and moose)). FDA responsible for all non-specified birds appropriate. including wild turkeys, wild ducks, and wild geese	The Federal Meat Inspection Act regulates the inspection of the following amenable species: cattle, sheep, swine, goats, horses, mules or other equines, including their carcasses and parts. It also covers any additional species of livestock that the Secretary of Agriculture considers Mandatory Inspection of Ratites and Squab (including emu) announced by USDA/FSIS April 2001	The Poultry Products Inspection Act (PPIA) defines the term poultry as any domesticated bird. USDA has interpreted this to include domestic chickens, turkeys, ducks, geese and guineas. The Poultry Products Inspection Act States poultry and poultry products shall be exempt from the provisions of the FD&C Act to the extent they are covered by the PPIA. Mandatory Inspection of Ratites and Squab announced by USDA/FSIS April 2001	The Egg Products Inspection Act defines egg to mean the shell egg of domesticated chicken, turkey, duck, goose or guinea. Voluntary grading of shell eggs is done under USDA supervision. (FDA enforces labels/labeling of shell eggs)
Products with 3% or less raw meat; less than 2% cooked meat or other portions of the carcass; or less than 30% fat, tallow or meat extract, alone or in combination	Products containing greater than 3% raw meat; 2% or more cooked meat or other portions of the carcass; or 30% or more fat, tallow or meat extract, alone or in combination[a]	Products containing 2% or more cooked poultry; more than 10% cooked poultry skins, giblets, fat and poultry meat in any combination[a]	Egg products processing plants (egg breaking and pasteurizing operations) are under USDA jurisdiction
Products containing less than 2% cooked poultry meat; less than 10% cooked poultry skins, giblets, fat and poultry meat (limited to less than 2%) in any combination[a]	Open-face sandwiches		
Closed-face sandwiches			

FDA JURISDICTION	USDA JURISDICTION		
FDA is responsible for shell eggs and egg containing products that do not meet USDA's definition of "egg product." FDA also has jurisdiction in establishments not covered by USDA; e.g. restaurants, bakeries, cake mix plants, etc. Egg processing plants (egg washing, sorting, packing) are under FDA jurisdiction			Products that meet USDA's definition of "egg product" are under USDA jurisdiction. The definition includes dried, frozen, or liquid eggs, with or without added ingredients, but mentions many exceptions. The following products, among others, are exempted as not being egg products: freeze-dried products, imitation egg products, egg substitutes, dietary foods, dried no-bake custard mixes, egg nog mixes, acidic dressings, noodles, milk and egg dip, cake mixes, French toast, sandwiches containing eggs or egg products, and balut and other similar ethnic delicacies. Products that do not fall under the definition, such as egg substitutes and cooked products, are under FDA jurisdiction
Cheese pizza, onion and mushroom pizza, meat flavored spaghetti sauce (less than 3% red meat), meat flavored spaghetti sauce with mushrooms, (2% meat), pork and beans, sliced egg sandwich (closed-face), frozen fish dinner, rabbit stew, shrimp-flavored instant noodles, venison jerky, buffalo burgers, alligator nuggets, noodle soup chicken flavor	Pepperoni pizza, meat-lovers stuffed crust pizza, meat sauces (3% red meat or more), spaghetti sauce with meat balls, open-faced roast beef sandwich, hot dogs, corn dogs, beef/vegetable pot pie	Chicken sandwich (open face), chicken noodle soup	

Jurisdiction for products produced under the School Lunch Program, for military use, etc. is determined via the same algorithm although the purchases are made under strict specifications so that the burden of compliance falls on the contractor. Compliance Policy Guide 565.100, 567.200 and 567.300 provide additional examples of jurisdiction. IOM 3.2.1 and 2.7.1 provide more information on our interactions with USDA and Detention Authority

[a] These percentages are based on the amount of meat or poultry product used in the product at formulation

1.6.2   The Impact of the Food Safety Modernization Act (FSMA)

The Food Safety Modernization Act of 2011 is the first substantive change to US food law since 1938. Yet, FSMA only amends the Food Drug and Cosmetic Act and thus impacts the FDA alone. The USDA and its enabling Acts were not part of the legislation passed. Here again one can see the impact the two-agency paradigm creates. Now the FDA will begin to modernize its approach to food law and safety, while the USDA is under no mandate to do the same. An arbitrary decision to pass separate legislation in 1906 now means an artificial boundary exists in grocery stores and markets. This unseen boundary means some products are produced under one standard, while others are not.

FSMA makes a number of fundamental changes to the FD&C. As will be discussed in the chapters to follow FSMA's changes are sweeping in scope and depth of coverage. While it modernizes many areas it keeps intact every definition from the 1906–1938 Act. It will have no bearing on what is "food" for the FDA. The classification question, however, takes on new significance under FSMA. As the rules enabling the legislation seek to expand the reach of the FDA it will become important to determine whether FSMA applies to "food" operations.

1.6.3   The Tale of Two Agencies

The question of jurisdiction was not by design but by historical happenstance. Here again the history of food law and regulation plays a role. Congress passed both the Federal Meat Inspection Act and the Pure Food and Drug Act, the precursors to our modern statutes, on the same day in 1906. Many historians conclude the Pure Food Act passed on the coattails of the Federal Meat Inspection Act. Public outcry focused on the conditions of meat slaughter houses as reported by Upton Sinclair, President Roosevelt's investigators and others. The drive was to restore confidence in the meat market. Passed as a set the two pieces of legislation were initially sent to one agency to implement and enforce—the USDA. Not only was the USDA a natural fit since it already focused on inspecting meat for export, it was also best suited for the job.

Initially the two Acts were implemented and enforced by two branches within the USDA. The Pure Food Act was assigned to the Division of Chemistry. The Division of Chemistry was created in 1837 in the US Patent Office to conduct chemical investigations of "agricultural matters" (Symposium 1990).[17] Later moved to the USDA, the Division of Chemistry received new funding and was elevated to bureau status. Its primary role before the 1906 Pure Food Act and for many years to follow was investigating drug composition. Perhaps, it was this function which made it the best governmental body to implement and enforce the Pure Food and Drug Act. It alone would have the expertise and equipment to determine whether drugs were adulterated.

The Bureau of Animal Industries took charge of the Federal Meat Inspection Act. Congress created the BAI in 1884 for the purpose of monitoring imported animals and preventing diseased animals from being used as food (FSIS History)[18] A few years later in 1890 the BAI also assumed functions to inspect meat products intended for export. In 1891 the law was expanded to include inspections of live animals intended for export. The export functions arose out of an outcry from Europe. Many European nations refused to import US beef without assurances from the federal government that the meat was prepared in sanitary slaughterhouses (Johnson 1982).[19] The BAI was no forerunner to modern food safety. Instead it provided what many call an "illusion inspection" since its authority only extended to slaughtering and not post-slaughter production.

As the two branches implementing the Acts drifted a two-agency system was created (see Fig. 1.7). The Bureau of Chemistry became the Food Drug and Insecticide Administration in

---

[17] Hutt (1990).

[18] FSIS History, http://www.fsis.usda.gov/wps/portal/informational/aboutfsis/history (last visited June 10, 2014).

[19] Johnson (1982, pp. 5–9).

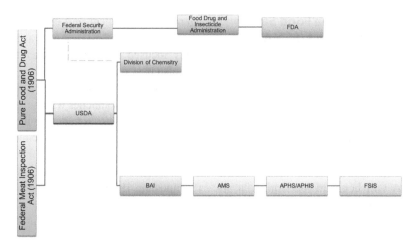

Fig. 1.7 Evolution of Modern Day FDA and FSIS from their Origins in 1906

1927. It remained as a branch within the USDA. Later in 1931 it was renamed the Food and Drug Administration. In 1938 the FDA left the USDA and was moved as a branch within the Federal Security Agency, a forerunner of the modern Department of Health and Human Services. Meanwhile the Bureau of Animal Industries, a branch within the USDA, experiences its own evolution, but always remained within the USDA.

1.7   Comparative Law—Food Agencies Around the World

Nearly every nation makes protecting the food supply a top priority. Differences in history and culture shape each country's approach to food regulation. Despite these differences food travels the world from farms and factories in one country to a wide and variable list of foreign and domestic markets. The USA, for example, imports 80% of its seafood, 50% of its fresh fruit, and 20% of its fresh vegetables (Global Engagement Report).[20] This creates an overlapping system of laws, regulations, and enforcement. Where possible this text will address the comparative elements of food law and regulation. As a start is a list of foreign agencies tasked with food safety.

Sampling of Governmental Agencies Focused on Food Safety and Enforcement	
Argentine	Secretarat of Agriculture, Livestock, Fishing and Food; National Food Safety and Quality Service
Canada	Minister of Health, Health Products and Food Branch
China	State Food and Drug Administration
India	Food Safety and Standards Authority of India
Philippines	Food and Drug Administration
South Korea	Ministry of Food and Drug Safety
European Union	European Food Safety Authority
	Committee on the Environment, Public Health and Food Safety
Germany	Federal Ministry of Food, Agriculture and Consumer Protection
Netherlands	Ministry of Economic Affairs, Agriculture and Innovation
Norway	Ministry of Agriculture and Food, Food Safety Authority
Spain	Catalan Food Safety Agency
United Kingdom	Food Standards Agency
Australia	Department of Agriculture
New Zealand	New Zealand Food Safety Authority
Japan	Administration of Food Safety

1.8   Chapter Summary

This chapter covers a lot of ground. It attempts to provide a view of the full landscape of food law and regulation. The structure and sources of

---

[20] FDA, Global Engagement Report (2013).

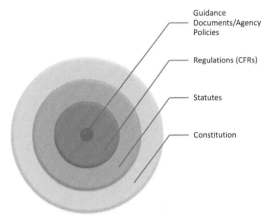

Guidance Documents/Agency Policies

Regulations (CFRs)

Statutes

Constitution

Fig. 1.8 Sources of law and limits of each source (e.g. a statutory

federal law emerge. The US government consists of three branches and a division of authority between States and the federal government. There are also three primary sources of federal law. The US Constitution provides the boundaries all statutes or regulations (CFRs) must abide within (see Fig. 1.8). In addition the chapter explored introductory topics of common food law statutes and definitions.

Overview of Key Points:

- The primary and secondary agencies regulating food safety and quality standards
- The minimal role of judicial opinions in agency enforcement
- The five components of a judicial opinion
- Three-tiered structure of federal courts
- Intended use under *Nutrilab*
- Dual classification and the "food exception"
- Definition of meat and poultry
- Jurisdictional boundaries between the two primary food agencies—FDA and FSIS
- The history of why two agencies were created
- Introduction to international food regulatory bodies

1.9 Discussion Questions

1. Outbreaks seem more frequent in the headlines, allergens more present than ever before, and our awareness of what we eat is at its apex. Does this mean food production has become more or less safe since 1906? Explain why using current headlines, regulations, or examples.
2. Identify a food with a label that could be subject to dual classification, both as a food and a drug. Explain how it would be classified as a food or drug using the decision in *Nutrilab*. Does that classification change using the decision in *American Health Products*?

References

FDA (2006) The FDA at work—cutting-edge science promoting public health, FDA Consumer magazine, The Centennial Edition/January-February, By Philip J. Hilts. http://www.fda.gov/aboutfda/whatwedo/history/overviews/ucm109801.htm. Accessed 12 March 2014

FDA, CFSAN (2014) Scope of responsibility. http://www.fda.gov/AboutFDA/CentersOffices/OfficeofFoods/CFSAN/WhatWeDo/. Accessed 4 Feb 2014

Friedman LM (1985) A history of american law, 2nd edn. Simon & Schuster, Inc, New York

Hutt PB (1990) Symposium on the history of fifty years of food regulation under the federal food, drug, and cosmetic act: a historical introduction. Food Drug Cosmet Law J 45:17

Johnson DR (1982) The history of the 1906 pure food and drugs act and the meat inspection act. Food Drug Cosmet Law J 37:5–9

Kroger C (2014) Market uneasy after cantaloupe outbreak. The Packer. http://www.thepacker.com/fruit-vegetable-news/Listeria-in-cantaloupe-horrific-but-markets-still-favorable-131237344.html?view=all Accessed 26 May 2014 (October 6, 2011)

Orin K (2007) How to read a legal opinion: a guide for new law students. Green Bag Entertain J Law 11(1):51

Federal Inspections and Enforcement

Abstract

This chapter begins with the substantive discussion of food regulation by looking at the enforcement authority of the two primary food agencies in the US. By beginning with the enforcement authority, later subjects on prohibited acts will be placed in the context of risks and consequences. Introduction of the enforcement powers also enables instructors to test students using practical scenarios from Form 483s, warning letters, and regulatory control actions. This chapter covers the full suite of enforcement actions available to the FDA and FSIS. It includes a detailed examination of Constitutional and Statutory defenses to inspections and enforcement actions.

2.1 Introduction

This chapter will explore the inspection models used by the USDA/FSIS and the FDA. Inspections are the chief mechanism for the agencies to either remove troubled products or proactively prevent their release into the stream of commerce. As will be discussed in the sections below, the two agencies follow radically different models of inspection. Comparing and contrasting the two approaches will highlight the advantages and disadvantages of both agencies' enforcement mechanisms. From this viewpoint, a better understanding of the risks of non compliance along with strategies to solving enforcement issues will emerge.

When considering the regulatory landscape, activities of the primary agencies can be classified in two ways. On the one hand are the activities needed to bring new ingredients or additives on the market and on the others, the activities conducted to ensure products are safe, and consumers are not intentionally deceived. Typically, the dichotomy is referred to as pre market approvals and post market surveillance. This text begins with enforcement and inspection, post market surveillance activities, both because it constitutes the bulk of regulatory work and it provides insight into the statutory standards for ingredient and label integrity.

2.2 FSIS Inspection Authority and Enforcement Tools

2.2.1 Statutory Authority and Early Origins of Inspection Authority

Initial authority for meat inspections came in the Federal Meat Inspection Act (FMI) of 1906. It authorized the USDA to conduct continuous inspections of all domestic meat intended for human consumption. The FMI required the USDA to inspect all animals covered under the Act brought into a plant for slaughter or a processing facility. Processing plant activities included boning whole carcasses or creating meat products like sausages or ham. Poultry inspections were not included in the 1906 FMI. Prior to World War II, poultry production in the USA remained a small farm activity with sales limited to neighbors and local markets. Poultry sales were intrastate rather than interstate, and thus outside of any federal legislative awareness or authority. It was after the 1957 Poultry Products Inspection (PPI) Act that poultry inspections were also made mandatory. FSIS conducted all meat and poultry inspections from 1957 to 1995. Beginning in May 1995, the authority to inspected processed eggs under the Egg Products Inspection Act (EPI) was transferred from the USDA's Agricultural Marketing Services (AMS) to FSIS. For the past 20 years, FSIS has operated as the sole arm of the USDA with authority to conduct egg inspections.

FSIS enabling Acts are among the most stringent. The statutes governing the safety of meat, poultry, and eggs are designed to prevent contaminated (adulterated) or mislabeled (misbranded) food from reaching the market. This in part requires FSIS to ensure all regulated foods are slaughtered and processed under sanitary conditions. FSIS enjoys unfettered continuous accesses to facilities and the power to prevent uninspected or condemned products from entering the market. Understanding the scope, process, and coverage of FSIS, inspection and enforcement authority proves crucial in managing the agency's reach.

2.2.2 Continuous Mandatory Inspection Requirement

The 1906 FMI required the continuous presence of inspectors in all establishments providing meat for interstate commerce. This edict applied to both the slaughter and processing of meat intended for domestic sale and human consumption. What constitutes "continuous," however, varies between slaughtering and processing. FSIS personnel inspect all meat and poultry animals at slaughter with at least one federal inspector per slaughter-line during all hours the plant is operating. No slaughter or dressing can occur without an inspector on-site and on the slaughter-line. The system even accounts for instances of

overtime or holiday shifts by utilizing a system allowing plants to pay a user-fee to bring an inspector on duty (CRS Meat and Poultry).[1] Inspectors at processing facilities in contrast remain on-site daily but do not require an FSIS inspector to monitor each product or process. Inspectors are on-site daily to ensure meat is processed in sanitary conditions, and regulations for ingredient levels, packaging, and labeling are followed. Processing plants are also considered under continuous inspection because of the daily visits and the presence of inspectors on-site at all times.

The 1906 crisis that sparked Congress action provides important context in evaluating the inspection model used by FSIS. In 1891, the USDA conducted limited ante- and post mortem inspections, but no inspection of processing plants. Upton Sinclair exposed the horrifying conditions in slaughter and processing plants in his book *The Jungle*. The FMI passed in 1906 largely because of the outcry from Sinclair's stories. It reflects not only a knee-jerk reaction to the crisis, but also a heavy emphasis on enforcement in order to restore public confidence.

In light of the strict prohibition against selling uninspected meat and poultry, the contours of FSIS jurisdiction become important. FSIS legal inspection responsibilities begin when animals arrive to slaughterhouses and end once products leave processing plants. The enabling acts provide the USDA and FSIS no further authority to engage in inspections of any type. This raises important questions about what happens to meat products when they leave the facility. Who monitors the shipping, storage, or preparation of these products? Such questions need not be rhetorical, but can often be the central issues of outbreak litigation.

2.2.3 Inspection Methods

Inspection with FSIS remained largely unchanged for 90 years. Meat inspection programs initially relied on organoleptic methods, namely sight,

---

[1] Congressional Research Service, "Meat and Poultry Inspection Issues" (Jean M. Rawson, 2003).

Fig. 2.1 Three marks of inspection utilized by the USDA/FSIS

touch, and smell, to determine the quality and presence of diseases. Inspectors stamp a mark of approval on each carcass and major cuts of meat passing their organoleptic inspection (see Fig. 2.1). Without the mark, the carcass cannot move on for further processing or enter the market. The purpose of this carcass-by-carcass inspection was originally aimed at reducing the potential for the transmission of diseases from sick animals to humans. This could arise either from a diseased animal brought to slaughter or via poor sanitary conditions in the slaughter and processing plants.

The processing plants experienced a similar inspection. Processing initially involved cutting and boning whole carcasses along with the production of meat products like ham or bacon. These functions were usually completed in a facility adjacent to the slaughtering facility. The focus for FSIS in processing was on the overall production line, not the individual products. The emphasis was on sanitary conditions, which would contaminate meat previously inspected and approved.

For nearly 90 years, the FSIS inspected for disease using organoleptic methods. As with the original passage of the 1906 Act only crisis compelled changes to how the USDA conducted inspections. The 1990 *E. Coli O157:H7* outbreak linked to the fast-food chain "Jack In the Box" brought a new inspection concept to FSIS. Prior to the outbreak FSIS explored ways to modernize its inspection system. The surge in establishments, the increasing range of products, and the emergence of new technologies, ingredients, and processes proved too complex for FSIS. FSIS simply was overwhelmed and the 1990 outbreak highlighted the extent of the gaps in its surveillance. In response, FSIS underwent structural changes and developed a new rule for inspectors known as "Pathogen Reduction/Hazard Analysis

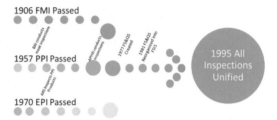

Fig. 2.2 Evolution of FSIS and unification of USDA inspections

and Critical Control Point System" (HACCP). More than two decades later, HACCP remains the industry standard for FSIS inspections.

2.2.4 Organization and Evolution of FSIS

FSIS did not start out with a clear name or mission. As discussed in Chapter 1 and shown in Fig. 1.7, FSIS began as the Bureau of Animal Industries. Inspection functions were housed in this sub agency from 1906 to 1953. President Eisenhower kicked off a lengthy series of changes; first, moving inspection functions to the Agricultural Research Service (ARS). In 1968, when poultry inspections were added, the sub agency was named Consumer and Marketing Services, a sub division within the ARS. In a short two-year-span, inspections were first moved to the Animal & Plant Health Service in 1971, renamed APHIS in 1972, and then moved to a new sub agency in 1982 called Food Safety & Quality Service (FS&QS). The final move came in 1981 when FS&QS was reorganized into FSIS (see Fig. 2.2). A great deal of upheaval for any organization, the sub agency experienced over five large organizational shuffles in less than 30 years.

Although FSIS's exercises a narrower scope of authority, it utilizes a complex organizational

Fig. 2.3 US map broken into the ten OFO regions

structure. Visiting the FSIS's website and explaining its organization, one can become easily lost. The main office to focus on is the Office of Field Operations (OFO) and Office of Investigation, Enforcement and Audit (OIEA). The OFO manages all FSIS inspections and initiates the corresponding enforcement actions. FSIS deploys approximately 8000 FSIS inspectors and staff to about 6200 meat slaughtering and/or processing plants nationwide. The OFO, like the FDA's ORA, organizes its inspectors into districts (see, Sect. 3.2 below). FSIS operates ten districts (see Fig. 2.3). Each district is overseen by a district manager (DM) and deputy district manager (DDM). Both would be involved in serious enforcement actions.

The OIEA supports the OFO by conducting both criminal violations and investigating investigating outbreaks. While OFO is focused on in-plant activities OIEA casts its attention toward in-commerce products. In particular, it investigates criminal violations and instances of intentional contamination. It will also play a key role in investigating foodborne illness outbreaks. Since OFO is limited to inspecting domestic facilities OIEA also verifies imported meat, poultry, and egg products meet applicable standards. This is a small sampling of the primary activities charged to the OIEA.

2.2.5 Enforcement Toolkit

Overview of Types of FSIS Enforcement Actions

FSIS enforcement options can be divided into three groups. The three groups or classes of enforcement actions as defined in the regulations are: regulatory control action, withholding action, and suspension. Each is defined in 9 CFR 500.1 as provided below. As can be seen in the excerpt of 500.1 below the enforcement actions escalate in severity. There is a final enforcement action, which is irreversible and in many ways the culmination of FSIS enforcement. That is the Withdrawal of Inspection under 9 CFR 500.6. Once inspectors are withdrawn from a facility, the facility cannot operate or re-open.

---

9 CFR 500.1

  a. A "regulatory control action" is the retention of product, rejection of equipment or facilities, slowing or stopping of lines, or refusal to allow the processing of specifically identified product.

b. A "withholding action" is the refusal to allow the marks of inspection to be applied to products. A withholding action may affect all product in the establishment or product produced by a particular process.

c. A "suspension" is an interruption in the assignment of program employees to all or part of an establishment.

Regulatory Control Actions

Regulatory controls actions are the most commonly used by FSIS inspectors. Regulatory control actions function as a low-level enforcement action that allows inspectors to correct an issue before a product leaves the facility or an equipment is reused. The violations are minor and the enforcement action is taken immediately. There are four scenarios provided in 9 CFR 500.2 when a regulatory control action may be taken. The four scenarios are provided below.

---

9 CFR 500.2(a)(1)-(4)

1. Insanitary conditions or practices;
2. Product adulteration or misbranding;
3. Conditions that preclude FSIS from determining that product is not adulterated or misbranded; or
4. Inhumane handling or slaughtering of livestock.

The focus for regulatory control actions centers on preventing non compliant products from leaving the facility. This includes potential contamination or adulteration as well as misbranding. The USDA FSIS Rules of Practice Regulation (RPR) provides examples of each of the four conditions (Rules of Practice).[2] The first three focus

on ensuring that the products are safe and wholesome and the facility's compliance program fully functioning. For example, ensuring equipment is clean (500.2(a)(1)), water does not collect on or around meat (500.2(a)(2), or the facility is well lit in order to allow inspectors to assess products and processes (500.2(a)(3)).

Regulatory control actions only result in a temporary delay. Typically product is retained and potentially reinspected. In other cases, equipment or facilities are closed until cleaned or repaired. In most instances, slaughter or processing lines are slowed or stopped temporarily.

The final basis for a regulatory control action finds its roots in a second enabling act. Congress originally passed the Humane Methods of Slaughter Act in 1958 (7 U.S.C. 1901 *et. sEq.*)[3] The HMSA was updated in 1978 and provided the USDA FSIS authority to stop a slaughtering line until the abuses were corrected. The HMSA and 500.2(a)(4) do not apply to the slaughter of chickens or other poultry, only to livestock such as sheep, pigs, or cattle. The USDA/FSIS have issued a number of regulations, directives, and guidance to industry on how to humanely slaughter and handle livestock (9 C.F.R 313; FSIS Compliance Guide).[4] A word of caution, if a reader is new to FSIS inspections, then be aware of enforcement reports and regulations in this area, in particular on inhumane handling and slaughter, can often be unsettling and graphic.

Although regulatory control actions are immediate, the facility still must be notified. The RPR makes clear the notification, typically via a Noncompliance Record (NR), may be provided to the facility after the action is taken. This allows the hazard to be contained and the facility notified of a potential gap in its compliance program. In some cases, the facility may seek an appeal of the enforcement action. This appeal is taken to the next level of FSIS supervision.

---

[2] USDA FSIS Rules of Practice: Inspection Methods (June 23, 2013).

[3] 7 U.S.C. 1901 *et seq.*

[4] See e.g., 9 CFR 313; FSIS Compliance Guide for a Systematic Approach to the Humane Handling of Livestock—to support the Humane Methods of Slaughter Act (2013).

Withholding Action or Suspension Without Notification

The remaining two categories of enforcement action can occur under two scenarios. Withholding actions refer to withholding the marks of inspection, which every product requires to enter the market legally. Suspension of inspection activities as the name suggests involves suspending inspectors and effectively stopping all production. Suspension differs from the most severe enforcement action—Withdrawal of Inspection. The Withdrwal of Inspection terminates FSIS inspections permanently and shutters the facility. Suspensions and withholding actions are similar, but a suspension will be in effect for longer than a withholding action.

There are certain violations FSIS deems as requiring enforcement action in these two categories immediately and without any prior notification to the facility. Subsection 500.3 provides four triggers for a withholding or suspension action without prior notice. Those are provided below.

---

9 CFR 500.3(a)(1)-(4)

a. FSIS may take a withholding action or impose a suspension without providing the establishment prior notification because:

1. The establishment produced and shipped adulterated or misbranded product as defined in 21 U.S.C. 453 or 21 U.S.C. 601;

2. The establishment does not have a HACCP plan as specified in Sec. 417.2 of this chapter;

3. The establishment does not have Sanitation Standard Operating Procedures as specified in Secs. 416.11–416.12 of this chapter;

4. Sanitary conditions are such that products in the establishment are or would be rendered adulterated;

5. The establishment violated the terms of a regulatory control action;

6. An establishment operator, officer, employee, or agent assaulted, threatened to assault, intimidated, or interfered with an FSIS employee; or

7. The establishment did not destroy a condemned meat or poultry carcass, or part or product thereof, in accordance with part 314 or part 381, subpart L, of this chapter within 3 days of notification.

The most common basis for withholding or suspension actions involves a serious and imminent threat to public health. Protecting the public health provides the primary rationale for taking a significant enforcement step without notification. In the RPR, FSIS directs inspectors to document the imminent threat when taking action under 500.3(a). Inspectors are also required to notify the facility orally and in writing "as promptly as the circumstances permit..." (Rules of Practice).[5]

The decision to take withholding and suspension actions come from higher levels of authority. The decision to take a withholding action originates with inspectors in the plant, but must be made by the inspector in charge (ICC) or the frontline supervisor. In some cases, the decision is made by the district office. Suspension decisions on the other hand may only be made by the district office.

There are other grounds for taking enforcement action without prior notification that lack an urgency to protect public health. For instance, if any regulatory control action is not corrected or is repeated, then FSIS may take a withholding or suspension action without notification. In a sense, the facility already received notification through the regulatory control action and the regulations.

It is important to highlight instances where notification may be withheld that do not directly relate to food safety. Namely, the ability to withhold notification where FSIS personnel are

---

[5] *Id*. US FSIS Rules of Practice at 6.

confronted and possibly assaulted. Needless to say, it can be a contentious environment operating a facility with constant regulatory supervision. FSIS relies on the ability to work continually and freely in a facility. If an environment is created where inspectors do not feel comfortable to perform their duties, then enforcement action without notification works to restore the trust between FSIS and the host facility.

Withholding Action or Suspension with Prior Notification

If there is no immediate threat to public health then withholding or suspension actions require notification. Subsection 500.4 provides the criteria for withholding or suspension actions that require notification. Those are provided below. Prior to withholding the marks FSIS must provide written notice it intends to either withhold the marks of inspection or suspend inspections.

> ### 9 CFR 500.4(a)-(e)
> FSIS may take a withholding action or impose a suspension after an establishment is provided prior notification and the opportunity to demonstrate or achieve compliance because:
> a. The HACCP system is inadequate, as specified in § 417.6 of this chapter, due to multiple or recurring noncompliances;
> b. The Sanitation Standard Operating Procedures have not been properly implemented or maintained as specified in §§ 416.13 through 416.16 of this chapter;
> c. The establishment has not maintained sanitary conditions as prescribed in §§ 416.2–416.8 of this chapter due to multiple or recurring noncompliances;
> d. The establishment did not collect and analyze samples for Escherichia coli Biotype I and

> record results in accordance with § 310.25(a) or § 381.94(a) of this chapter;
> e. The establishment did not meet the Salmonella performance standard requirements prescribed in § 310.25(b) or § 381.94(b) of this chapter.

Enforcement actions taken under 500.4 involve notification largely because it involves repeated non compliance. Unlike 500.3 where there is no HACCP or standard operating procedures (SOPs), 500.4 involve deficiencies in the compliance program. If these were one-off errors in the program, then they would most likely be caught in a regulatory control action. Section 500.4 instead aims for the gaps in the compliance program that result from inadequate procedures or processes. As such, the RPR directs inspectors to compile "extensive information" to provide both a factual basis for the facility to analyze and challenge and to demonstrate a pattern or history of failed corrective or preventative actions (Rules of Practice).[6] Once presented with the notification and supporting evidence a facility is given an opportunity to respond by identifying areas of disagreement or share an interpretation of the regulations. This is in many ways similar to the Form 483 used by the FDA, which will be discussed in the Section 3.4 below.

Withdrawal of Inspection

Withdrawal of FSIS inspectors represents the pinnacle of the agency's enforcement powers. There are several bases for withdrawing inspectors, which includes all of the previous actions that lead to withholding or suspension actions.

> ### 9 CFR 500.6
> The FSIS Administrator may file a complaint to withdraw a grant of Federal inspection in accordance with the Uniform

---

[6] Id. US FSIS Rules of Practice at 7.

Rules of Practice, 7 CFR subtitle A, part 1, subpart H because:

a. An establishment produced and shipped adulterated product;

b. An establishment did not have or maintain a HACCP plan in accordance with part 417 of this chapter;

c. An establishment did not have or maintain Sanitation Standard Operating Procedures in accordance with part 416 of this chapter;

d. An establishment did not maintain sanitary conditions;

e. An establishment did not collect and analyze samples for Escherichia coli Biotype I and record results as prescribed in § 310.25(a) or § 381.94(a) of this chapter;

f. An establishment did not comply with the Salmonella performance standard requirements as prescribed in §§ 310.25(b) and 381.94(b) of this chapter;

g. An establishment did not slaughter or handle livestock humanely;

h. An establishment operator, officer, employee, or agent assaulted, threatened to assault, intimidated, or interfered with an FSIS program employee; or

i. A recipient of inspection or anyone responsibly connected to the recipient is unfit to engage in any business requiring inspection as specified in section 401 of the FMIA or section 18(a) of the PPIA.

The slaughter and processing of meat is uniquely viewed as a privilege not a right. Unlike other food facilities, FSIS regulated facilities must apply for a grant of inspection. Think of it as applying for a driver's license. And like a driver's license can be revoked, so can a grant of inspection. Abuse the privilege, lose the privilege. The process to take revoke the grant of inspection can be lengthy. It not only involves a documented history of non compliance, but a hearing before an Administrative Law Judge (ALJ). A hearing preserves due process (see, Sect. 2.2.6 below). This process, and truly the entire grant of inspection feature, is unique to FSIS. The FDA while requiring a facility register prior to beginning operations is unable to bar a facility from beginning operations like FSIS can. This level of control requires careful checks to ensure the privilege is properly revoked.

2.2.6 Lessons from FSIS's History— Food Law Is a Floor Not a Ceiling

There are a number of lessons to take away from the history of FSIS. In particular, note the slow pace of change in the inspection methods. Despite rapid changes in technology, demand, and the range of products, FSIS clung to outdated inspection criteria. There were rumblings of change prior to adopting HACCP, but it still took crisis to create change. If compliance is seen as a floor, then this can lead a facility into a false sense of security. As facilities struggle with the balancing marketability and compliance, it is important to look at the headlines. No facility wants to be associated with the crisis that leads to new rules or regulations.

Case Study: New Poultry Inspections Rule; First Change in Over 50 Years

An excellent example of the pace of change comes in a recent rule change announced by the USDA/FSIS. In the summer of 2014, the USDA/FSIS announced a new rule to poultry inspection. The new rule replaced the inspection model used when the PPI was adopted in 1957. For nearly 60 years, FSIS did not require facilities to test for *Salmonella* and *Campylobacter*. Under the new rule, known as the New Poultry Inspection System (NPIS), facilities will be required to take preventative measure against *Salmonella* and *Campylobacter* contamination (NPIS Final Rule).[7] This will include mandatory testing at two points in the production process. The NPIS will leave unchanged the maximum line speeds, which are currently 140 birds per minute. Although the speed sounds dizzying, pilot programs set maximum line speeds of 175 birds per minute.

Startling to consider how much the scientific knowledge of these two pathogens changed over 60 years or to think how the industry developed in that timeframe. Yet rather than elect for incremental changes the current model of regulation waits and issues sweeping regulations in an attempt to catch-up.

The history of FSIS also provides insight into political priorities. As governmental agencies, the USDA and FDA are only as effective as properly funded. Shifting an agency around, renaming and altering responsibilities does not speak to a well-regarded agency. The numbers support the notion. Not only are meat and poultry inspections deemed suitable for shuffling, but also for basic funding. Numerous studies of budget appropriations conclude the meat inspection budget either remains stagnant or contracts even in the face of a swelling mandate. Simply, consumers are demanding more meat and poultry, but the budget appropriations to ensure the safety of that meat and poultry is not gaining approval.

If the goal of an agency is consumer confidence than the agency name matters. When one hears the name Food Safety and Inspection Service a clear mandate emerges without knowing anything else about the agency. A name like Agricultural Research Service or Animal and Plant Health Service signals little about what the agency does. Would one really expect an agency named Agricultural Research Service to ensure the safety of meat products? The US Federal government is vast and names matter. Names provide clarity to the public about an agency's mission and primary functions.

2.3 Overview of FDA Inspection Process and Enforcement Tools

2.3.1 Evolution of Inspection Authority

The FDA conducts warrantless inspections of the premises of regulated industries. The inspections may be "for cause" such as an inspection during a recall or adverse event or simply "surveillance" inspections as required by the Act. In either case, the FDA inspectors arrive unannounced and request total access to a facility for a period of four or more days. The FDA did not always enjoy the authority to inspect facilities. The 1906 Act, at only five pages, made no explicit reference to an ability to inspect facilities. An agent could arrive at a facility and request entry. If refused the FDA would need to go to court and obtain a warrant. With the 1938 Act Congress worked to patch-up this oversight. The 1938 Act created Section 704 to authorize the FDA to inspect facilities and added refusal to consent to inspect to the list of prohibited acts in Section 331. As a prohibited Act, a refusal became a crime, typically a misdemeanor (see, Chap. 7).

The Supreme Court struck down the penalty in the 1938 Act. The court determined the provision in Section 704 which allowed consent, but penalized for withdrawing consent, as too vague to be enforceable.

UNITED STATES v. CARDIFF, 344 U.S. 174 (1952)
MR. JUSTICE DOUGLAS delivered the opinion of the Court.

Respondent was convicted of violating 301 (f) of the Federal Food, Drug, and Cosmetic Act, 52 Stat. 1040, 21 U.S.C. 331 (f). That section prohibits "The refusal to permit entry or inspection as authorized by

[7] USDA/FSIS, Modernization of Poultry Slaughter Inspection, Docket No. FSIS-2011-0012 (Final Rule 2014).

section 704." Section 704 authorizes the federal officers or employees "after first making request and obtaining permission... of the owner, operator, or custodian" of the plant or factory "to enter" and "to inspect" the establishment, equipment, materials and the like "at reasonable times."

Respondent is president of a corporation which processes apples at Yakima, Washington, for shipment in interstate commerce. Authorized agents applied to respondent for permission to enter and inspect his factory at reasonable hours. He refused permission, and it was that refusal which was the basis of the information filed against him and under which he was convicted and fined... The Court of Appeals reversed, holding that 301 (f), when read with 704, prohibits a refusal to permit entry and inspection only if such permission has previously been granted.

The Department of Justice urges us to read 301 (f) as prohibiting a refusal to permit entry or inspection at any reasonable time. It argues that that construction is needed if the Act is to have real sanctions and if the benign purposes of the Act are to be realized. It points out that factory inspection has become the primary investigative device for enforcement of this law, that it is from factory inspections that about 80 % of the violations are discovered, that the small force of inspectors makes factory inspection, rather than random sampling... of finished goods, the only effective method of enforcing the Act.

All that the Department says may be true. But it does not enable us to make sense out of the statute. Nowhere does the Act say that a factory manager must allow entry and inspection at a reasonable hour. Section 704 makes entry and inspection conditioned on "making request and obtaining permission." It is that entry and inspection which 301 (f) backs with a sanction. It would seem therefore on the face of the statute that the Act prohibits the refusal

to permit inspection only if permission has been previously granted. Under that view the Act makes illegal the revocation of permission once given, not the failure to give permission. But that view would breed a host of problems. Would revocation of permission once given carry the criminal penalty no matter how long ago it was granted and no matter if it had no relation to the inspection demanded? Or must the permission granted and revoked relate to the demand for inspection on which the prosecution is based? Those uncertainties make that construction pregnant with danger for the regulated business.

The alternative construction pressed on us is equally treacherous because it gives conflicting commands. It makes inspection dependent on consent and makes refusal to allow inspection a crime. However we read 301 (f) we think it is not fair warning... to the factory manager that if he fails to give consent, he is a criminal. The vice of vagueness in criminal statutes is the treachery they conceal either in determining what persons are included or what acts are prohibited. Words which are vague and fluid...may be as much of a trap for the innocent as the ancient laws of Caligula. We cannot sanction taking a man by the heels for refusing to grant the permission which this Act on its face apparently gave him the right to withhold. That would be making an act criminal without fair and effective notice...

This Supreme Court opinion led Congress to pass the Factory Inspection Amendment in 1953. The Factory Inspection Amendment remains a critical provision some 60 years later. The Amendment introduced several new concepts. First, it removed the consent requirement from the FDA's inspection authority. Instead, it required the FDA to present credentials and a written notice of inspection. This written notice is known as Form 482. The procedure created, and still used today, allows the FDA entry to any facility during a reasonable time simply by showing their credentials, typically an FDA badge, and the Form 482. The criminal penalties were retained under the Amendment. It remains a criminal penalty to refuse any FDA agent making a request for entry under a Form 482. Second, the Amendment added a report of significant violations observed during the inspection. This is known as Form 483 "Inspectional Observations."

The Form 483 stands as the most visible and widely recognized aspect of FDA investigations. It states the investigator's opinion on any significant violations found in the facility during their visit. The Form 483 is reviewed by a compliance officer who makes a determination whether further enforcement actions, like a warning letter, are warranted. Prior to the Amendment following a facility inspection, owners were given no summary of violations which could foreshadow potential enforcement actions. This not only blindsided facilities when enforcement actions were taken many months later, but also risked misbranded or adulterated products leaving the facility.

Investigator or Inspector?

The terms are often used interchangeably when referring to the US FDA agent conducting a civil investigation. In either case, it will refer to the field personnel conducting the facility inspection. Within the FDA, however, there is a meaningful distinction between an "investigator" and an "inspector." The FDA assigns an inspector to inspect "technologically non complex" facilities like bakeries and candy manufacturers and to collect samples or investigate routine consumer complaints (*DeBell & Chesney*).[8] An investigator can complete those same tasks, but carries more training, a technical background, or advanced degree, which allows them to inspect more complex facilities or complaints like

dietary supplement manufacturing. A facility may be visited by either type depending not only on the facility, but also the overall availability of personnel in the field office.

2.3.2 Where We Find the FDA

Unlike FSIS, the FDA is not in the facility on a daily basis. This raises the question, where do we find the FDA? The answer depends on the circumstances. Typically regulatory matters can be divided into two camps. The first camp, commonly called "pre market," relates to the approval process for new ingredients and additives for example. It is important to note straightaway the FDA adopts the stance that it is not a consultative agency. Pre market submissions will be a discussion in a later chapter (see, Chap. 6). For now, be cognizant of the limitation to approach the FDA freely with questions and/or advice. The second camp, referred to as "post market surveillance," encompasses all of the enforcement activities conducted by the agency. This includes inspections at foreign and domestic facilities, inspections of imports, and criminal matters.

The locus of pre market submissions is CFSAN. CFSAN is one of the six product-oriented centers that comprise the FDA. The other five are: CVM, Center for Drug Evaluation and Research (CDER), Center for Biologics Evaluation and Research (CBER), Center for Device and Radiological Health (CDRH), and Center for Tobacco Products (CBT). This organization intuitively informs industry and consumers the personnel within the agency focused on their regulated product. The bulk of interactions with CFSAN occur at the FDA's campus in College Park, Maryland. CFSAN also operates research facilities in Laurel, Maryland, Bedford, IL, and in Dauphin Island, Alabama, but it is uncommon to interact with those facilities (About CFSAN).[9]

---

[8] The FDA Inspections Process (Lee E. DeBell and David L. Chesney 1982).

[9] About the Center for Food Safety and Applied Nutrition (FDA) available at http://www.fda.gov/aboutfda/centersoffices/officeoffoods/cfsan/default.htm (last visited June 18, 2014).

Fig. 2.4 US map broken into the five FDA regions

The FDA relies on its Office of Regulatory Affairs (ORA) to manage post market surveillance activities. The ORA operates all FDA field activities, including facility inspections, foreign and domestic, as wells reviews of imported products. It works for all six centers, not just CFSAN. Thus, it conducts investigations of drug, medical device, cosmetic, biologics, veterinary products, and food, beverage, and dietary supplements. The result of limited budgets and personnel can lead to an investigator conducting inspections in five or six areas. It is not uncommon for such an overlap to lead to errors. The FDA organizes all of its functions, including the ORA, into regions (see Fig. 2.4). The five regions are further divided into 16 districts. Within each district there is both a district office and several field offices. Overall, there are approximately 1900 FDA offices throughout the US.

2.3.3   FDA Criminal Division

In addition to routine civil investigations the FDA also conducts criminal investigations. This function is conducted by the Office of Criminal Investigations (OCI). Its scope of authority includes all FDA-regulated products, not just food, beverages, and dietary supplements. Criminal investigations are uncommon with civil enforcement actions occurring the most frequently. OCI concentrates its resources to four primary areas:

1. Unapproved, counterfeit, and substandard medical devices and drugs;

2. Escalation of civil enforcement actions where "the normal regulatory process has been unable to remedy the problem…";

3. Violations that post a significant risk to public health and the only remedy is criminal enforcement; and

4. Impeding the FDA from properly performing its regulatory functions, such as providing "false statements to the FDA during the regulatory process and obstruction of justice" (OCI Mission).[10]

In terms of food law and regulation OCI becomes involved when repeated civil enforcement actions fail to correct a violation. For example, following repeated investigations and warning letters from ORA the case will be transferred to OCI. Sans any significant risk to public health, this process usually follows several warning letters over a period of three or more years. OCI will then conduct its own investigations, including utilizing under-cover agents. Criminal actions will be discussed in detail in Chapter 7.

2.3.4   The Role of Form 483

The Factory Inspection Amendment gave birth to Form 483. The same Amendment which gave the FDA authority to inspect and the written notice of inspection known as Form 482 also required Form 483. Section 374(b) states:

> Upon completion of any such inspection … and prior to leaving the premises, the officer or employee making the inspection shall give to the owner, operator, or agent in charge a report in writing setting forth any conditions or practices observed by him which, in his judgment, indicate that any food, drug, device, tobacco product, or cosmetic in such establishment (1) consists in whole or in part of any filthy, putrid, or decomposed substance, or (2) has been prepared, packed, or held under insanitary conditions whereby it may have become contaminated with filth, or whereby it may have been rendered injurious to health. A copy of such report shall be sent promptly to the Secretary.

---

[10] OCI Mission (FDA) available at: http://www.fda.gov/ICECI/CriminalInvestigations/ucm123027.htm (last visited June 18, 2014).

Form 483 at its simplest is a notice of potential violations found during an inspection. Form 483 is issued at the conclusion of a facility inspection by the inspector conducting the audit. It contains their opinion on significant violations, but is not reviewed by a compliance office, district director or any other FDA officer prior to being issued. Section 374(b) requires the reporting of adulteration observations, but the more common Good Manufacturing Practices (GMPs) violations are listed by FDA policy. The IOM provides further guidance to the investigator on what observations to cite. In Section 5.2.3 the IOM directs investigators to adhere to two general principles.

IOM 5.2.3   General Principles of Reporting Observations

Observations which are listed should be significant and correlate to regulated products or processes being inspected.

Observations of questionable significance should not be listed on the FDA-483, but will be discussed with the firm's management so that they understand how uncorrected problems could become a violation. This discussion will be detailed in the EIR.

There are some observations considered violations that are not reported in Form 483. These violations could serve as the basis for a warning letter or other enforcement action. The policy holds that the violations should not be reported because they require further review before citing. The IOM provides a list of these non-reportable violations in Section 5.2.3.3.

IOM 5.2.3.3   Non-Reportable Observations
  1. Label and labeling content with some exceptions.
  2. Promotional materials.
  3. The classification of a cosmetic or device as a drug.
  4. The classification of a drug as a new drug.

  5. Non-conformance with the New Drug Regulations, 21 CFR 312.1 (New Drugs for Investigational Use in Human Beings: Exemptions from Section 505(a)) unless instructed by the particular program or assignment.
  6. The lack of registration required by Section 415 and 510 of the FD&C Act. The lack of registration per 21 CFR 1271 Subpart B Procedures for Registration and Listing, promulgated under Section 361 of the PHS Act.
  7. Patient names, donor names, etc. If such identification is necessary, use initials, code numbers, record numbers, etc.
  8. Corrective actions. Specific actions taken by the firm in response to observations noted on the FDA 483 or during the inspection are not listed on the FDA 483, but are reported in the EIR. Except as described in IOM 5.2.3.4.
  9. The use of an unsafe food additive or color additive in a food product.

Form 483 is used by an investigator to write an Establishment Inspection Report (EIR). Together the EIR and Form 483 inform a compliance officer or district director of potential violations. The EIR will experience substantial review to determine whether a warning letter or other enforcement action is required. As a matter of policy a company may submit a response to the FDA within 15 days of receiving Form 483. A response is not required, but if not given the EIR and Form 483 will only provide the FDA compliance officer one side of the story. An appropriate response can avoid or mitigate further enforcement action while a fumbled response can exacerbate or hasten additional enforcement activity.

As the formal name of Form 483 suggests its application is limited to facility inspections.

It will be used for any facility inspection, foreign or domestic, for-cause or surveillance, but not in any other situation. It is not utilized, for example, in the case of import refusals or detentions. Imports will involve a separate set of forms and notices. An import refusal may only trigger an inspection, which could lead to a Form 483. This attenuated example is as close as a Form 483 comes to imports.

The primary purpose of adding Form 483 to the Factory Inspection Amendment was notification of potential errors. Prior to the Amendment an inspection would occur with the FDA leaving no indication of the inspection findings. This left formal enforcement as the only means for facilities to engage in a discussion about minor violations that could be corrected quickly, potential errors made by the investigator, or adopt a proactive stance to get ahead of major violations. Formal enforcement actions are time and cost intensive along with being incredibly public. Formal enforcement actions can also be slow to initiate and create a lengthy gap between when the violation was observed and when corrective actions are sought. In short Form 483 stepped in to fill an important policy shortcoming.

How to Respond to a Form 483

As a pre warning of enforcement action Form 483 offers facilities an opportunity to avoid more burdensome enforcement actions. Here are some tips in responding to a Form 483:

- **Avoid the Three Stooges response**: Rather than firing off a response before reviewing the regulations, observations, and corrective actions be calm, clear, cogent addressing each observation separately;
- **Avoid the Sue Sylvester anger-first response**: An inspection can feel deeply personal, like an attack, especially for small and medium-sized facilities. Drop the anger-first response. Verbally

assaulting the FDA, the investigator, or jumping to broad illegalities will not resolve Form 483 and likely serve to make the enforcement matter worse.

- **Avoid the Urkel "Did I Do That?" Response**: Some facility owners adopt the idea that playing dumb will garner pity from the agency and make the whole matter disappear. This typically involves a response with a series of "Did I Do That?" and statements disclaiming any knowledge of the regulations. Instead of pity the FDA quickly becomes concerned by a facility that appears to lack control of its operations. Show command of the issues, the facility, and the potential crisis at hand. Confidence rather than ignorance will grease the wheels in quickly resolving a Form 483.

In many instances Form 483 serves as the precursor to additional FDA enforcement action. Although not required by the regulations, it often works in tandem with a warning letter. The common escalation of enforcement actions, for facility inspections, at least begins with Form 483 followed by a warning letter and potentially additional formal enforcement actions (see Fig. 2.5). A Form 483 will escalate to a warning letter where the observations are too serious to be resolved at the Form 483 stage or the response indicates further FDA involvement is required.

2.3.5 Warning Letters as Enforcement Instruments

Warning letters are unofficially the first enforcement tool used by the FDA. This is not necessarily how the FDA views warning letters. Under FDA policy warning letters are merely considered "informal enforcement actions" simply intended

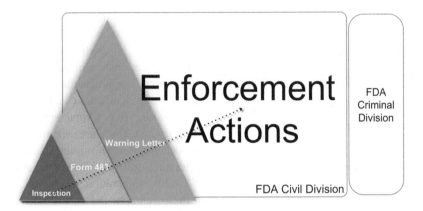

Fig. 2.5 Inspection progressing to a Form 483, warning letter and its relationship with the FDA criminal division

to provide a recommendation on the steps need to achieve compliance. These informal recommendations elicit real changes that impact facilities in much the same way as formal enforcement actions. Yet, for the agency the warning letters simply function as "prior notice."

Prior notice serves both as notice and an ultimatum. The Regulatory Procedures Manual (RPM) describes prior notice in Chapter 10. There the agency states the practicality of prior notice. Practical because it compels or induces voluntary compliance and acts a place-holder or evidence of violative behavior all without the time or cost of formal enforcement actions. Prior notice is not a statutory requirement but an administrative agency's own policy. The FDA can and does take formal enforcement action without warning letters.

Prior notice is not appropriate in all circumstances. In some cases immediate formal enforcement action must be taken. In others an accumulation of warning letters or Form 483s serve as enough notice. Outside of the mundane ordinary cases prior notice does not work well. Where there is evidence of criminal or intentional violations, for example, the agency will take formal action first. Likewise where there is a reasonable and real threat to the public, the agency pounces rather than provide prior notice. The RPM in subchapter 4-1-1 provides other examples on when a warning letter may be inappropriate.

RPM 4.1.1   Examples Where Enforcement Act May Precede a Warning Letter

Examples of situations where the agency will take enforcement action without necessarily issuing a warning letter include:

- The violation reflects a history of repeated or continual conduct of a similar or substantially similar nature during which time the individual and/or firm has been notified of a similar or substantially similar violation;
- The violation is intentional or flagrant;
- The violation presents a reasonable possibility of injury or death;
- The violations…are intentional and willful acts that once having occurred cannot be retracted. Also, such a felony violation does not require prior notice. Therefore, Title 18 U.S.C. 1001 violations are not suitable for inclusion in warning letters; and,
- When adequate notice has been given by other means and the violations have not been corrected, or are continuing. See Chapter 10, Prior Notice, for other methods of establishing prior notice.

Fig. 2.6 Five formal FDA enforcement actions

Warning letters may also flow concurrently with formal enforcement actions. Warning letters are not a statutory creature, but one created by the agency. Thus, it is entirely discretionary and should not be viewed as a prerequisite to formal enforcement action. Neither does it commit the agency to act. In some scenarios provided in the RPM the FDA holds formal enforcement actions will occur prior to or concurrently with warning letters.

RPM 4.1.1 Examples Where Enforcement Acts are Concurrent with Warning Letters
In certain situations, the agency may also take other actions as an alternative to, or concurrently with, the issuance of a Warning Letter. For example:
- The product is adulterated under Section 402(a)(3) or 402(a)(4) of the Act;
- There is a violation of CGMP;
- The product contains illegal pesticide residues; or
- The product shows short contents, subpotency, or superpotency.

2.3.6 Other Enforcement Mechanisms

Depending on the risks and the facility involved, the FDA can select from a wide variety of enforcement options. All of the options focus on protecting the public from hazardous or fraudulent products. Some options offer an ability to remove products from the market or in limited cases stop the products from entering the market. Other options are both punitive and protective. All of the options are meaningful. They are not paper tigers. This is also the first place FSMA begins to make significant changes.

There are five formal enforcement actions (see Fig. 2.6). Those are seizures and administrative detentions, recalls, import refusals and alerts, restraining order or injunctions, and suspension of facility registration. Each will be discussed in the paragraphs below and will describe how FSMA enhanced the enforcement action.

Seizure and Administrative Detentions
The first formal enforcement power requires court approval. The FDA will only seize product when its safety is in question. This can be done in issues of undeclared allergens, a form of misbranding, or contamination. The FDA cannot seize products on its own but must obtain a warrant from a Federal district court. When issued the warrant will direct US Marshals, not FDA agents, to take possession or seize the food items identified in the warrant. The FDA will be involved indirectly in the actual seizure of the product.

Only a few options can be taken by the owner of seized product. In some cases the owner washes their hands of the situation and takes no action to resolve the seizure. In this case the judge will issue a default order directing the product to be destroyed. Alternatively, an owner could review the seizure and challenge the action as invalid on the basis of constitutional or statutory violations. Common challenges would involve questions about the validity of the warrant. Confronted with no basis to overturn the seizure an owner could enter into negotiations with the FDA on how to recondition the product. If reconditioning is available, such as relabeling product to properly declare an allergen, the FDA will enter a Consent Decree outlining the settlement. Those three options are the only paths to resolving a seizure.

An administrative detention is similar to a seizure but less severe (see Fig. 2.7). Unlike a seizure that acts to permanently bar the product from the market an administrative detention acts

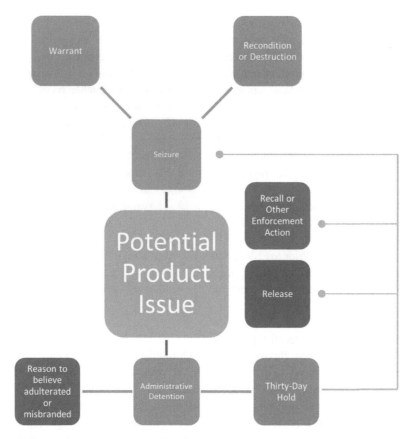

Fig. 2.7   Comparing administrative detention and seizures

as a temporary hold. The hold allows the FDA to assess the product and determine if violations exist that would trigger more severe enforcement actions like a seizure or a recall. An administrative detention does not require a court order. Thus, it makes it an attractive alternative to a formal seizure. The temporary hold under the regulations cannot exceed 30 days. During this window, the administrative detention acts like a seizure barring the product from leaving FDA control. The the FDA assessment tools used at this time include product sampling and facility inspections.

FSMA dramatically impacted the administrative detention regulations. Prior to FSMA the statute required "credible evidence or information" of serious health hazard. This limited the administrative detention tool to scenarios of repeat violators or glaring violations. FSMA amended the statutory standard for administrative

detention. Now FDA agents only need a "reason to believe food is adulterated or misbranded" (FSMA 2011).[11] Gone is the objective stringent standard. Now any notion of adulteration or misbranding can lead to a 30-day hold by the FDA.

Recalls

An FDA recall is perhaps the most widely publicized and therefore best known enforcement action. As a news media favorite, readers likely come to this text with some knowledge about recalls. The basic function at least is well-known – a recall serves to remove the product from the market. In most cases the product removed is involved in some type of outbreak of foodborne illness. A review of the popular press revels the most widely covered recalls are those linked to

---

[11] Bioterrorism Act of 2002, Section 303; Section 207 FSMA 2011.

outbreaks. Recalls, however, are not indelibly linked to foodborne illnesses. There are a wide variety of scenarios that warrant recalls. Recalls also represent the pinnacle of product removal. Other less public options are available, such as a stock recovery. Recalls are more nuanced characters than the news media may portray.

In most cases a recall is a voluntary action. A facility must act to remove a potentially harmful product from the market not only for regulatory compliance reasons but also to protect its consumers in order to protect its brand. For much of the FDA's history voluntary recalls were not only a facility's first impulse, but also the agency's only option for a timely recall. Prior to FSMA if a firm was reluctant to conduct a recall the agency would need a court order to compel a recall. In most instances time is of the essence and going to court could squander precious time. Now under FSMA the FDA can compel a recall without a court order. FSMA added section 206 to the Act which provided the FDA authority to order a recall under certain circumstances after first giving a firm the opportunity to initiate a voluntary recall (FD&C).[12] A mandatory recall is now in itself a threat to coerce action.

Recalls are classified based on the risks associated with the product issue. Chapter 7 of the IOM outlines the criteria for classifying a recall as Classes I, II, or III. Class I recalls are those seen often in the news media where food is contaminated with a foodborne illness like *Salmonella* or botulinum toxin. It also includes instances where food enters the market with undeclared allergens. Class III recalls by comparison involve minor container defects or imported products lacking an English label.

IOM 7.1.1   Recall Classification Criteria
7.1.1.2—Recall Classification

Recall Classification is the numerical designation, i.e., I, II, or III, assigned by the FDA to a particular product recall to indicate the relative degree of health hazard presented by the product being recalled.

Class I Recall is a situation in which there is a reasonable probability that the use of, or exposure to, a violative product will cause serious adverse health consequences or death.

Class II Recall is a situation in which use of, or exposure to, a violative product may cause temporary or medically reversible adverse health consequences or where the probability of serious adverse health consequences is remote.

Class III Recall is a situation in which use of, or exposure to, a violative product is not likely to cause adverse health consequences.

In limited circumstances a recall may be replaced by a stock recovery or market withdrawal. A stock recovery involves products still in the facility warehouse or distributor's storage. The regulations are specific in authorizing stock recovery only where the product is not on the market and still in the facility's direct control (21 C.F.R. 7.3).[13] A market withdrawal acts the same as a recall by removing product from the market but addresses only minor violations. A market withdrawal conducted to address a product issue is only appropriate if the issue is not one the FDA would take action against (21 C.F.R. 7.3).[14] Facilities should evaluate stock recoveries and market withdrawals before opting for a recall (see Fig. 2.8).

Import Refusal and Import Alerts
The Basics of Importing
Managing and enforcing imports serves an important public health role. The FDA estimates there are more foreign facilities producing food products than domestic facilities. The exact number can be difficult to determine because

---

[12] Amending the FD&C Act to be codified at 21 U.S.C. § 423(a).

[13] See, 21 CFR 7.3(k).
[14] See, 21 CFR 7.3(j).

Fig. 2.8 Progression of options for product removal

Fig. 2.9 Import enforcement process

some foreign facilities are not required to register with the FDA. For example, suppliers to foreign manufacturers who do not otherwise export may not be registered with the FDA. It is not only the quantity of imports that concern the FDA, but the agency's limited ability to inspect foreign facilities. The agency only reaches a scintilla of foreign food facilities. In some countries where robust government systems, infrastructure, and legal systems exist this may not matter. For other countries lacking such systems or sufficient food safety laws and enforcement imports pose a unique threat. This is the important context through which import refusals and alerts must be viewed.

Imported products must pass through both customs and FDA hurdles to enter the US market (see Fig. 2.9). All shipments must make certain declarations to the customs officials. Once an FDA-regulated product is declared, customs officials will request the FDA's assistance in determining whether to allow the shipment entry. One of the primary tools in assessing whether the shipment meets applicable FDA regulations is to conduct an inspection. Given the massive scale of imports not every shipment can be inspected. Prior to 2010 the FDA agent on duty would randomly identify containers for inspection based on their experience or perhaps simply on a hunch. In 2010 the FDA launched a new system named Predictive Risk-based Evaluation for Dynamic Import Compliance Targeting (PREDICT) that utilizes a complex algorithm to determine the likelihood of violations. This risk assessment provides FDA agents an objective means to focus its attention.

Imports and the Constitution

Imports stand at a unique crossroads that domestic food products do not experience. Imported products are the singular instance where food law intersects directly with constitutional law. That is to say the FDA not only exercises statutory authority when it conducts import inspections but also as a constitutional authority. In no other instance is the FDA's authority as broad as at the border. The US Constitution enumerates what is known as the "foreign commerce power." In Article 1 the Constitution provides Congress the power to "regulate commerce with foreign Nations" (US Constitution).[15] The result leaves only a superior constitutional right as the only means to curb the FDA's import inspection authority (*Butterfield v. Stranahan*).[16] Given the few options to challenge the FDA's authority, facilities must be administrative experts in order to smoothly navigate and resolve imports and the corresponding refusals, detentions, and alerts.

---

[15] U.S. Const. art 1, sec. 8, cl. 3.

[16] See, e.g., *Buttfield v. Stranahan*, 192 U.S. 470, 479 (1904) (upholding Tea Act of 1897 under Congress' power to regulate foreign commerce).

Imports and the FD&C

As with many areas of the FDA law the 1938 Act continues to provide the FDA its authority to enforce the importation of food, beverages, and dietary supplements. The Act sets an easily cleared threshold for detaining imported product in Section 801. The statute simply requires the imported product appear to violate the Act (FD&C).[17] This is at par with the new FSMA standard for administrative detentions. A detention is permitted on the basis of an FDA agent's unsubstantiated opinion. Again note the unique contours of food law for imported food products. Section 801 only applies to imported products meaning it establishes two standards for assessing violations.

Imported products must clear a higher level of compliance than their domestic counterparts. The "appears" standard subjects foreign facilities to a stricter interpretation of the Act. Domestic products in most cases must actually be proven to be adulterated or misbranded. Based on this "appears" standard imported products can be detained, refused, returned, destroyed, and even blacklisted. This is a severe litany of consequences flowing from a stricter interpretation of the Act. Domestic products would not be subject to a similar set of consequences because they merely appeared to violate the Act. This is the unfortunate and immutable reality of importing FDA-regulated products. As with the constitutional constraints on imports the result requires importers to be particularly keen on understanding and complying with the regulations.

The threshold for import detentions proves nearly inescapable for importers. Section 801 provides the entirety of the criteria for detaining imported products. No additional regulations, guidance documents, or policy statements clarify when a product will "appear" to violate the Act. The breadth of the authority requires careful attention to due process. Due process requires notification, hearing, and typically an appeal in order to ensure fairness and equal treatment under the law. Due process is particularly important when depriving a regulated party of property interests. Once a product is detained a notification and hearing are provided to the importer.

Notice of FDA Action to Import Alerts

The FDA provides a notice of its decision to refuse admittance. Once it refuses, the FDA issues a Notice of FDA Action also known as an FDA Import Detention Notice. The notice broadly states the reason for refusal, typically by citing the section of the Act defining adulteration or misbranding. Seemingly aware of the dearth of information in the notice, many compliance officers overseeing a detention will provide more information on the basis of the refusal. The notice is issued to the "owner of consignee" of the shipment. This can be the foreign manufacturer's shipping carrier, such as DHL or FedEx, or its US distributor. There is no obligation to notify the foreign manufacturer, which can complicate issues for the foreign entity.

The notice opens a small window for a hearing. The notice provides a ten-day window to request a hearing or respond with testimony or evidence indicating the basis for refusal is in error (21 C.F.R. § 1.94).[18] Although called a hearing, this is not what one may imagine for a hearing. It is neither adversarial nor formal. Instead the importer can introduce oral testimony or submit written evidence to an FDA compliance officer. There will be no cross-examination and little questioning. The most common evidence introduced is counter-laboratory analysis or proof of proper facility registration.

Following a refusal the product will be released, returned, reconditioned, or destroyed. The detained product cannot leave the border facility until a resolution is reached on the refusal. If an importer is able to prove the product is in compliance the product will be released. An importer can also request to recondition the product, such as by relabeling, under bond. Once reconditioned and approved the product will be released. If there is no response to the notice or the

---

[17] 21 U.S.C. § 381 (originally codified as § 801) § 381(a) (the standard states a detention is permitted if a product "appears from the examination of such samples or otherwise" adulterated or misbranded).

---

[18] See, 21 C.F.R. § 1.94.

testimony of evidence falls short then a second notice is issued known as the FDA Import Refusal Notice. This notice will direct the importer to either return the product to its country of origin or destroy it. Destruction will automatically occur if the product is not exported within days of the refusal notice.

The notices of detention and refusal can escalate one level further. Section 801 contains one additional ambiguous term used to broaden the FDA's import authority. The statute states the FDA may refuse entry to a product that "appears" from the examination of samples "or otherwise" to be adulterated or misbranded. The use of samples is explicitly authorized. Yet, this is not the only basis the FDA will use to determine if a product "appears" adulterated or misbranded. The FDA interprets the term "otherwise" to provide it limitless discretion to develop criteria for refusing shipments.

The most powerful outgrowth of the "otherwise" language is the import alert. The import alert acts essentially as a blacklist, a ban on a foreign manufacturer. The import alert is formally known as detention without physical examination (DWPE). A facility on an import alert will be detained and typically refused for every shipment simply because it appears on the list. There may be no product issue, but because the facility is on the list it faces scrutiny.

The FDA develops an import alert based on a number of factors. The most common reason for listing a facility on the import alert is a prior refusal. Listing on the import alert is the most important reason for properly resolving a Notice of Detention. Other reasons for a facility qualifying for listing may have nothing to do with the specific facility or its products. Criteria for listing includes the conditions of a particular country or region, known product category issues, such as particular dietary ingredients, or any other factor. The RPM in Subchapter 9-6 provides examples of when an import alert may be issued.

No Obligation for Notification

The entire import process can quickly leave a foreign manufacturer totally unaware of a refusal or its listing on an import alert (see Fig. 2.9). All notifications, Notice of Detention, refusal, and import alert are directed to the "owner or consignee." Notification of the import alert is not mandated by policy or law but often given only as a courtesy. The FDA policy reasons that the publication of the import alert on its website serves as notification to the effected parties. It does rank high on search engine algorithms making it difficult to miss. It is still highly unusual to first learn of legal action via a web search. A foreign manufacturer would not be able to look at a later Notice of Detention or refusal to determine if it is on an import alert. These notices will not state an import alert is the basis for detention or refusal. Strong contractual arrangement and proactive compliance are the best tools to stay in the loop.

Once listed on an import alert there is a process for removal. This involves a lengthy petitioning process. The entire process can take over a year. The petition-for-removal must show how the violation was corrected. The FDA assess the petition to determine the likelihood of the same violation reoccurring. The petition may involve a number of objective factors such as evidence of sampling, microbial testing, relabeling, registration, ingredient approval, and implementing new GMPs.

The Expansion of Agency Import Authority under FSMA and FSVP

FSMA creates new hurdles for importers. FSMA aimed not only to reform domestic food production, but also ensured foreign foods were produced to the same new standards of safety and quality. At first blush this appears as an innocuous seeming goal to preserve parity between domestic and foreign producers as domestic producers comply with new rules and regulations under FSMA. Yet, as noted above foreign facilities are held to a more stringent application of the Act. Thus, all FSMA rules and regulations will be more strictly applied to foreign facilities.

One rule in particular will pose new challenges. FSMA mandated the creation of a new rule known as the Foreign Supplier Verification Program (FSVP). The FSVP rule introduces new standards for domestic importers of foreign food products, including dietary supplements. The dense rule can be summarized as requiring domestic importers to identify and verify the control of hazards associated with the imported product. Supplier verification may require in some cases on-site inspections of the foreign facility. Failure to document the activities required under the rule will offer FDA agents a new basis to deem a shipment "appears" adulterated. In time it will not be surprising to see the creation of a new import alert focused solely on the FSVP. Facilities which fail to demonstrate FSVP compliance will be added to the import alert list with all subsequent shipments detained and potentially refused until a petition for removal lifts the alert.

Restraining Order or Injunctions

In some cases the FDA needs an enforcement tool to quickly control a hazard. A seizure warrant requires demonstration of probable cause, which often can lead to unacceptable delays. The administrative detention under FSMA may begin to fill the role of restraining order. Currently the judicial injunction or restraining order is the preferred tool to control a hazard when a seizure is impractical.

There are several options ranging from temporary actions to permanent injunctions. The FDA may seek a temporary restraining order from a judge with little showing of potential of harm. The temporary restraining order will go into effect for ten days. Following the ten-day window the FDA must either release the product or seek a preliminary injunction. A preliminary injunction requires a hearing and evidence of harm posed by the product. It is at the judge's discretion how long to impose a preliminary injunction. The preliminary injunction may be put in place until certain criteria are met, such as the FDA conducting inspection or a firm implementing corrective actions. When a preliminary injunction begins to expire and the agency wants to continue to hold the product it can request a permanent injunction. The permanent injunction can be the result of a settlement, where the FDA may promise to withhold other civil penalties in exchange for the permanent injunction, or a hearing before a judge. A permanent injunction is a court order that requires a separate hearing and order to revoke.

Suspension of Registration

Prior to FSMA, suspension of a facility's registration was not an option. All FDA-regulated facilities are required to complete some form of registration. This allows the FDA to monitor what facilities to inspect. It also helps it identify the root causes of outbreaks, consumer complaints, and other issues. FSMA granted the FDA new authority to suspend a facility's registration. The practical effect of the suspension is to stop a facility's operations. It cannot sell, produce, or market its products while the suspension is in effect.

The FDA quickly utilized its new authority. Shortly after the statutory provision went into effect the FDA suspended two facilities, registrations (FDA Notices).[19] The statute sets strict criteria on when the enforcement tool can be used. Under Section 415(b) of the FD&C, two primary criteria must be met for the FDA to suspend a facility's registration.

Section 415(b) Suspension Criteria
1. "Reasonable probability of causing serious adverse health consequences or death to humans or animals;" and
2a. "Created, caused or was otherwise responsible for such reasonable probability;" or
2b. "Knew of or had reason to know of such reasonable probability AND packed, received or held such food."

Reacquiring a facility's registration involves a lengthy judicial proceeding. As with other areas of enforcement the FDA's main goal is to ensure corrective actions are taken to address the

---

[19] FDA Notice of Opportunity for Hearing, Roos Foods Inc., March 11, 2014 (http://www.fda.gov/Regulatory-Information/FOI/ElectronicReadingRoom/ucm388921. htm); FDA Notice of Opportunity for Hearing, Sunland Inc., November 26, 2012 (http://www.fda.gov/AboutF-DA/CentersOffices/OfficeofFoods/CFSAN/CFSAN-FOIAElectronicReadingRoom/ucm329370.htm).

underlying causes leading to the suspension. This is not a process the FDA undertakes alone. It involves the US Attorney, who is an officer of the Department of Justice (DOJ), seeking judicial approval of the suspension. In both examples of registration suspension, the court also approved the reinstatement of the registration.

Facility registration suspension matches an FSIS enforcement tool. As mentioned above, one of the tools available to FSIS is suspension or withdrawal of inspectors. Both the new FSMA provisions and the decades old FSIS enforcement mechanism effectively shut down a facility until compliance objectives are met. FSMA suspension still does not go as far as the FSIS authority. Unlike the FDA, FSIS retains the ability to execute a permanent withdrawal. The two agencies have long taken divergent paths leading to the disparate treatment of certain classes of food products. FSMA, while only impacting the enabling Act for the FDA, begins to unify and match FSIS authority. The two agencies are likely never to carry the same enforcement tools, but some parity does help facilities cope with the regulations.

2.3.7   Impact of FSMA

FSMA introduced a number of direct and indirect changes to formal enforcement mechanisms. FSMA changes some provisions, such as lowering the administrative detention threshold, while also creating entirely new ones, like facility registration suspension. The goal was often to unencumber the agency from seeking judicial approval prior to acting. Thus, allowing the agency to become more nimble and responsive. This was the case in providing the agency new authority to mandate recalls. FSMA represents sweeping changes many of which will be enforced under the new and enhanced enforcement authorities.

Informal enforcement mechanisms are also impacted by FSMA. New standards and regulations mean fresh areas for inspection observations in Form 483s or warning letters. The industry will not doubt experience growing pains as the rules and provisions take effect. The top observations cited in Form 483 will likely reflect this period of adjustment with FSMA observations cited

more frequently than any other provision. This is as the drafters of FSMA intended it to be. The goal of food law since its inception was to pressure industry to self-regulate and self-enforce. Citations in Form 483s and warning letters along with formal enforcement actions will motivate compliance.

2.3.8   Comparing and Contrasting Inspection Models

There are a number of fundamental differences between the USDA/FSIS and the FDA/ORA. Many can be explained by the legislative history of the two enabling acts. The FMI represents a strong reaction to a crisis resulting in the most pervasive inspection and enforcement ability in the food industry. The FMI also bars any facility from operating without FSIS approval and inspectors on-site. Not only does FSIS operate with more authority but also with a smaller scope of responsibility. Whereas the FDA regulates all food, FSIS oversees only limited categories of meat, poultry, and eggs. This background sets the stage for two fundamentally different approaches to inspections and enforcement actions.

The two agencies operate with a different inspection model and emphasis during inspections. The USDA utilizes a continuous mandatory inspection model focused on preventing contaminated meat products from reaching the market. The FDA elected a random inspection paradigm aimed at removing problematic products from the market. The FDA approach can lead to gaps of 7–10 years between inspections (CRS 2007).[20] Often the gaps are much larger allowing issues to billow and build up. Two models can create trouble.

A number of examples highlight the issues spawned by this split approach. Perhaps, the best example can be seen in a processing plant producing pepperoni and cheese frozen pizzas. Recall from Chapter 1, the USDA's preventative model covers meat, poultry, and some egg products while the FDA's reactive approach covers all

---

[20] Congressional Research Service Report To Congress: The Federal Food Safety System: Primer (2007).

other foods. The pepperoni is a meat product and any line producing the pepperoni pizza will experience continuous inspection from the USDA/FSIS. The cheese pizza lacks any meat or poultry leaving it to the FDA to inspect. The FDA inspection will occur randomly, perhaps as often as every 10 years or so. One facility two product lines, one line subject to continuous inspection the other inspected rarely and randomly.

The bifurcation is random and not based on any logical approach. The food products under the USDA are of no greater risk to the public than the products under the FDA's purview. Take high-risk foods such as seafood as an example. Seafood falls outside the enabling acts for the USDA and is subject only to random inspection by the FDA. One could also ask whether pepperonis foster a higher risk of foodborne illness or hazards than cheese in the pizza example. Different hazards no doubt exist between pepperonis and cheese but not in a meaningful way to justify different inspection treatment. In nearly every example the answer lies not in science or risk evaluations but in legislative history.

The agencies also wield a different array of enforcement tools. Different missions and approaches to enforcement require a different set of tools. The FDA, for example, can compel a recall, but FSIS can only recommend recalls. FSIS can escalate enforcement actions to the point of permanently shuttering a facility, whereas the FDA can only make temporary suspensions. For facilities solely producing an FDA or FSIS product, the difference in enforcement mechanisms matters little. For facilities producing both FDA and FSIS products, the difference can be frustrating and burdensome.

2.4 Constitutionality of Warrantless Inspections

2.4.1 Introduction to Challenging Enforcement Actions

There are two types of challenges to federal enforcement actions. Enforcement actions may be challenged using either the US Constitution or the enabling statute. Recall from Chapter 1, the role of the US Constitution. It enumerates both the limits of federal power with the states and with individuals. There are a number of constitutional challenges to raise in enforcement actions. Among the more common challenges is the constitutionality of warrantless searches or the restrictions of free speech on product labels (see Chapter 4 and 7 for a discussion on the freedom of speech). The enabling statute serves a similar purpose. Each statute delegates a specific authority from the federal powers to a federal agency. The agency is bound by the explicit and implied terms of that statute. For example, Congress among numerous acts, but namely through the FD&C delegated authority to the FDA to manage food regulation and safety. An egregious example would be the FDA promulgating rules under the Federal Clean Air Act to reduce the carbon emissions emitted by coal-powered electric plants. Any coal plant which the FDA attempted to subject to such a rule could initiate a statutory challenge. In more nuanced examples, one can imagine the FDA taking enforcement actions against 1.89% cooked meat. Using the IOM Exhibit provided in Chapter 1, does this technically fall outside the FSIS's agreed scope of authority? Does it fall outside the FMI? Each valid question to invoke in a statutory challenge.

2.4.2 Constitutional Questions of Warrantless Inspections Conducted by the FDA

At the onset, it should be noted that many facilities acquiesce to an inspection rather than challenge the FDA's authority. Facilities possess only so much regulatory capital. Regulatory capital captures the concept of maintaining a certain amount of good will with the regulatory agencies charged with overseeing a facility. This good will can grease the wheels and create a partner in enforcement actions and inspections rather than creating an antagonist. There are a number of examples where regulatory capital can be used from requesting more information on statutory requirements to finding flexibility in implementing corrective measures. Ultimately by demonstrating openness to the regulations, a facility avoids building a negative reputation with the agency, especially the district of field office, which could make enforcement actions more burdensome to overcome.

There remain valid constitutional questions about warrantless inspections conducted by the FDA. Carefully selecting those issues and challenges preserves a facility's regulatory capital, while maintaining its constitutional rights. This section will explore the background case law on warrantless administrative searches before turning to those open constitutional questions, such as photos and cellphones.

The constitution only confers rights to US citizens and domestic companies. Foreign facilities subject to an inspection will need to raise statutory challenges to the questionable FDA inspection activities.

2.4.3   The Fourth Amendment

The constitutionality of warrantless inspections follows a progeny of several cases. As discussed above in passing the 1953 Factory Inspection Amendment, Congress made it abundantly clear that it intended the FDA to conduct inspections without a warrant. Congress wanted to avoid giving offending facilities any opportunity to commit acts of subterfuge like hiding or destroying evidence. The Fourth Amendment of the US Constitution, however, restricts the federal government's ability to conduct warrantless searches. Its limits are aimed at abuses of power and protecting personal privacy and security from arbitrary or oppressive government intrusions. The full text of the Fourth Amendment reads:

> The right of the people to be secure in their persons, houses, papers, and effects, against unreasonable searches and seizures, shall not be violated, and no warrants shall issue, but upon probable cause, supported by oath or affirmation, and particularly describing the place to be searched, and the persons or things to be seized (U.S. Constitution).[21]

The Fourth Amendment does not act as a prohibition against a search. It merely requires a judge to issue a warrant to federal officials based on a probable cause prior to any search. This is the so-called warrant requirement. Therefore, under the Fourth Amendment, if a federal official conducts a search without a warrant it is presumed "unreasonable" unless one of the generally recognized exceptions to the warrant requirement. Already one can see a constitutional challenge teed-up under the FD&C—the Factory Inspection Amendment removed the consent requirement while insisting on the remaining warrantless requirements.

2.4.4   Waiving a Challenge by Consenting to a Search

The chief exception to the warrant requirement is consent. If a party consents to an inspection, then they effectively waive any ability to raise a Fourth Amendment challenge. The Supreme Court addressed the consent-waiver in three cases involving additional Fourth Amendment issues that developed other doctrines explored in this section (*Camara, See,* and *Barlow's Inc.*).[22] The consent-waiver poses a substantial barrier to the challenging inspections since the majority of facilities consent to searches.

The issue of consent can become a contentious point of debate. The question of whether consent was in fact given can become a factual issue to argue in court. Factual issues appear mundane especially compared to what one imagines of constitutional challenges. Still factual challenges remain a potent instrument in raising a constitutional challenge under the Fourth Amendment. Typically, the issue is brought via a pretrial motion to suppress evidence obtained through the warrantless and allegedly unconsented search. The Factory Inspection Amendment criminalizes a facility refusing consent. Thus, the factual issue of consent involves a number of questions about who gave consent, how and when it was refused, and whether criminal penalties should be imposed.

---

[21] Amendment IV, U.S. Constitution.

[22] See, *Camara v. Municipal Court, 387 U.S. 523 (1967),* See *v. City of Seattle, 387 U.S. 541 (1967), and Marshall v. Barlow's Inc., 436 U.S. 307 (1978).*

As developed by courts, administrative consent is a unique form of consent. The consent for administrative searches differs from the consent developed in criminal cases involving Fourth Amendment questions. It developed thorough its own history of case law. The administrative consent is much broader than the criminal consent. In the case excerpt below, we see a number of factors demonstrating the sheer breadth of the administrative consent-waiver.

United States v. Thirftmart, Inc. 429 F.2d 1006 (9th Cir.), cert. denied, 40 U.S. 926 (1970)

Appellants have been convicted of violations of the Federal Food, Drug & Cosmetic Act... Upon inspection, food in four company warehouses had been found to be infested with insects...Fines were imposed on [warehouse manager and supervisor].

The principal issue on appeal relates to the constitutionality of searches of appellants' warehouses conducted by Food & Drug Administration (FDA) inspectors. The inspections were routine and similar ones had been conducted periodically in the past. The inspectors testified that on arrival at the warehouses they approached the managers, filled out and presented their notices of inspection, requested permission to inspect and in each case were told, *"Go ahead"* or words of similar import. The inspection notices contained a recitation of 21 U.S.C. § 374(a), which authorizes FDA inspectors to enter at reasonable times to inspect food warehouses. The inspectors did not have search warrants nor did they advise the warehouse managers that they had a right to insist upon a search warrant.

[The facility] contend that this warrantless inspection was unconstitutional under...

The precise issue raised is whether the informal and casual consent to search given by the warehouse managers made it unnecessary to secure a search warrant. Appellants argue that a waiver of search warrant "cannot be conclusively presumed from a verbal expression of assent. The court must determine from all the circumstances whether the verbal assent reflected an understanding, uncoerced, and unequivocal election to grant officers a license which the person knows may be freely and effectively withheld..." Since the managers were not warned that they had a right to refuse entry and since there was no proof that they knew they had such a right, [they] argue that the consent was not effective to remove the need for a search warrant.

In a criminal search the inherent coercion of the badge and the presence of armed police make it likely that the consent to a criminal search is not voluntary. Further, there is likelihood that confrontation comes as a surprise for which the citizen is unprepared and the subject of a criminal search will probably be uninformed as to his rights and the consequences of denial of entry...

These circumstances are not present in the administrative inspection. The citizen is not likely to be uninformed or surprised. Food inspections occur with regularity. As here, the judgment as to consent to access is often a matter of company policy rather than of local managerial decision. FDA inspectors are unarmed and make their inspections during business hours. Also, the consent to an inspection is not only not suspect but is to be expected. The inspection itself is inevitable. Nothing is to be gained by demanding a warrant except that the inspectors have been put to trouble... an unlikely aim for the businessman anxious for administrative good will.

Here, the managers were asked for permission to inspect; the request implied an option to refuse and presented an opportunity to object to the inspection in an atmosphere uncharged with coercive elements. The fact that the inspectors did not warn the managers of their right to insist upon a warrant and the possibility that the managers were not aware of the precise nature of their rights under the Fourth Amendment did not render their consent unknowing or involuntary. They, as representatives of Thriftimart, Inc., were presented with a clear opportunity to object to the inspection and were asked if they had any objection. Their manifestation of assent, no matter how casual, can reasonably be accepted as waiver of warrant.

In *Thirftmart*, a wide range of factors did not negate the validity of the administrative consent-waiver. The court does not raise any concern about who gave the consent. The consent was given by my three employees, but remains valid. The pressure of criminal penalties under Section 331, which ordinarily could impact criminal consent, is held not to constitute a form of coercion or forced consent. In the case excerpt, the consent is casual, almost flippantly given, yet still upheld. The court further finds the administrative consent valid even if a party giving consent is unaware of their right to refuse a warrantless inspection. A facility is not required to give consent, but once given it is broadly construed against it.

The issue of who can give consent draws attention to the important matter of employee training. Any FDA-regulated facility can readily expect an inspection. The case law makes it clear that the administrative consent is equally binding whether given by a manger or a facility owner. Proper training on managing FDA inspections can help mitigate a number of issues, including waiving Fourth Amendment challenges before the facility owner even evaluates their merits.

The administrative consent not only differs from the criminal consent, but also from consent to a warrant. There are instances where the FDA will seek a warrant prior to conducting an inspection. Typically, this occurs when the FDA anticipates a refusal or there is a history of non compliance. Consent to a warrant carries its own case law and meaning. Consenting to an inspection under a warrant will still waive any Fourth Amendment challenge, but cannot abrogate a challenge to the validity of the warrant. If the warrant is challenged and found invalid, then the search itself is considered invalid. Plaintiffs often challenge the probable cause serving as the basis of the warrant. Probable cause for an administrative warrant will be discussed next.

2.4.5 Administrative Warrant Requirement under the *Camara-See* Principle

Courts initially developed a doctrine requiring administrative agencies that obtain a search warrant when confronted with a refusal to inspect. In companion cases, *Camara* and *See* established the doctrine for administrative search warrants.

Camara and *See* stand for the principle requiring an administrative agency to obtain a search warrant only after entry is refused. *Camara* involved a private residence and local statute. In *Camara,* a San Francisco resident was charged with a criminal violation of the housing code after refusing a warrantless inspection by the city of their home. The Supreme Court held the home owner who enjoyed a constitutional right under the Fourth Amendment to insist inspectors to obtain a warrant for the search. It also held the criminal conviction inappropriate on the basis of refusing consent to a warrantless inspection. In *See*, the issues revolved around the inspection of a commercial facility. There the court held administrative agencies that could not coerce or compel an owner of a business to permit inspection. The only means to override consent was through a valid search warrant.

An administrative search warrant requires a different level of proof then a criminal search warrant. As mentioned above, the probable cause standard is incredibly relaxed in the administrative warrant context. In the criminal context, a warrant is only issued upon specific knowledge that a crime has or will occur. Not so for an administrative warrant. Probable cause could be based on "reasonable legislative or administrative standards" such as the amount of time since the last inspection. Therefore under the *Camara–See* principle if consent is denied an administrative agency, like the FDA, must obtain an administrative search warrant based on a lowered standard of probable cause.

The decisions in *See* and *Camara* changed little in how the FDA conducted its investigations. In part, this is because most facility owners did not refuse consent. The only change following *See* and *Camara* came through new guidance to inspectors requiring they obtain a warrant if refused access to the facility following a presentation of their credentials and Form 482 (FDA Press Release).[23] The agency also adopted the policy of applying for an inspection warrant when the district office anticipated refusal to inspect or if living quarters were on the premises.

2.4.6　*Collonade–Biswell* and the Pervasively Regulated Business Doctrine

Following *See*, the Supreme Court developed a new doctrine over the course of two cases decided two years apart. The cases did not involve the FDA's inspection authority but are largely thought to apply to the agency. In 1970 and again in 1972, the Supreme Court decided on two cases which together established an exception to the warrant requirement for administrative inspections. The cases did not involve the FDA but some courts interpret the exception to apply to the FDA inspection. In *Colonnade Catering Corp. v. United States*, the Internal Revenue Service (IRS) searched a catering business and holder of a "federal retail liquor dealer's occupational tax stamp" (*Colonnade*).[24] A federal agent and member of the Alcohol and Tobacco Tax Division attended a party catered by Colonnade and developed a

suspicion; the liquor bottles were being refilled in violation of the law. Essentially, the violation involved avoiding taxes. IRS agents later returned and inspected the cellar where they found the liquor storeroom locked. The manager and later president of the company refused to unlock the storeroom. Thus, an agent broke the lock, entered the storeroom, and seized several bottles of liquor. The court held that *See* did not apply because there was a reduced expectation of privacy for heavily regulated industry, like the liquor industry. In reaching this, holding the Court emphasized the long history and breadth of regulation and overview for the liquor industry. Therefore, this part of the doctrine holds the warrant requirement that does not apply to heavily regulated industry because there is a reduced expectation of privacy. *Colonnade* also stands for a secondary principle. The court held the forcible entry without a warrant that was precluded by the statute authorizing the IRS to inspect. Thus, the inspection was illegal.

The second half of the exception came when the court decided *Biswell v. United States* in 1972 (*Biswell*).[25] *Biswell* involved a pawn shop that also operated as a federally licensed gun dealer. A police officer and federal treasury agent visited the pawn shop to inspect the operator's records. During the inspection, the agent requested entry into a locked gun storeroom. When asked if the agent had a search warrant, the operator was supplied a copy of the Gun Control act which authorized inspection. The operator replied, "Well, that's what it says, so I guess it's okay." He proceeded to unlock the storeroom where the agent found and seized two prohibited sawed off rifles. On appealling, the court of appeals found the statute authorizing the search unconstitutional and the consent invalid. The Supreme Court reversed the finding of the lawfulness of search, independent of consent. It established the principle for warrantless inspections of "pervasively regulated" businesses where the inspection is authorized by a valid statute and carefully limited in "time, place, and scope." The court analogized the valid statute to a valid search warrant save for

---

[23] FDA Press Release, June 18 1967.

[24] 397 U.S., 72 (1970).

[25] *406 U.S. 311 (1972).*

the ability to gain forcible entry under statutory authority. Both provide a right to inspect regardless of consent and penalize refusal to inspection with possible criminal prosecution.

The FDA adopted a new policy following *Biswell and Colonnade*. The FDA maintained the authority of Section 704 to conduct warrantless inspections and to penalize refusals to admit agents. It sets the caveat that the inspection must be reasonable, a requirement of Section 704, and the inspectors must present their credentials and Form 482.

Despite the policy change, challenges under *Biswell* and *Colonnade* remain viable. Recall neither case that involved a review of Section 704 nor a challenge to the FDA inspection authority. Neither did *Biswell*, decided 2 years after *Colonnade*, overturn the holding in *Colonnade*. This means some courts may apply the rationale of *Colonnade, Biswell,* or *Colonnade–Biswell* in reviewing challenges to the FDA inspections. Under a *Colonnade* challenge, once an owner or operator of a facility refuses to consent any agency action risks being construed as forceful in violation of the authority in Section 704 and the Fourth Amendment. A court reviewing a challenge through the lens of *Biswell*, however, could hold the issue of consent is irrelevant. In either review the courts would agree Section 704 stands in place of a search warrant, but the role of consent, similar to *See* and *Camara*, could remain a point of contention.

See and *Camara* could also remain a viable standard to measure the constitutionality of FDA inspections. Early examples emerged following *Colonnade* and *Biswell* where courts applied the standard in *See* rather than the *Colonnade–Biswell* doctrine. The question became more poignant following a Supreme Court decision in 1978.

2.4.7 The Evolution of Colonnade– Biswell Doctrine

A key development in the *Colonnade–Biswell* pervasively regulated exception occurred when the court decided *Marshall v. Barlow's, Inc. (Barlow's).*[26] *Barlow's* involved an electrical

and plumbing installation company inspected by the Occupational Safety and Health Administration (OSHA). An OSHA inspector arrived at the Barlow facility to make an inspection, but was refused admission. OSHA sought a district court order compelling *Barlow's* to allow the inspection. Unmoved Barlow again refused to allow OSHA inspectors into its facility. It sued the federal agency and sought injunctive relief against the warrantless search. A three-judge panel held a warrant that was required by the Fourth Amendment and cited to *Camara* and *See* to support that is holding. The Supreme Court agreed and affirmed the lower court's holding. The court reasoned the general rule is that all warrantless searches are generally unreasonable and thus violate the Fourth Amendment. The issue then becomes whether there is a recognized exception to the warrant requirement. OSHA argued the "pervasively regulated" business exception articulated in *Colonnade–Biswell* applied to the inspection. The court, however, held the rationale behind *Colonnade–Biswell* lie in the doctrine of consent. It found there must be a "long tradition of close government supervision of which any person who chooses to enter such a business must already be aware." Unlike *Colonnade* and *Biswell*, the court found no such history for OSHA inspections.

The decision in *Barlow's* provides more detail to the *Colonnade–Biswell* doctrine but no greater clarity of its application to FDA inspection. *Barlow's* informs federal agencies the "pervasively regulated" business exception is narrow only applying in "relatively unique circumstances." For it to apply to a federal inspection, the regulation must be both extensive and in effect for a lengthy period of time. One way to know if these two criteria exist is whether a person entering that regulated field would reasonably be expected to be aware of government supervisions.

On the surface, *Barlow's* appears to strike a blow to the FDA's warrantless inspection practice, but the status quo remains. The FDA did not change its policy following *Barlow's*. The question becomes whether the FDA viewed in its entirety, namely as a regulator of drugs, medical devices, food/dietary supplements, cosmetics, and

---

[26] 436 U.S. 307 (1978).

animal products, is pervasively regulated under *Colonnade–Biswell* or whether each product center is assessed separately. Arguably drugs and medical devices are pervasively regulated, but can the same be said of food or cosmetics?

The FDA adopts the view that all its inspections are covered by *Collonade–Biswell*. The question has largely fallen away because the majority of facilities agree to inspections. Only two judicial challenges focused on the question with the Eighth Circuit applying *Colonnade–Biswell* to a drug facility and two district courts applying the doctrine to food facilities (8th Cir. 1981; D. Mass 1980).[27] Thus, there is no definitive judicial answer on the question.

2.4.8   Outstanding Constitutional Challenges and Open Questions

Not every constitutional challenge involves an outright refusal to allow the FDA to inspect. Refusing other requests during an inspection, sampling or photos for instance, present more subtle shades of refusal and consent. An added wrinkle comes when applying the FDC&A to foreign facilities.

The FDA adopts a broad stance on its authority to conduct inspections. It bases its authority to conduct extra statutory activities during the inspection in part on *Colonnade–Biswell*. It also extrapolates a right to take photos on two other judicial opinions that raised the issue of photography in public spaces. The FDA's position on photography can be found in Subchapter 5.3 of the IOM (IOM).[28]

The FDA policy rests on a theory of implied consent and minimal expectation of privacy. The IOM cites two cases, *Dow Chem. Co. v. United States*[29] and *United States v. Acri Wholesale Grocery Co.*[30] as the moorings for its right to take photos. *Dow Chemical* involved a challenge to the EPA's use of a commercial aerial photographer to photograph its facilities. The aerial photographer at all times flew in navigable, or public, airspace. Immediately, a factual contrast emerges with the FDA's position since there is a greater expectation of privacy within a facility than compared to photos from public airspace. The key to the *Dow* stands in the courts finding that although the aerial photography was not specifically authorized in the EPA's enabling acts it was permitted. Thus, the FDA likewise adopts the policy that its investigations are not bound by explicit statutory authorizations.

Excerpt from Dow Chem. Co. v. United States

The use of aerial observation and photography is within EPA's statutory authority. When Congress invests an agency such as EPA with enforcement and investigatory authority, it is not necessary to identify explicitly every technique that may be used in the course of executing the statutory mission. Although 114(a) of the Clean Air Act, which provides for EPA's right of entry to premises for inspection purposes … does not authorize aerial observation, that section appears to expand, not restrict, EPA's general investigatory powers, and there is no suggestion in the statute that the powers conferred by 114(a) are intended to be exclusive. EPA needs no explicit statutory provision to employ methods of observation commonly available to the public at large…

---

[27] See, *United States v. Jamieson-McKames Pharmaceuticals, Inc.*, 651 F.2d 532 (8th Cir. 1981). *United States v. New England Grocers Supply Co.* 488 F. Supp. 230 (D. Mass. 1980; appeal from conviction by magistrate following the decision in *Barlow*); *United States v. Gel Spice Co. Inc.*, 601 F.Supp. 1214 (E.D.N.Y 1985; appeal of magistrate's denial of motions to suppress certain evidence and dismiss some of the ten count charges).

[28] IOM at Subchapter 5.3 "Evidence Development" available at http://www.fda.gov/ICECI/Inspections/IOM/ucm122531.htm (last visited June 29, 2014).

[29] *476 U.S. 227 (1986) (held warrantless aerial photographs taken by EPA using a commercial aerial photography were within the EPA's statutory authority because a regulatory and enforcement agency requires no explicit authorization to employ methods of observation available to the public.).*

[30] *409 F. Supp. 529 (S.D. Iowa 1976).*

The *Acri Wholesale* case involves a question of implied consent. There the FDA conducted an inspection of the grocery store's warehouse. It found a pervasive rodent infestation and proceeded to take photographs. The facility consented both to the search and the photos. Consent abrogated any subsequent challenge to the FDA using the photos in its criminal case.

The cases raise serious questions about the soundness of the FDA's position. The Supreme Court in *Dow* made it clear the EPA did not need explicit authority to utilize images the public could take. The district court in *Acri* made the failure to object to photos central to its holding. Together the opinions leave open considerable room for a challenge where the facility owner objects to photography in parts of the facility closed to the public.

The FDA outlined the rationale behind its photography policy on photos in 2013. The FDA issued a guidance document aimed at the drug industry titled, "Guidance for Industry Circumstances that Constitute Delaying, Denying, Limiting, or Refusing a Drug Inspection." Rather than assert new authority to photos, the agency expounds on the rationale for using photographs.

Excerpt from Guidance

Photographs are an integral part of an FDA inspection because they present an accurate picture of facility conditions. Not allowing photography by an FDA investigator may be considered a limitation if such photographs are determined by the investigator(s) to be necessary to effectively conduct that particular inspection. Examples of conditions or practices effectively documented by photographs include, but are not limited to: evidence of rodents or insect infestation; faulty construction or maintenance of equipment or facilities; product storage conditions; product labels and labeling; and visible contamination of raw materials or finished products.

As with many areas of FDA, regulation facilities find little benefit in challenging the agency. The majority of domestic facilities find it futile to refuse a reasonable request or stop an inspector from taking photographs. The FDA agent will simply return with a warrant and perhaps a more aggressive posture. Foreign facilities confront a similar conundrum. If the objection to photographs is considered a refusal then it faces a number of penalties, including losing the ability to export to the US typically via an import alert. Yet challenges will surely come. Technology continues to outpace the law. The ability to capture high-resolution video easily, and in some cases covertly like Google Glass, raises new challenges. While courts find that the Act provides a flexible standard there must be some meaningful limits. Recall we are considering an Act from 1938 long before the technology of today. These challenges open the viability of the *Camara–See* principle for refusals to take photos or release data.

The ability to capture other data will also raise questions. In particular, consider the role of smartphones and tablets. The wealth of information contained on these devices is copious and diverse. In most of the food industry, there are low barriers of entry allowing entrepreneurs to quickly begin importing, producing, or selling food items. Those entrepreneurs may rely on devices to serve both personal and professional purposes. Surely Section 704 is not so flexible or broad to allow the FDA full access or to penalize an objection.

The Supreme Court is beginning to shape Fourth Amendment law in light of new technologies. In the past two years, the Court visited the issue in the criminal context. It began by looking at the use GPS tracking (*Jones*)[31] followed by a closely watched case involving the warrantless search of smartphones (Riley).[32] In both cases, the Court required warrants to gain access to the data.

The question now becomes whether an obsequious mindset will stall challenges to FDA inspections. The FDA is yet to take such invasive measures, but it is showing signs of interest. The

---

[31] *United States v. Jones*, 32 S. Ct. 945 (2012).

[32] *Riley v. California*, 134 S.Ct. 999 (2014)

FDA, for instance, began routinely searching for public metadata and social media posts for labeling violations. How long before the interest extends to requests for access to smartphones, tablets, or data stored in the cloud?

2.4.9   Broader Lessons from *Camara, See, Colonnade,* and *Biswell*

It can be deflating to review the case law on constitutional challenges. Broad powers are met with the practical questions of the benefits of challenging. Although the answer is often to acquiesce, there are still broader lessons in the case law. We see the right to object and require the agency to obtain a warrant in order to ensure fairness and objectivity. Important limits are also set like the prohibition on using force. There are also standards which act as a sentential guarding against unnecessary or arbitrary intrusions. The 1938 Act could not imagine many aspects of our modern world and the growth of the food industry. The current case law provides a backdrop for future challenges.

2.5   Statutory Challenges to Inspections and Enforcement.

The statutory challenge offers an effective alternative to the Constitutional challenge. Often broader in scope statutory challenges can take two forms. A statutory challenge can be raised under the Administrative Procedures Act (APA) or under the enabling statute. The APA applies to all Federal agencies working to ensure fairness in rulemaking and agency adjudication.

2.5.1   Administrative Procedure Act Challenges

The APA serves many functions. It identifies four main purposes that can be distilled into two primary functions. The APA governs agency rulemaking setting standards to keep industry and the public at large aware and opportunities to engage in agency rulemaking. It also establishes standards for agency adjudication. As seen with the discussion a great deal of law, regulation, and policy is set within the FDA. The APA acts as a layer of enforcement over the agency to ensure that adjudication is fair and uniform. It also provides thresholds on when judicial review would be appropriate.

The APA offers an administrative cudgel for facilities challenging inspections and enforcement actions. The vast number of cases involves what is known as an "arbitrary and capricious" challenge. Section 706 of the APA sets a standard for setting aside or overturning agency actions. At the surface the arbitrary and capricious challenge questions whether the enforcement action is based on facts, existing policy, and the conclusion is rational. The FDA often makes decisions on a case-by-case basis many times with different regions or compliance officers involved in the decision making. The arbitrary and capricious challenge pauses the process to ask whether there is a basic consistency and fairness between the immediate enforcement matter and past Agency actions.

> **5 U.S.C. § 706(a) (APA)**
> The reviewing court shall hold unlawful and set aside agency actions, findings, and conclusions found to be arbitrary, capricious, an abuse of discretion, or otherwise not in accordance with law.

The arbitrary and capricious standard varies by federal agency. For the FDA, the standard requires a uniform application of the FD&C, CFR and Guidance documents across industry. As the agency takes enforcement actions it builds a series of "case law" similar to what a court may do. The FDA is bound by this "case law" unless it can articulate a rationale supported by facts for a different outcome. Thus, the agency not only required to uniformly apply the law but consistently follows its own enforcement precedent. Give the sparingly few judicial challenges there is a morass of enforcement history.

The arbitrary and capricious standard applies in courts but can be raised at any time. Section 706 makes it clear the arbitrary and capricious standard is intended for a "reviewing court." Ultimately the judicial venue is the only one to offer full relief from an enforcement decision. Still the FDA remains sensitive to arbitrary and capricious challenges. This makes traipsing through the jungle of enforcement history important. The intent to pursue an APA challenge can be raised with a compliance officer, but it must be well articulated and supported.

APA and Reviewability of Warning Letters
Troubling hurdles emerge when considering a judicial challenge to a warning letter using the APA. A judicial concept known as ripeness proves an unmovable obstacle. Warning letters are currently deemed "unripe" for judicial review. Courts will only review cases when certain criteria are met, including ripeness. Ripeness refers to the timing of hearing the case. If a court hears a case too soon, there is a risk that not all of the facts will be in place. In the context of the APA, ripeness bears several features.

Courts developed three unique hallmarks to administrative ripeness. In the context of the APA, courts hold ripeness to mean fitness, hardship to the plaintiff and finality (Example Cases).[33] Warning letters continually cannot clear the final criteria—finality (Example Cases).[34] Cases considering APA challenges to warning letters concluded that warning letters do not constitute final agency action. Since there is the possibility of further agency decision-making, the case is unripe for adjudication. Two factors gird the finding that warning letters lack finality. First, finality means the challenged actions mark the

"consummation of the agency's decision making process" and second, the action is "one by which right or obligations have been determined, or from which legal consequences will flow" (*Holistic Candlers*).[35] Recall the FDA in its description of warning letters described it as informal notice by a single agent that does not commit the FDA to any course of action.

From this description of warning letters and the two-prong test for finality courts dismiss APA challenges. A variety of reasons are given for dismissing the challenge. In some courts that agree with the FDA and hold warning letters are tentative actions because the letters contain conclusions of subordinate officials (Biotics Research Corp.).[36] In other cases, a court points to the FDA's policy that warning letters are not a commitment to take enforcement action as support that no rights or legal consequences are determined as a reason to find finality lacking (*Clinical Reference Lab*).[37] At this point in the case law, APA's challenges to warning letters in court are a futile affair. As FSMA begins to take effect, however, the question becomes whether there are new grounds for challenging the current precedent.

FSMA likely impacts the reviewability of warning letters in a way not anticipated by the FDA. One unintended effect of FSMA may be its impact on the reviewability of warning letters. As seen in the section above, FSMA dramatically enhanced the FDA enforcement capabilities. Whether it was augmenting the FDA's ability to initiate formal enforcement actions or creating new statutory requirements, rules, and penalties, the FDA is more capable than before. Facilities are now confronted with real consequences. Consequences, like high-risk facility designation, that question the finality of warning letters. A fresh judicial review will be required to determine whether warning letters citing FSMA standards or imposing FSMA penalties are at last ripe for review.

---

[33] See, e.g., *McKart v. United States*, 395 U.S. 185, 193-94 (1969); Pub. *Water Supply Dist. No. 10 v. City of Peculiar*, 345 F.3d 570, 573 (8th Cir. 2003); *Farm-to-Consumer Legal Def. Fund v. Sebelius*, 734 F. Supp. 2d 668, 695 (N.D. Iowa 2010) (citing O'Shea v. Littleton, 414 U.S. 488, 494 (1974)).

[34] See, e.g., *Dietary Supplemental Coal., Inc. v. Sullivan*, 978 F.2d 560, 563 (9th Cir. 1983); *Cody Labs, Inc. v. Sebelius*, No. 10-CV-00147-ABJ, 2010 U.S. Dist. Lexis 80118, at *32-33 (D. Wyo. July 26, 2010); *Summit Tech. v High-Line Med. Instruments Co.*, 922 F. Supp. 299, 306 (C.F. Cal. 1996).

[35] *Holistic Candlers and Consumers Ass'n v. FDA*, 664 F.3d 940, 943 (D.C. Cir. 2012 (quoting *Bennett v. Spear*, 520 U.S. 154, 177–78 (1997)).

[36] See, e.g., *Biotics Research Corp. v. Heckler*, 710 F.2d 1375, 1378 (9th Cir. 1983).

[37] See, e.g., *Clinical Reference Lab., Inc. v. Sullivan*, 791 F. Supp. 1499, 1503-04 (D. Kan. 1992).

2.5.2   APA and Import Alerts

Import alerts are particularly susceptible to APA challenges. Rather than an arbitrary and capricious challenge, import alerts are open to challenges under Section 553 of the APA. Section 553 of the APA establishes a notice-and-comment period for new agency rules or regulations. The rationale underlying the notice-and-comment period is that industry participation in rule brings not only fairness to the process, but also improves the regulations promulgated. Not every agency action is subject to Section 553. Section 553(b)(3)(A) creates a series of exceptions to the notice-and-comment procedures. The list is lengthy including general statements of policy and case-by-case adjudications. The question becomes what are import alerts?

---

5 U.S.C. § 553(b)(3)(A) Exceptions to Notice-and-Comment

(b) General notice of proposed rulemaking shall be published in the Federal Register, unless persons subject thereto are named and either personally served or otherwise have actual notice thereof in accordance with law. The notice shall include—

1. a statement of the time, place, and nature of public rule making proceedings;
2. reference to the legal authority under which the rule is proposed; and
3. either the terms or substance of the proposed rule or a description of the subjects and issues involved.

Except when notice or hearing is required by statute, this subsection does not apply—

a. to interpretative rules, general statements of policy, or rules of agency organization, procedure, or practice; or
b. when the agency for good cause finds (and incorporates the find-

---

ing and a brief statement of reasons therefor in the rules issued) that notice and public procedure thereon are impracticable, unnecessary, or contrary to the public interest.

The FDA deems import alerts as guidance documents. Although the import alerts contain binding requirements affecting a wide range of facilities, the FDA holds the documents that are merely "guidance." As can be seen by this point, the FDA can be adept in using the APA to find strategic interpretations to its actions. Labeling import alerts guidance, the FDA avoids the time and cost of case-by-case adjudication and the time and cost of notice-and-comment for each new alert.

There have only been three cases challenging the FDA's issuance of a specific import alert using the APA. All three cases involved drugs offered for import (*Bellarno Int'l*).[38] Each challenge focused on whether the new import alert was a substantive rule of general applicability or simply as the FDA claimed an interpretative rule. In order for an import alert to be an interpretive rule, there must be a supporting regulation or statutory provisions. Absent a legislative basis for the enforcement action the Import Alert is not interpreting a statute or regulation, but issuing a new regulation. APA challenges are fertile, but often overlooked as a basis for challenging import alerts.

2.5.3   Statute Specific Challenges

Interstate Commerce Presumption
The FDA may only exercise authority over food products sold outside of a state. This is known as interstate commerce or the commerce between states. Interstate commerce is the constitutional

---

[38] See, *Bellarno Int'l, Ltd. v. Foo & Drug Admin.*, 678 F. Supp. 410 (E.D.N.Y 1988); *Community Nutrition Inst. v. Young*, 818 F.2d 943 (D.C. Cir. 1987); *Syncor Int'l Corp. v. Shalala*, 127 F.2d 90 (D.C. Cir. 1997).

basis that allows Congress to pass many of its federal laws. Rather than allowing every producer to raise a constitutional challenge questioning the FDA's authority to regulate Congress, passed a statutory presumption in Section 379a. The interstate commerce presumption states all FDA-regulated products are in interstate commerce. The burden then shifts off the FDA of proving interstate commerce to the facility seeking to raise the defense. The facility must establish evidence showing the regulated products are not sold outside the state it is located in.

The FDA and Section 704

In addition to APA challenges enabling statutes, such as the FD&C, provide limitations on authority that can be the basis for challenging enforcement actions. Foreign facilities in particular, which are not protected by the US Constitution, benefit immensely from understanding and utilizing statutory challenges. For FDA-regulated facilities, the statutory section that proscribes the FDA's inspection authority is Section 704. This section provides ample options to challenge an inspection.

The heart of Section 704 is the reasonableness standard. The word "reasonable" appears over a dozen times in the statute. Section 704 sets the FDA inspection authority. It authorizes the FDA agents to enter and inspect a facility at "reasonable times" and "within reasonable limits and in a reasonable manner" audit the facility, equipment, and other materials (FD&C).[39] It also sets the timeframe for an inspection requiring agents to "commence and complet[e]" the inspection with "reasonable promptness" (FD&C).[40] Despite its prevalence in the statute there are no additional regulations or guidance documents defining "reasonable." Also as one might expect given the number of facilities consenting to inspections, there are no meaningful cases interpreting reasonableness in the Act. The IOM provides an unhelpful circular definition stating reasonableness means "…what is reasonably necessary to

achieve the objective of the inspection" (IOM).[41] Reasonableness is than context dependent and if analyzed carefully by a facility, clear boundaries can be placed on an inspection using the "reasonableness" standard.

What Is Reasonable?

The reasonableness standard depends on the type of facility. The Act covers every type of food facility, both in terms of the range of products and activities. From seafood to dietary supplements and from farmers and manufacturers to grocery stores and distributors, the Act covers a wide range of facilities. Reasonableness will not carry the same meaning for all facilities.

The timing of an inspection is a basic starting point for measuring reasonableness. Through its compliance programs, the FDA carries the singular policy of conducting inspections when a facility is operating. In other words, the inspection occurs during normal business hours, not covertly when no one is around. The policy of inspecting an operating facility also informs a facility about whether other requests are reasonable. For instance, a facility with high-risk production lines, like switching between allergen and non allergen containing products, an agent may request this operation to run if off-line during an inspection. Other requests, no matter how burdensome or inconvenient, will emerge. The central question must always be, "Are the requests reasonable?" In some cases, the answer will be no. For example, a facility with living quarters on-site may deem a request to access the living quarters unreasonable. Each request requires carefully balancing risks and benefits, which makes planning for an inspection a key part of assessing reasonableness.

A facility may be reluctant to impose its own limits on the inspection. If there are limits, a facility would impose on any visitor or employee than those requests that are reasonable. For example, a facility may have a policy to only allow an inspection when the inspector is accompanied by a facility or a quality manager. Another example is requiring an inspector to follow safety or good

---

[39] 21 U.S.C. § 374(A) and (B).

[40] *Id. at* § 374(B).

[41] See, IOM at § 2.2.1.1.

manufacturing practices also followed by personnel. The inspection is an opportunity to interact with the FDA agent and express areas of concern. If there is an activity, the agent begins to perform that facility can raise the reasonableness question.

Foreign Facilities and Advanced Notice

Inspecting foreign facilities presents unique challenges to the FDA. Given the logistics of coordinating a foreign inspection, advanced notice is typically given. This is a matter of practicality rather than adherence to regulatory requirements. Every facility should conduct a mock inspection, but the advanced notice provides foreign facilities an opportunity to assess reasonableness prior to an inspection. Foreign facilities may not be sure what to expect or what constraints it can impose on the inspection. A smooth inspection process can be secured by understanding the FDA's policy on sampling, photographs, and how to raise reasonableness challenges.

2.6 Comparative Law—International Inspection Methods

A wide variety of inspection models are utilized by various countries. A number of factors dictate how a country may verify food safety and compliance, such as resources and the number of imports. Imports in particular from high-risk countries with known safety and quality issues can become a strong driving force. Japan offers not only one of the most robust and complex food safety system, but also one that takes a proactive stance. The EU is similar to the US FDA relying on post market surveillance.

Japan utilizes a hybrid approach of strong premarket controls along with post market inspections. Japan conducts food sanitation inspections for all fruits, vegetables, and seafood products (The World Bank 2005).[42] Also, like the FDA, it conducts customs inspections on all incoming shipments. Japan imports more food from China

than the USA. China as a notorious food safety violator earned a poor reputation following several high-profile incidents. Japan relies heavily on imports with 60 % of its food supply coming from foreign sources (Fackler, NY Times 2007).[43] This makes it particularly sensitive to safety and quality compliance. Japan not only utilizes high standards, but also subjects imported foods to stringent random testing. The Japanese Health Ministry estimates its test samples from over 10 % of its imports (Fackler, NY Times 2007) . In comparison, the US FDA estimates its tests less than 1 % of its imports. If one of the chief goals of inspections is not only to verify compliance, but act as a deterrent, than arguably a high rate of random inspections serves an equally effective purpose.

The EU in 2013 began consideration of a new proposal for unannounced inspections. Here again, crisis drives change. Following the horse meat scandal, where horse-meat was sold across the EU without disclosure on the label, new rules are under consideration. The rules would set a minimum number of announced inspections (Reuters 2013).[44] EU member states under their own agencies already conduct some level of inspections. The proposal aims to strengthen gaps in the food safety net which lead to the horse-meat scandal and related incidents. The unannounced inspection model used by the EFSA is closest to what the US FDA utilizes.

Although the inspection methods and regulations vary, the driving purpose remains the same. Despite this, the regulatory body's goal is to ensure safe food and accurate labeling. Each country takes its own approach to enacting laws and developing an inspection model to verify compliance. Examining different models allows us to assess the effectiveness of our own approach and find potential new solutions to compliance and regulation.

---

[42] The World Bank 2005).

[43] Fackler 2007.

[44] Reuters 2013.

2.7 Chapter Summary

The primary lesson from this chapter is to introduce the reader to different enforcement models used by the FDA and FSIS. Understanding how the models work and the criteria for formal and informal enforcement actions provides a strong foundation for the remaining chapters. It also allows the reader to compare the effectiveness of the two models. This chapter also examined the grounds to raise challenges to enforcement actions and inspections. This includes the constitutional defenses to the FDA's warrantless inspections under the Fourth Amendment and the concept of due process. Statutory relief was also explored using the APA and enabling Acts.

Overview of Key Points:

- FSIS's use of a compulsory continuous inspection model
- Three classes of FSIS enforcement action
- History of FSIS and its inspection techniques
- FDA use of random inspection model
- FDA use of "for cause" and "surveillance" inspections
- The use of Form 483s for facility inspections
- The role and reviewability of Warning Letters
- The five types of FDA enforcement tools
- Constitutional challenges under the Fourth Amendment
- Statutory challenges under the APA
- Section 704 challenges to FDA inspections

2.8 Discussion Questions

1. Are warrantless searches (i.e. facility inspections) an appropriate tool for a federal agency to use? Are warrantless inspections an integral component to ensuring safe food production? Explain your position using current headlines, regulations, or examples.
2. Does on-site necessarily mean better protection? Explain whether you think the USDA enforcement model protects the public from harm using current headlines, regulations or examples.
3. Find examples of warning letters and regulatory control actions. How easy are the notices to find? Does this act as an effective deterrent?

References

Fackler M (2007) New York Times, Safe Food for Japan, 11 Oct 2007

Reuters (2013) New York Times, Europe Seeks Crackdown on Food Fraud, 6 May 2013

World Bank (2005) Food Safety and Quality Standards in Japan

Adulteration

3

Abstract

This chapter presents the first of two prohibited acts under US food law—adulteration. The concept of adulteration will be traced from its origins in the 1906 Act through the adulteration amendments of the 1950s and 1960s. It will define regulatory definitions and standards of proof required for regulatory agencies to take enforcement actions. The chapter expounds on the idea of economic adulteration, intentional adulteration, indirect adulteration, and the relationship to FSMA, GMPs, tolerances, action levels, and standards of identity. It introduces the adulteration amendments, covering topics such as food additives, animal drug residues, and the Delaney Clause. It leaves for a later chapter a deeper discussion on food and color additives.

3.1 Introduction

With the concept and contours of regulatory enforcement and inspection in hand, the text now turns to one of two statutory triggers for enforcement actions. This section looks at the concept of adulteration and the history of regulation beginning with the original 1906 Act. It will explore the evolution of adulteration as a regulatory and legal concept from its inception in 1906 to the 1938 Act. The 1938 Act continues to provide the central regulatory framework and definitions for adulteration. This chapter will also look at early amendments to the 1938 Act enacted to address specific forms of adulteration. In particular the text will discuss the 1954 Pesticide Residues Amendment, the 1958 Food Additives Amendment, the Delaney Clauses, the Color Additive Amendments, and the 1968 Animal Drug Amendments. This complex system requires a careful understanding of definitions and classification.

3.1.1 Defining Adulteration

Adulteration takes on a specific statutory meaning. One may be quick to draw conclusions about what food adulteration may look like. Experience within a facility may suggest a rodent or insect infestation, while experience as a consumer likely points to the obscene stories of a band-aid, rodent, or some other foreign object found in food. Both would roughly be right, but in the absence of an understanding of the regulations one could not say why the product was adulterated or under what scheme from the Food Drug and Cosmetic Act (FD&C or "the Act").

Comparing the 1906 and 1938 Act

The definition of adulteration begins in 1906. Under the 1906 Act any food that contained "any added poisonous or other added deleterious ingredient which may render such article injurious to health"

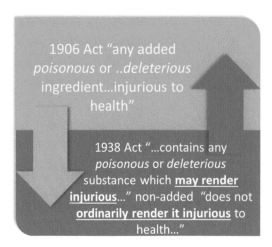

Fig. 3.1 Changes between 1906 and 1938 Act

was adulterated (1906 Act).[1] Terms like "poisonous," "deleterious," and "injurious to health" carried clear meaning. Adulteration referred to a food product that would make you ill or hurt you in some way, such as with glass fragments. Congress, however, did not define the term "added" which could mean any number of things. For instance was Congress referring to intentionally added ingredients or naturally occurring contaminants?

Congress repealed the 1906 Act and expanded the 1938 Act's control over adulterants. Under the 1938 Act any food that "…bears or contains any poisonous or deleterious substance which may render it injurious to health; but in case the substance is not an added substance such food shall not be considered adulterated under this clause if the quantity of such substance in such food does not ordinarily render it injurious to health…" was adulterated (FD&C).[2] This definition, absent the caveats from subsequent Amendments, remains the definition of adulteration. The Food Safety Modernization Act (FSMA) did not alter the definition of adulteration in the 1938 Act. As will be discussed it only added a new rule on intentional adulteration.

The 1938 Act bifurcated the adulteration decision tree without clarifying the term "added." The Act now distinguished between "added" and non-added ("not an added substance"). If the substance is added it must pass the "may render injurious standard," whereas if it is non-added it must meet the "ordinarily render injurious" standard. Thus, two new standards emerged in the 1938 Act (see Fig. 3.1). Yet there was no corresponding clarity in the Act on when one standard applied over the other. The resulting ambiguity left the FDA wide latitude to exercise discretion on which added substances would be evaluated as "ordinarily injurious" or "may render injurious."

When drafting the 1938 Act Congress wanted to provide the FDA greater control over certain added substances. As will be discussed in detail in Section 4 the "may render injurious" standard is less rigorous and permits a wide range of additives into food compared to the "ordinarily injurious" test. Congress enacted a type of license scheme in Section 406 of the 1938 Act. Section 406 provides the FDA to establish tolerances for added "poisonous or deleterious" substances only when the substance is "required" or "cannot be avoided" (FD&C).[3] Thus, for substances that are useful or difficult to eliminate, the FDA can regulate tolerances to ensure the additives remain safe.

The food adulteration definition was originally codified in Section 402. All subsequent amendments follow in sequence. For instance the Pesticides Residue Amendment was codified as Section 408. The adulteration section is currently codified as Section 342 with all Amendments codified in sequential order beginning with Section 346. When searching for definitions and case law be aware of the original codification, which is still often used as a reference, and the current location of the adulteration regulations.

Overview of the Amendments

The 1938 Act provides a triad of controls that functioned in the market for nearly 20 years. It lists adulteration as a prohibited act in Section 331. A substance under the 1938 Act was adulterated if an added substance failed the "may render injurious" test, if a non-added substance fumbled the "ordinarily injurious" standard, or in limited cases where added substances regulated under a tolerance exceeded the maximum

---

[1] 34 Stat. 768 (repealed 1938).

[2] 21 U.S.C. § 342(a)(1) (2014).

[3] 21 U.S.C. § 346.

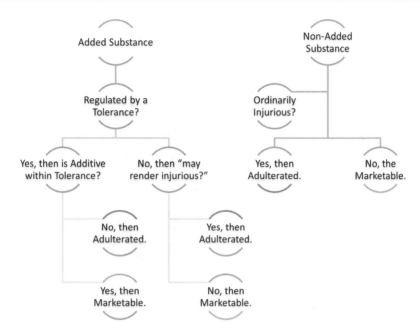

Fig. 3.2 Initial decision tree to determine adulteration

threshold (see Fig. 3.2). This framework sufficed from 1938 until the first Amendment was passed in 1954.

Beginning in 1954 Congress began considering and passing a series of adulteration amendments, a process that would last over the course of six years. It began with the Pesticides Residues Amendment in 1954. The Pesticides Amendment changed the criteria for finding raw agricultural commodities (RACs) adulterated. Originally enacted as Section 408 it is currently codified as Section 346a. The Pesticides Amendment specifically amended Section 406, which authorized the agency to set tolerances. The Amendment deemed a RAC adulterated if it contains a pesticide residue in excess of the tolerances established in the Amendment. The Amendment provides a complex system for determining the tolerances of various pesticides.

Two years later Congress passed the Food Additives Amendment. The Amendment was originally codified as Section 409 but is currently found in Section 348. The Amendment applies to a wide range of food substances, including food packaging materials. It is a complex system that will be addressed as a separate topic in Chapter 6.

For our purposes in Chapter 3 it suffices to say any food additive not approved under the food additive licensure scheme or exceeding the limits of its approval is adulterated.

The Food Additives Amendment also added a new definition to the Act. Prior to 1958 the Act did not need and did not define the term "food additive." Adulteration pertained only to foods as defined in the Act. The food additive definition is found in Section 201(s). The definition sweeps in both direct and indirect additives. Food packaging is an example of an indirect food additive. The main criteria for determining whether the added substance qualifies as a food additive under the Act is whether it exercises a technical effect on the food it is added to.

Food Additive Definition in Section 201(s)
"food additive" means any substance the intended use of which results or may reasonably be expected to result, directly or indirectly, in its becoming a component or otherwise affecting the characteristics of

any food (including any substance intended for use in producing, manufacturing, packing, processing, preparing, treating, packaging, transporting, or holding food; and including any source of radiation intended for any such use)…

The final amendment came in 1960 when Congress passed the Color Additive Amendment. The Color Additive Amendment split apart color additives from the 1958 Amendment. The result was a stricter system of approval and use of color additives. Whereas food additives, those that are not used primarily to impart color, can find approval through a series of exemptions (see Chap. 6), color additives are not subject to any safe harbors. Color additives must be approved and listed by the FDA. Otherwise the color additive will be deemed adulterated. Chapter 6 will discuss color additive approval in more detail.

Color Additive Amendments broke from the sequential numbering model of prior Amendments. The Color Additive Amendment was originally codified as Section 721 far from Section 409. This was in part because color additive regulations applied to all FDA regulated products, not just foods bearing color additives. The 1960 Amendment is now codified as Section 379e.

Although not a separate Amendment the Delaney Clauses deserve a separate discussion. The Delaney Clauses are named after Congressman James "Jim" Delaney (see Fig. 3.3). The first Delaney Clause is found in the Food Additives Amendment. It precludes the FDA from approving any additive that may be found to induce cancer in humans or experimental animals. The Delaney Clause is an addition to the Food Additives Amendment and only applies to substances governed by Section 409. The Delaney Clause was also added to the Color Additive Amendment.

The final and most recent adulteration Amendment to the 1938 Act came 30 years later when Congress *passed* the Animal Drug Amendment.

Fig. 3.3 Image of Congressman Jim Delaney

Following the passage of the Food Additives Amendment, the FDA regulated drugs administered to food-producing animals under Section 409 and the appropriate drug approval provisions. Section 409 applies to drugs administered directly to animals that could "reasonably be expected" to leave residues in human food. The 1968 Amendment simplified the process. Under the Amendment a single approval system for animal drugs was implemented. The Amendment also prohibited the sale of drugs likely to leave residues and the sale of food containing residues absent FDA approval.

The new system of original definitions and Amendments created a complex system (see Fig. 3.4). In the majority of cases it can be challenging to define exactly why a product is adulterated. Even the seemingly straightforward example layers of exemptions and statutory criteria require verification and analysis. It is important to understand the web of regulations both for enforcement purposes and when developing a new product.

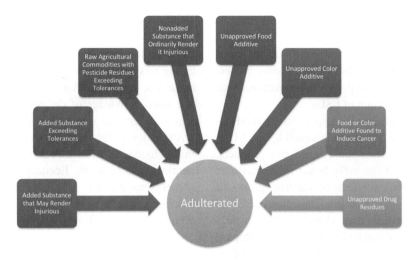

Fig. 3.4 Adulteration framework

The USDA and Adulteration and Food Additives

Notably absent from the discussion above is any mention of the USDA and FSIS. Each of the enabling acts controlled by FSIS contains a definition of adulteration. FSIS remains in primary control for inspecting and enforcing adulterated meat, poultry, and eggs. The definitions of adulteration are identical to the definitions used by the FDA.

Poultry Products Inspection Act Section 453(g)(1)
Poultry product is adulterated " … if it bears or contains any poisonous or deleterious substance which may render it injurious to health; but in case the substance is not an added substance, such article shall not be considered adulterated under this clause if the quantity of such substance in or on such article does not ordinarily render it injurious to health…"

Federal Meat Inspection Act Section 601(m)(1)
Meat or meat food products are adulterated " … if it bears or contains any poisonous

or deleterious substance which may render it injurious to health; but in case the substance is not an added substance, such article shall not be considered adulterated under this clause if the quantity of such substance in or on such article does not ordinarily render it injurious to health…"

Egg Products Inspection Act Section 1033(a)(1)
Egg or egg products are adulterated " … if it bears or contains any poisonous or deleterious substance which may render it injurious to health; but in case the substance is not an added substance, such article shall not be considered adulterated under this clause if the quantity of such substance in or on such article does not ordinarily render it injurious to health…"

The FDA remains the primary agency for regulating the use of food additives even in meat and poultry. The FDA and FSIS share responsibility for the safety of food additives used in meat, poultry, and egg products. The FDA, however, sets the minimum safety regulations which FSIS must enforce. Under the Food Additives Amendment all proposed additives are first evaluated by the FDA. The FDA will determine whether or not the additives may be used. A secondary review may be conducted by the Risk, Innovations, and Management division of FSIS.

FSIS can only elect to apply stricter food additive standards than those adopted by the FDA. As experts in meat, poultry, and egg products, FSIS believes it can make better determinations about the technical effects of additives proposed for FSIS regulated products. Thus, it may elect to set higher standards than what the FDA sets for a food additive. When a higher standard is adopted an additive must be approved first by the FDA, then by FSIS. FSIS may never set a lower standard than the one set by the FDA. FSIS provides the example of sorbic acid on its website (FSIS Additives).[4] The FDA approved sorbic acid as a food additive. But in order to use this product in meat, any applicant seeking to use it would need approval from FSIS who adopted a more stringent sorbic acid standard. FSIS set a higher standard for sorbic acid in meat out of concern that the additive could "mask spoilage caused by organisms that cause foodborne illness" (FSIS Additives).[5]

3.2   Types of Adulteration

The definition of adulteration provided in Section 1 above was intentionally incomplete. It addressed only the poisonous and deleterious standard for added or naturally occurring (non-added) substances. It omitted the remaining provisions addressing other means by which food becomes adulterated. In Subsections 3 and 4 the Act deems adulterated facility conditions that could indirectly contribute to adulteration. Subsection (b) regulates the concept of economic adulteration. A final type of adulteration considered by the agency is intentional adulteration or contamination. This can take several forms, including acts of terrorism, disgruntled employees, consumers, or competitors. It captures an act done with knowledge of the potential injury or harm to others. As will be discussed below the FDA's policy is for facilities to adopt food defense or food security measures to protect against intentional adulteration. Economic adulteration is often swept into this concept since it is clearly an intentional act, but for purposes of this text will be treated separately.

---

21 U.S.C. 342(a)(3)-(5);(b)

...or (3) if it consists in whole or in part of any filthy, putrid, or decomposed substance, or if it is otherwise unfit for food; or (4) if it has been prepared, packed, or held under insanitary conditions whereby it may have become contaminated with filth, or whereby it may have been rendered injurious to health; or (5) if it is, in whole or in part, the product of a diseased animal or of an animal which has died otherwise than by slaughter...

(b) Absence, substitution, or addition of constituents

(1) If any valuable constituent has been in whole or in part omitted or abstracted therefrom; or (2) if any substance has been substituted wholly or in part therefor; or (3) if damage or inferiority has been concealed in any manner; or (4) if any substance has been added thereto or mixed or packed therewith so as to increase its bulk or weight, or reduce its quality or strength, or make it appear better or of greater value than it is.

---

[4] United States Department of Agriculture, Food Safety Inspection Service, *Additives in Meat and Poultry Products*, available at http://www.fsis.usda.gov/wps/portal/fsis/topics/food-safety-education/get-answers/food-safety-fact-sheets/food-labeling/additives-in-meat-and-poultry-products/additives-in-meat-and-poultry-products.

[5] *Id.*

The 1938 Act encompassed a comprehensive view of adulteration. Its provisions regulated scenarios where substances were added, naturally occurring, and posed a risk of contamination during production. The Act ensured the public's safety from harmful products and from products whose only harm was economic. The sections below will look closer at the new concepts of intentional, indirect, and economic adulteration.

3.2.1 Economic Adulteration

Economic adulteration involves the selling of a product containing inferior ingredients but marketed as containing superior ingredients. It is an attempt to confuse the consumer and profit from the fraud. This type of pass-off offense finds ancient roots in China, Greece, and Rome. Examples in the early 1900s when the Pure Food and Drug Act was passed include milk diluted with water or maple syrup diluted with cane sugar. In most cases there is no risk of harm.

Examples of Ancient Economic Adulteration (Hart 1952)[6]

Law of Moses

The fat of an animal found dead or torn by wild animals may be used for any other purpose, but you must not eat it. (Leviticus, 7:24.)

Ancient Chinese Literature and Legal Code

"The Supervisor of Markets had agents whose duty it was to prohibit the making of spurious products, and the defrauding of purchasers."

When dried or fresh meats cause men to become ill, all the leftover portions should be speedily burned. The violator will be flogged 90 strokes. He who deliberately gives or sells it to another will be banished for a year, and if the person to whom it has been given or sold dies, the offender will be hanged...

Roman—Pliny the Great

"...the dealers have set up regular factories where they give a dark hue to their wine by means of smoke, and, I regret to say, employ noxious herbs, in as much as a dealer actually used aloe for adulterating the flavor and color of his wine."

Pliny also provides examples of wine diluted with water and flour mixed with chalk to increase its whiteness and shortness.

The lack of harm makes economic adulteration unique. In general a food is deemed adulterated when it contains a "poisonous or deleterious" substance that could pose a health risk. In many cases economically adulterated products lack this risk. There are early examples, such as ink dyes used as coloring agents, which were injurious to both body and purse. Arguably such a case today would be treated as intentionally adding an adulterant rather than economic adulteration. In fact, it is a struggle to find cases where the FDA uses its limited resources to pursue economic adulteration. This is because the resources are directed to the products posing a risk of physical injury, which economic adulteration lacks.

Sole Economic Harm Not Classified as Misbranding

Economic adulteration may be a misnomer. Typically when the Act enforces provisions related to economic harm, deception, or fraud where there is no physical risk, it does so as misbrandings. Misbrandings will be discussed in the next chapter, but can generally be described as labeling offenses. The distinction between misbranding and adulteration lies in the enforcement tools the agency can use.

Adulteration allows the agency to move quickly with little proof of adulteration. The bulk of adulteration refers to areas of potential injury, which requires immediate and swift action. Thus, the FDA can make multiple seizures or administrative detention of adulterated foods. Misbranded items in many cases lack the immediacy of adulteration. Some cases like labels failing to declare an allergen may be an exception. When the concept is limited to fraudulent claims, imposing only economic harm it is much harder for the FDA to act.

---

[6] See F. Leslie Hart, *A History of the Adulteration of Food Before 1906*, 7 Food Drug Cosmetic Law Journal 5 (Jan. 1952).

Economic adulteration shares the characteristics of misbranding but carries the enforcement triggers of adulteration. It is a hybrid creature born of the circumstances of the 1906 Act. Economic adulteration bears little, if any, difference between other fraudulent statements made on a label. It relates to ingredients, whereas other claims may relate to health benefits or product features. Yet for no reason other than historical happenstance the use of inferior ingredients passed off as superior ones is treated as adulteration. Only when considering the history of the 1906 Act does this make sense.

The 1906 Act arouse out of a tumultuous period of food production. This text in previous chapters described the findings of Upton Sinclair. It was not only in meat packing plants that deplorable conditions and acts were found. It was fairly pervasive throughout the food industry. It is likely that this lingering fear and revulsion of abuse of trust in food production is what continues to keep these acts classified as adulteration.

Concurrent Jurisdiction—FDA and Federal Trade Commission

Economic adulteration is the first area of law where we encounter concurrent jurisdiction with a federal agency that is unaffiliated with food regulation. In 1914 Congress provided the Federal Trade Commission (FTC) to enforce deceptive commercial practices. This came in the form of the Wheeler Lea Act. The FTC interprets its enabling acts to provide it authority to stop the sale of food products that deceive the public or provide the seller an unfair advantage in the market.

The FDA exercises primary control over economic adulteration. As will be seen with misbranding, the FTC largely defers to the FDA on economic adulteration. The agencies operate under a working agreement which outlines the boundaries of authority (FDA-FTC MOU).[7] Under the working agreement the two agencies

will either act concurrently or through case referrals. The FTC claims jurisdiction over advertising and marketing and the FDA is charged with regulating the labeling. Labeling leaves the FDA in control of the standards on what constitutes economic adulteration. The reason being the front of the label will declare the product's identity, such as blueberry muffins, and the ingredient panel will declare ingredients, like blueberry bits (see Modern Examples on p. 78). This is the only area of adulteration where the FTC can exercise authority. In every other place the FTC can act in relation to marketing and claims. This again raises the question why economic adulteration is not a misbranding offense. The two agencies are moving towards cooperation rather than competition, as will be discussed in Chapter 4, and are even issuing joint warning letters.

Problematic 1906 Act

The 1906 Act attempted to regulate economic adulteration. The Act began with a general definition of adulteration. Language like "substitutions," however, could be broadly interpreted to preclude any modern developments in food production. Seemingly aware of the impediment Section 7 the drafters added two important exemptions.

Section 7 of the Pure Food and Drug Act (1906)

First. If any substance has been mixed and packed with it so as to reduce or lower or injuriously affect its quality or strength.

Second. If any substance has been substituted wholly or in part for the article.

Third. If any valuable constituent of the article has been wholly or in part abstracted.

Fourth. If it be mixed, colored, powdered, coated, or stained in a manner whereby damage or inferiority is concealed…

---

[7] MOU 225-71-8003, Memorandum of Understanding between the Federal Trade Commission and the Food and Drug Administration, available at: http://www.fda.gov/AboutFDA/PartnershipsCollaborations/MemorandaofUnderstandingMOUs/DomesticMOUs/ucm115791.htm.

The exemptions were necessary to make the economic definition work but were also prone to abuse. Section 8 on misbranding provided two labeling requirements which, if followed, allowed

Fig. 3.5 Examples from the FDA of distinctive names

modern processed foods. The first exemption is known as the "distinctive name" exemption. The most infamous example is "Bred Spread" which is neither peanut butter nor jam, but its own distinctive name (see Fig. 3.5). Another early example was "Grape-Smack" which was imitation grape juice. Section 8 allowed a fabricated food on the market if it was labeled with a distinctive name. A second provision allowed compounds, imitations, or blends if proper labeled. Thus, rather than "butter" a label could bear the statement "imitation butter." Under the distinctive names and imitation exemptions honest manufacturers could develop and sell new innovative foods. It also worked to effectively gut the economic adulteration provisions.

> **What is Hummus?**
> Sabra Dipping Co., LLC submitted a citizens petition requesting a standard of identity for hummus. Currently a variety of products sell under the common name "hummus" without needing to adhere to a standard of identity. Sabra seeks to identify hummus as consisting of chickpeas and no less than 5% hummus. The 11-page petition can be found using Docket FDA-2014-P-0259.

Enforcing the distinctive names exemption proved challenging. Manufacturers quickly learned to adopt distinctive names and claim compliance with Section 8. Grape-Smack, for example, escaped enforcement for economic adulteration. The imitation grape juice contained calcium acid phosphate and corn starch, but was labeled with the initials of a more expensive brand, "C.A.P." (*100 Barrels of Calcium Acid Phosphate*).[8] The courts were soon flooded with cases focused not on the inferior ingredients or alleged fraud, but on whether names likes "Macaroons" or "Maple Flavo" were distinctive names. This was an abstruse and bizarre consequence of the Section 8 distinctive names exemption.

The second exemption did not fare any better in court. The exemption was written broadly to provide safe harbor to any product labeled "compound," "imitation," or "blend." Some courts interpreted this language plainly giving shelter to any product appropriately labeled. Other courts read Sections 7 and 8 together and required the ingredients of the compound to be listed. For instance one court found "Fruit Wild Cherry Compound" (*Weeks*)[9] not economically adulterated because it was labeled "compound," but a second court found "Compound Ess Grape" (*Schider*)[10] economically adulterated despite complying with Section 8 because it listed a single ingredient imitation grape essence. Litigation was unpredictable and the ease of challenging the FDA left few cases resolved administratively.

Improvements in the 1938 Act

For over 30 years courts, industry, and the FDA suffered under the 1906 economic adulteration fiasco. The 1938 Act directly addresses all of the concerns raised, enforcing the economic adulteration provisions between 1906 and 1938. It resolves all of the confusion and abuses of the 1906 Act. It began by striking the distinctive names and imitation exemption, then added new provisions to take their place.

---

[8] *United States v. 100 Barrels of Calcium Acid Phosphate*, White & Gates 58 (N.D. Cal. 1909).

[9] *Weeks v. United States*, 224 Fed. 64 (2 s Cir. 1915), *aff'd on other grounds* 245 U.S. 618 (1918).

[10] *United States v. Schider*, 246 Y, S, 519 (1918).

The 1938 Act replaced distinctive names with standards of identity. The FDA gained new authority to created standards of identity for each food product following public notice and hearing. No longer would "Bred Spred" escape enforcement as economically adulterated. If it did not meet the standard of identity for peanut butter or jam then it was economically adulterated. It could continue to be its own non-standardized product, but it would need to label itself imitation and state its ingredients on the label. Thus a product either meets a standard of identity or must be labeled imitation and list all its ingredients. It could also petition the FDA under 21 CFR 130.5 to create a new standard of identity for its product. The petition is not unique to standards of identity but is called a citizens petition, which can be used for a variety of purposes.

The same definition of economic adulteration from the 1906 Act continues today. The 1938 Act did not change the overall definition of economic adulteration. It remains as broad and unworkable today as it was when enacted over a hundred years ago. Absent the exemptions to provide some boundaries defining economic adulteration, the provision is ambiguous. Many commentators describe economic adulteration as too broad to interpret plainly and too ambiguous to be interpret intelligently.

Modern Examples

Economic adulteration is not a relic of the past. It remains as vibrant and attractive cost-saving measure as it was in ancient Rome, Greece, or China. Many modern examples highlight the use and lack of enforcement over economic adulteration. For instance, blueberries are an expensive but highly marketable ingredient. Read your labels carefully. Does that package list real blueberries? In 2011 a consumer advocacy group found several products claiming blueberries, such as blueberry bagels, muffins, and cereals that did not contain actual blueberries. Many labels instead claimed "blueberry bits" which included sugar, food dye, and corn syrup. Unsuspecting consumers would buy the product believing it contained a superior ingredient, blueberries, when it actually contained an inferior ingredient, blueberry

bits. This lead to several labeling lawsuits as discussed in Chapter 7. Blueberry bits are a classic example of economic adulteration and one that went unenforced by the FDA.

3.2.2   Indirect Adulteration—Filth and Insanitary Conditions

Adulteration by filth or insanitary conditions is made for television material. It involves exactly the worse we can imagine—facilities infested with mice or cockroaches, or food contaminated with objects, insects, or worse. This type of adulteration echoes *The Jungle* in finding the worst of food production.

FSIS and the FDA share the same definition for indirect adulteration. Indirect adulteration, it should be noted, is distinct from indirect food additives. Rather than discussing Section 409, food additives, here the provisions for indirect adulteration are found in Sections 402(a)(3) and (4). As with the all aspects of adulteration the FMI, PPI, EPI mirror the FD&C. Therefore both agencies are operating under the same concept of adulteration by filth and insanitary conditions.

The FDA organizes indirect adulteration under 402(a)(3) and (4) into three subcategories. When evaluating adulteration the FDA will look for health hazards, indicators of insanitation, and natural or unavoidable defects. The health hazard category includes physical, chemical, and microbiological hazards associated with extraneous materials. Insanitary conditions include criteria like visibly objectionable contaminants and evidence of infestations. The health hazard category encompasses criteria from the Hazard Analysis and Critical Control Point (HACCP) system. All three categories are subject to control.

Defect Action Levels for Filth

Most filth is unavoidable. As unsettling as it may seem food will never be defect-free. Insect parts for example are an accepted element of harvesting everything from grapes to cocoa. Similar to the tolerances for added substances, the FDA sets "action levels" for unavoidable poisonous or deleterious substances. Tolerances and action levels are similar in that both set a maximum level for a substance that could pose harm. Tolerances, however, as intentionally added substances are subject to a formal rule. The FDA will provide public notice and hold hearings before setting a tolerance. The tolerance is then giving the full force of law as if it were part of the Act. Action levels are informal judgment on the quantities of a particular contaminant that will not pose a health risk. Action levels, therefore, represent a promise by the FDA not to enforce any level of contamination from unavoidable sources but only commit to formal enforcement when a threshold or action level is crossed.

Action levels only apply to the unavoidable or natural defect category. The FDA publishes all of the action levels in a guidance document. If the contaminant is not subject to an action level, the enforcement discretion will not apply. This includes blending a food product containing a substance in excess of an action level with another food product in an attempt to dilute the contaminant (FDA Guidance).[11] Action levels can also be found in Parts 109 and 509 of the CFR.

Examples of FDA Action Levels (FDA Guidance)[12]

LEAD		
Commodity	Action level (μg/ml leaching solution)	Reference
Ceramicware		
Flatware (average of 6 units)	3.0	CPG 545.450
Small hollow-ware (other than cups and mugs) (any 1 of 6 units)	2.0	CPG 545.450

Commodity	Action level (μg/ml leaching solution)	Reference
Large hollow-ware (other than pitchers) (any 1 of 6 units)	1.0	CPG 545.450
Cups and mugs (any 1 of 6 units)	0.5	CPG 545.450
Pitchers (any 1 of 6 units)	0.5	CPG 545.450
Silver-plated hollowware		
Product intended for use by adults (average of 6 units)	7	CPG 545.450
Product intended for use by infants and children (any 1 of 6 units)	0.5	CPG 545.450

	MERCURY	
Commodity	Action level	Reference
Fish, shellfish, crustaceans, other aquatic animals (fresh, frozen or processed)	1 ppm methyl mercury in edible portion	CPG 540.600
Wheat (pink kernels only)	1 ppm on pink kernels and an average of 10 or more pink kernels/500 g	CPG 578.40

---

[11] See Guidance for Industry: Action Levels for Poisonous or Deleterious Substances in Human Food and Animal Feed (2000).

[12] See Guidance for Industry: Action Levels for Poisonous or Deleterious Substances in Human Food and Animal Feed (2000) available at: http://www.fda.gov/food/guidanceregulation/guidancedocumentsregulatoryinformation/chemicalcontaminantsmetalsnaturaltoxinspesticides/ucm077969.htm#merc.

Good Manufacturing Practices and Insanitary Conditions

Good Manufacturing Practices (GMPs, or cGMPs) are the primary control to maintain minimum sanitary conditions in a facility. The Good Manufacturing Practices are published in 21 CFR 110 for food and 111 for dietary supplements. The GMPs outline all of the sanitation controls and methods for equipment, grounds, shipping, and facility. An entire GMP is dedicated to pest control, the most prevalent citation for insanitary conditions.

GMPs are important for many reasons. The obvious value in GMPs is providing a uniform safety net to ensure all food products are produced, stored, and shipped in a way that protects the integrity and wholesomeness of the food. GMPs are also important because the procedures serve as invisible link in the bond of trust between consumer and producer. A consumer cannot visit every food facility that makes the product in their cupboard. The global economy is about production occurring away from our homes and communities by a number of actors we trust but cannot see. Neither can consumers see the effects of that facility—the rats near finished product, the cockroaches near production lines, the mold on the floor or ceiling, and so on. GMPs when followed provide the public confidence that the facility their food originated at is a clean facility. This is often why GMP violations are prevalent in Form 483.

FSMA and the Hazard Analysis Risk-Based Controls Rule

Identifying and controlling hazards ensures food is safe. Hazards can include sanitation practices, but also a variety of issues unique to particular product or class of products. Soft cheeses, for example, are prone to contamination with Listeria or seafood, which requires refrigeration at specific temperatures. Failing to control facility-wide and product-specific standards allows potential hazards into a food product, including chemicals, physical hazards, and microbiological contamination.

FSMA fundamentally changes how hazards are controlled. Prior to FSMA, facilities could voluntarily adopt a HACCP plan, which aims at identifying and controlling hazards. Otherwise there was a risk of adulteration, but there was no enforcement or mechanism to ensure that hazards were controlled. FSMA changed the existing paradigm by introducing a new rule known as the Hazard Analysis Risk-Based Control Rule, or Preventative Controls rule. The Preventative Controls rule implements two changes. It revises the GMP regulations with minor tweaks and more importantly introduces a mandatory hazard control system.

The Preventative Controls rule essentially introduces a HACCP-esque system for qualifying food facilities. The rule varies from HACCP in key areas, but the overall aim is the same. The rule requires science and risk-based preventive controls as necessary to prevent hazard associated with a facility and its products. Not all facilities are subject to the rule with exemptions for small and very small facilities. There are also modified requirements for certain low-risk activities. A facility must identify only those hazards "reasonably likely" to occur. It is then given flexibility in what preventative controls are used, but the FDA will assess the adequacy of the control. Hazard identification and control will soon be a mandatory means for ensuring against indirect adulteration.

3.2.3 Intentional Adulteration

Intentional adulteration involves any number of acts taken with the purpose of causing harm. This is not the typical case of adulteration. Instead, it is the rare case that requires vigilance to protect against the widespread harm that could result from such an act. It can involve a person slipping away in a transport vehicle, tampering with food in the grocery store, or introducing an adulterant in a production facility.

Interagency Approach

Intentional adulteration typically involves acts beyond a food facility's or the FDA's control. Acts of terrorism or tampering with products in a grocery store require an interagency approach. The FDA and USDA coordinate with a long list of agencies on food defense, including the Center for Disease Control and Prevention (CDC), the Department of Homeland Security (DHS), the Federal Bureau of Investigation (FBI), the Environmental Protection Agency (EPA), the Department of Defense, the Department of Energy, the Department of Commerce, and the Department of the Interior. The coordination also envelops state, local, tribal, territorial, and private sector partners to develop food safety plans. The FDA for its part completed vulnerability assessments of a variety of products and processes within the food and agriculture sector along with developing several guidance documents for industry.

Guidance Documents

Prior to FSMA the main tool for the FDA to address intentional adulteration was through Guidance Documents. The FDA issued five Guidance Documents and launched a new ALERT system to assist industry in protecting the food supply against intentional adulteration. All five Guidance Documents were issued in 2003. Each Guidance Document addressed food security preventive measure for specific types of facilities. It includes guidance for producer, processors, and transporters, importers, milk industry, retail food stores, and food service establishments, and the cosmetic industry. The Guidance Documents outline best practices in food defense and preventative measures to reduce the risk of intentional adulteration. The Guidance Documents were updated most recently in 2007. The FDA also developed a number of training programs and resources for State and industry partners.

FSMA and the Rule for Focused Mitigation Strategies to Protect Food Against Intentional Adulteration

FSMA introduced a new rule to mandate preventative controls aimed at mitigating the risk of intentional adulteration. The rule requires facilities covered by the rule to develop Food Defense Plans. The FDA identified four key activities most vulnerable to intentional adulteration.

> **Four Activities Identified in the Proposed Rule**
> 1. bulk liquid receiving and loading;
> 2. liquid storage and handling;
> 3. secondary ingredient handling (the step where ingredients other than the primary ingredient of the food are handled before being combined with the primary ingredient); and
> 4. mixing and similar activities.

Facilities engaged in any of the four activities would need to conduct a vulnerability assessment as part of a Food Defense Plan. The vulnerability assessment would identify specific mitigation strategies and steps to reduce a facility's risk of intentional adulteration. Together this forms the facility's Food Defense Plan, which would commit the facility to preventative actions to protect against intentional adulteration. It is a nebulous rule aimed at tackling a rare and ambiguous threat.

3.3 Added Substances and Adulteration

Determining if a product is "poisonous and deleterious" and thus adulterated depends on whether the substance is added or non-added (naturally occurring). The 1906 Act contained one standard, while the 1938 Act retained the standard for added substances and created a new standard for naturally occurring substances. This is in part what makes the adulteration framework difficult to work within. There are series of classifications and definitions to navigate before learning the appropriate statutory standard to apply. The process begins by determining the meaning of "added."

3.3.1 Defining Added for the FDA

The statute does not provide a definition for "added." Unlike most statutory provisions the term "added," both in the 1906 Act and the 1938 Act, remains undefined. As mentioned in the introduction such ambiguity invited the FDA to exercise its discretion in determining when a substance was added and when it was naturally occurring.

Not until 1977 did the FDA issue a regulation defining "added." In 21 CFR 109.3(c) and (d) it defined both added and non-added as used in Section 402. Under the 1977 definition, an added substance is defined by the absence of a naturally occurring or non-added substance. Essentially if a substance is not naturally occurring, defined as inherent in the substance added to food, then it is added. This mirrors the argument the FDA made in court prior to issuing the regulations (*An Article of Food ... Swordfish*).[13] Naturally occurring substances can migrate from subsection (c) to subsection (d) if increased to "abnormal levels." The definition brings new clarity and expands its authority to apply the more rigorous standard for added substances to naturally occurring substances in excess of a FDA set threshold. This raises new questions about when "abnormal levels" become added.

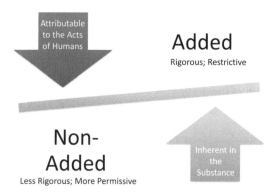

Fig. 3.6 Comparing added and non-added standards

21 CR 109.3(c) and (d) Non-Added and Added Definition

(c) A naturally occurring poisonous or deleterious substance is a poisonous or deleterious substance that is an inherent natural constituent of a food and is not the result of environmental, agricultural, industrial, or other contamination.

(d) An added poisonous or deleterious substance is a poisonous or deleterious substance that is not a naturally occurring poisonous or deleterious substance. When a naturally occurring poisonous or deleterious substance is increased to abnormal levels through mishandling or other intervening acts, it is an added poisonous or deleterious substance to the extent of such increase.

The few courts to review the 1977 definitions generally upheld the expanded definition. The leading case interpreting the "added" language of Section 402 is *United States v. Anderson Seafoods Inc.*[14] *Anderson* involved a seller of swordfish containing levels of mercury in excess of the FDA's action limit of 0.5 ppm of mercury for fish. The FDA sought an injunction and Anderson sought

declaratory judgment stating the fish was safe. The district court decided against Anderson and issued a declaratory judgment holding that fish tissue with mercury levels in excess of 1.0 ppm was adulterated under the Act. The district court held that all mercury was "added" under the Act regardless of its source (*Anderson*).[15] Anderson appealed to the Fifth Circuit.

The Fifth Circuit sustained the district court but clarified the meaning of "added" in Section 402. The court looked to the sparse legislative history for the drafter's intent in using the language. It concluded the legislative history showed "added" meant attributable to acts of man, and "not-added" meant attributable to events of nature. The court, however, held that the FDA was not required to differentiate between the portion of mercury found in the fish attributable to acts of humankind and the quantity of mercury from natural sources. It concluded:

> ...where some portion of a toxin present in a food has been introduced by man, the entirety of that substance present in the food will be treated as an added substance and so considered under the "may render injurious to health" standard of the Act (*Anderson*).[16]

Therefore, an added substance is any portion of a substance present in food which is attributable to some degree to the acts of humankind (see Fig. 3.6).

---

[13] *United States v. An Article of Food Consisting of Cartons of Swordfish*, 395 F. Supp. 1184 (S.D.N.Y. 1975) (held mercury in fish was an "added" substance even though present for centuries because it did not occur naturally).

[14] *622* F.2d 157 (5th Cir. 1980).

[15] *Id.* at 158–159.

[16] *Id.* at 162.

3.3.2   Added and Meat and Poultry

FSIS enforces identical language in its enabling acts requiring an analysis of the term "added" for meat and poultry. In its training materials FSIS adopts a definition of "added" similar to that used by the FDA. FSIS products contain a wide range of natural pathogens, like *Salmonella* on poultry products, which emphasizes the need for a clear definition of "added." FSIS, like the FDA, looks to determine if the presence of the contaminant is attributable to the acts of humans.

FSIS cites two examples of added adulterants under the Section 601(m)(1) of the FMI. The examples are the presence of *Listeria* and *E. Coli* O157:H7 in ready-to-eat (RTE) meat products. RTE products pose the greatest possibility of risk because unlike cuts of meats that are cooked or carry instructions on proper handling, RTEs will not be exposed to any kill-step prior to consumption. FSIS concluded based on scientific studies that the pathogens are only present "due to the way in which product is handled or produced" (FSIS Training).[17] For example, *E. coli* is deemed an added substance because it is spread from the digestive tract during slaughtering and processing. The RTE would not inherently carry the O157:H7 strain of *E.coli*. Thus, "added" for FSIS purposes under the FMI focuses on some act during slaughter or process that would introduce the adulterant in the food.

Courts have been less receptive of FSIS' attempts to expand its authority under the "added" standard. As FSIS began to roll out the new HACCP program in 1998, it increasingly relied on its own testing to determine whether a plant's HACCP plan was working properly. HACCP required microbial testing of *Salmonella* levels in finished meat and poultry. Under regulations adopted by the USDA if a plant's products repeatedly exceeded the *Salmonella* limits imposed by the regulations, the USDA could suspend or permanently withdraw inspectors.

A Circuit Court struck down FSIS' ability to control excess levels of *Salmonella*. Supreme Beef Processors Inc. operated as a meat processor and grinder. It was well known for supplying ground beef to public schools. In 1999 Supreme Beef failed three FSIS *Salmonella* tests in the course of eight months. One test found 47 % of ground beef samples contaminated with *Salmonella*. Consistent with its regulations FSIS notified the company of its intent to permanently withdraw its inspectors. Supreme Beef sought judicial relief from the decision and ultimately won (*Supreme Beef*).[18]

The Circuit Court found *Salmonella* was not an adulterant under the FMI. FSIS was not brining a traditional "added" adulterant case. Instead it was attempting to use the *Salmonella* test as a verification of HACCP. If a facility's HACCP plan worked effectively, then the *Salmonella* rates should be reduced. Otherwise, there was a risk the meat products were "produced or held" in insanitary conditions. FSIS was using the presence of *Salmonella* not as finding of adulteration but as a proxy for the potential presence of indirect adulteration. Since *Salmonella* was not only naturally occurring in meat products but also subject to a kill-step in the form of proper cooking and handling instructions, it was not controlled by the adulteration scheme. Hence, current FSIS rules focus on RTEs, which are neither subject to a kill-step nor expected to naturally contain adulterants like *Salmonella*.

3.3.3   May Render Injurious to Health Standard

May render injurious is a prohibitory standard and not a licensing system. Unlike some of the Amendments discussed below, the "may render injurious" standard does not set criteria or a framework for added substances. It is purely a prohibitory standard that plays a part in formal enforcement. To enforce the standard, as was the case in *Anderson*, the FDA must first test samples of potentially contaminated food. This can

---

[17] United States Department of Agriculture, Food Safety Inspection Service, Entry Training: FSIS Statutes And Your Role, available at http://www.fsis.usda.gov/wps/wcm/connect/b751f8c8-ed46-428b-8867-0e5f70c3e394/PHVt-Statutes_Role.pdf?MOD=AJPERES

[18] 275 F.3d 432 (5th Cir. 2001).

occur by seizure or through an inspection. This is one reason why sampling in facilities can be a highly contentious issue. Once a sample is tested the FDA must find witnesses, typically experts in their respective field, to testify why the added substance may be poisonous or deleterious to consumers. All of this must be proven in court. Straightaway it becomes apparent the evidentiary hurdles created by the "may render injurious" standard are enormous for the FDA. It is only once these hurdles are cleared and proven in court can action be taken on the adulterated food.

The leading case on the may render injurious standard was decided in 1914. Recall the 1938 Act did not change the standard for added substances from the 1906 Act. Thus, this one-hundred-year-old case still bears relevance on the 108-year-old statutory standard. The case is *United States v. Lexington Mill & Elevator Co.*[19] In *Lexington Mill* the FDA seized and sought to condemn flour treated with a process using nitrogen peroxide gas. Some quantity of the gas remained in the flour following the process. The district court provided the jury instructions for its deliberations informing it that any quantity of a poisonous substance would render the food adulterated (*Lexington Mill*).[20] The Supreme Court found the instruction an incorrect interpretation of the 1906 Act.

The Supreme Court's decision clarified many aspects of the "may render injurious" standard. It began by holding the standard applied to the food containing the added substance and not the added ingredient itself (*Lexington Mill*).[21] The FDA then held the burden of showing that the food "may be injurious" to consumers. The court noted this showing need not be conclusive, but the "may render injurious" standard was satisfied if it showed a significant possibility the food could be harmful (*Lexington Mill*).[22] Assessing the possibility of harm requires a detailed analysis. The FDA would need to consider a variety of factors such as the vulnerability of particular

segments of the population, like the sick, the young, and the aged, who may eat the food. If this analysis showed the food may pose a possibility of injury then the may render injurious standard was met. But if the analysis found the food could not after "reasonably consider[ation]" injure the health of any consumer than the food was not adulterated (*Lexington Mill*).[23] It did not matter that it contained a small addition of a poisonous or deleterious ingredient. If the analysis failed then the FDA could not take enforcement action based on adulteration.

A 1958 challenge to the survivability of *Lexington* failed. The Supreme Court ruled the language interpreted in *Lexington* survived the 1938 recodification in Section 402(a)(1) (*Florida Citrus Exchange*).[24] The language is identical in both acts. Thus, one hundred years later *Lexington* exercises an effect on the FDA. The precedent bars the FDA from concluding that a food is adulterated under the "may render injurious" standard absent a showing a significant possibility of harm to consumers, including vulnerable populations.

U.S. v. Lexington Mill & Elevator Co.
Mr. Justice Day delivered the opinion of the court:

The petitioner, the United States of America, proceeding under § 10 of the food and drugs act … sought to seize and condemn 625 sacks of flour in the possession of one Terry, which had been shipped from Lexington, Nebraska, to Castle, Missouri, and which remained in original, unbroken packages. The judgment of the district court, upon verdict in favor of the government, was reversed by the circuit court of appeals for the eighth circuit … and this writ of certiorari is to review the judgment of that court.

The amended libel charged that the flour had been treated by the 'Alsop Process,'

---

[19] 232 U.S. 399 (1914).

[20] *Id.* at 412.

[21] *Id.* at 410–11.

[22] *Id.* at 411.

[23] *Id.* at 411.

[24] See *Fleming v. Florida Citrus Exch.*, 358 U.S. 153, 161 (1958).

so called, by which nitrogen peroxide gas, generated by electricity, was mixed with atmospheric air, and the mixture then brought in contact with the flour, and that it was thereby adulterated under the fourth and fifth subdivisions of § 7 of the act; namely, (1) in that the flour had been mixed, colored, and stained in a manner whereby damage and inferiority were concealed and the flour given the appearance of a better grade of flour than it really was, and (2) in that the flour had been caused to contain added poisonous or other added deleterious ingredients, to-wit, nitrites or nitrite reacting material, nitrogen peroxide, nitrous acid, nitric acid, and other poisonous and deleterious substances which might render the flour injurious to health…

The Lexington Mill & Elevator Company, the respondent herein, appeared … admitting that the flour had been treated by the Alsop Process, but denying that it had been adulterated, and attacking the constitutionality of the act

A special verdict to the effect that the flour was adulterated was returned and judgment of condemnation entered. The case was taken to the circuit court of appeals upon writ of error. The respondent contended that, among other errors, the instructions of the trial court as to adulteration were erroneous and that the act was unconstitutional. The circuit court of appeals held that the testimony was insufficient to show that by the bleaching process the flour was so colored as to conceal inferiority, and was thereby adulterated… That court also held—and this holding gives rise to the principal controversy here—that the trial court erred in instructing the jury that the addition of a poisonous substance, in any quantity, would adulterate the article, for the reason that "the possibility of injury to health due to the added ingredient, and in the quantity in which it is added, is plainly made an essential element of the

prohibition." It did not pass upon the constitutionality of the act, in view of its rulings on the act's construction.

The case requires a construction of the food and drugs act…

Without reciting the testimony in detail it is enough to say that for the government it tended to show that the added poisonous substances introduced into the flour by the Alsop Process, in the proportion of 1:8 parts per million, calculated as nitrogen, may be injurious to the health of those who use the flour in bread and other forms of food.

On the other hand, the testimony for the respondent tended to show that the process does not add to the flour any poisonous or deleterious ingredients which can in any manner render it injurious to the health of a consumer.

On these conflicting proofs the trial court was required to submit the case to the jury. That court—after stating the claims of the parties, the government insisting that the flour was adulterated and should be condemned if it contained *any* added poisonous or other added deleterious ingredient of a kind or character which was capable of rendering such article injurious to health; the respondent contending that the flour should not be condemned unless the added substances were present *in such quantity* that the flour would be thereby rendered injurious to health—gave certain instructions to the jury… (emphasis added).

…It is evident from the charge given and refused that the trial court regarded the addition to the flour of any poisonous ingredient as an offense within this statute, *no matter how small the quantity*, and whether the flour might or might not injure the health of the consumer (emphasis added). At least, such is the purport of the part of the charge above given, and if not correct, it was clearly misleading, notwithstanding other parts of the charge seem to recognize that, in order to prove adulteration, it is necessary to show that the flour may be injurious to health. The testimony shows that the effect of the Alsop Process is to bleach or whiten the flour, and thus make it more marketable. If the testimony introduced on the part of the respondent was believed by the jury, they must necessarily have found that the added ingredient, nitrites of a poisonous character, did not have the effect to make the consumption of the flour by any possibility injurious to the health of the consumer.

The statute upon its face shows that the primary purpose of Congress was to prevent injury to the public health by the sale and transportation in interstate commerce of misbranded and adulterated foods. The legislation, as against misbranding, intended to make it possible that the consumer should know that an article purchased was what it purported to be; that it might be bought for what it really was, and not upon misrepresentations as to character and quality. As against adulteration, the statute was intended to protect the public health from possible injury by adding to articles of food consumption poisonous and deleterious substances which might render such articles injurious to the health of consumers. If this purpose has been effected by plain and unambiguous language, and the act is within the power of Congress, the only duty of the courts is to give it effect according to its terms. This principle has been frequently recognized in this court…

Applying these well-known principles in considering this statute, we find that the fifth subdivision of § 7 provides that food shall be deemed to be adulterated "if it contain any added poisonous or other added deleterious ingredient which may render such article injurious to health." The instruction of the trial court permitted his statute to be read without the final and qualifying words, concerning the effect of the article

upon health. If Congress had so intended, the provision would have stopped with the condemnation of food which contained any added poisonous or other added deleterious ingredient. In other words, the first and familiar consideration is that, if Congress had intended to enact the statute in that form, it would have done so by choice of apt words to express that intent. It did not do so, but only condemned food containing an added poisonous or other added deleterious ingredient when such addition might render the article of food injurious to the health. Congress has here, in this statute, with its penalties and forfeitures, definitely outlined its inhibition against a particular class of adulteration.

It is not required that the article of food containing added poisonous or other added deleterious ingredients must affect the public health, and it is not incumbent upon the government in order to make out a case to establish that fact. The act has placed upon the government the burden of establishing, in order to secure a verdict of condemnation under this statute, that the added poisonous or deleterious substances must be such as may render such article injurious to health. The word "may" is here used in its ordinary and usual signification, there being nothing to show the intention of Congress to affix to it any other meaning. In thus describing the offense, Congress doubtless took into consideration that flour may be used in many ways, in bread, cake, gravy, broth, etc. It may be consumed, when prepared as a food, by the strong and the weak, the old and the young, the well and the sick; and it is intended that if any flour, because of any added poisonous or other deleterious ingredient, may possibly injure the health of any of these, it shall come within the ban of the statute. If it cannot by any possibility, when the facts are reasonably considered, injure the health of any consumer, such flour, though having a small addition of poisonous or deleterious ingredients, may not be condemned under the act. This is the plain meaning of the words, and in our view needs no additional support by reference to reports and debates, although it may be said in passing that the meaning which we have given to the statute was well expressed by Mr. Heyburn, chairman of the committee having it in charge upon the floor of the Senate (Congressional Record, vol. 40, pt. 2, p. 1131): "As to the use of the term 'poisonous,' let me state that everything which contains poison is not poison. It depends on the quantity and the combination. A very large majority of the things consumed by the human family contain, under analysis, some kind of poison, but it depends upon the combination, the chemical relation which it bears to the body in which it exists, as to whether or not it is dangerous to take into the human system."

…We reach the conclusion that the circuit court of appeals did not err in reversing the judgment of the district court for error in its charge with reference to subdivision 5 of § 7.

…It follows that the judgment of the Circuit Court of Appeals, reversing the judgment of the District Court, must be affirmed, and the case remanded to the District Court for a new trial.

Affirmed.

3.3.4   Tolerances

Tolerances were created to ease the FDA's evidentiary burden under the may render injurious standard. Section 402(a)(2)(4) contains a carve-out stating that a food is deemed adulterated if it "bears or contains any added poisonous or added deleterious substance … which is unsafe within the meaning of Section 406." Section 406 provides the FDA authority to set tolerances.

Section 346 (Originally Codified as Section 406)

Any poisonous or deleterious substance added to any food, except where such substance is required in the production thereof or cannot be avoided by good manufacturing practice shall be deemed to be unsafe for purposes of the application of clause (2)(A) of Section 342 (a) of this title; but when such substance is so required or cannot be so avoided, the Secretary shall promulgate regulations limiting the quantity therein or thereon to such extent as he finds necessary for the protection of public health, and any quantity exceeding the limits so fixed shall also be deemed to be unsafe for purposes of the application of clause (2)(A) of Section 342 (a) of this title…

Section 406 experienced long periods of inactivity. Initially it was used to regulate pesticides. In 1958 Congress enacted Section 408 authorizing the FDA to establish tolerances for RACs. After Section 408 was enacted, Section 406 fell into disuse. It was only revived later when it needed a way to regulate the new contaminants like PCB or mercury.

Section 406 tolerances are limited to a small group of added substances. Only those added substances that are unavoidable trough GMPs or are "necessary in the production of food" may be subject to tolerances. The statute requires the FDA consider three criteria when establishing a tolerance.

Section 346 (Originally Codified as Section 406)

(b) A tolerance for an added poisonous or deleterious substance in any food may be established when the following criteria are met:

1. The substance cannot be avoided by good manufacturing practice.
2. The tolerance established is sufficient for the protection of the public health,

taking into account the extent to which the presence of the substance cannot be avoided and the other ways in which the consumer may be affected by the same or related poisonous or deleterious substances.

3. No technological or other changes are foreseeable in the near future that might affect the appropriateness of the tolerance established. Examples of changes that might affect the appropriateness of the tolerance include anticipated improvements in good manufacturing practice that would change the extent to which use of the substance is unavoidable and anticipated studies expected to provide significant new toxicological or use data.

Tolerances present a number of challenges to the FDA. The FDA is allowed to establish tolerances for substances that cannot be avoided through GMPs. The tolerance sets a level above which the substance poses health risk. Under this structure the FDA is provided no option to address added substances that pose a health risk at any level. Many scientists would agree, for example, no level of aflatoxin can be considered safe for humans. Yet, the FDA sets aflatoxin standards for both human food and animal feed. A fair reading of Section 406 is to set limits not bans on added substances. Thus, the FDA must set some level of tolerance.

Tolerances are difficult to square with subsequent amendments and are generally challenging to enforce. The FDA largely prefers informal action levels to tolerances. Tolerances were conceived of as a way to manage the FDA's evidentiary burden for added substances prior to the introduction of more than five amendments addressing a wide gamut of added substances. During this period of change and revision Section 406 was never repealed. It is a challenging section in its own right to make use of and presents the FDA with a number of dilemmas it prefers to ignore.

Section 109.30 Tolerances for polychlorinated biphenyls (PCB's)

(a) Polychlorinated biphenyls (PCB's) are toxic, industrial chemicals. Because of their widespread, uncontrolled industrial applications, PCB's have become a persistent and ubiquitous contaminant in the environment. As a result, certain foods and animal feeds, principally those of animal and marine origin, contain PCB's as unavoidable, environmental contaminants. PCB's are transmitted to the food portion (meat, milk, and eggs) of food-producing animals ingesting PCB-contaminated animal feed. In addition, a significant percentage of paper food-packaging materials contain PCB's which may migrate to the packaged food. The source of PCB's in paper food-packaging materials is primarily of certain types of carbonless copy paper (containing 3 to 5% PCB's) in waste paper stocks used for manufacturing recycled paper. Therefore, temporary tolerances for residues of PCB's as unavoidable environmental or industrial contaminants are established for a sufficient period of time following the effective date of this paragraph to permit the elimination of such contaminants at the earliest practicable time. For the purposes of this paragraph, the term "polychlorinated biphenyls (PCB's)" is applicable to mixtures of chlorinated biphenyl compounds, irrespective of which mixture of PCB's is present as the residue. The temporary tolerances for residues of PCB's are as follows:

1. 1.5 parts per million in milk (fat basis).
2. 1.5 parts per million in manufactured dairy products (fat basis).
3. 3 parts per million in poultry (fat basis).
4. 0.3 parts per million in eggs.
5. 0.2 parts per million in finished animal feed for food-producing animals (except the following finished animal feeds: feed concentrates, feed supplements, and feed premixes).

6. 2 parts per million in animal feed components of animal origin, including fishmeal and other by-products of marine origin and in finished animal feed concentrates, supplements, and premixes intended for food producing animals.

7. 2 parts per million in fish and shellfish (edible portion). The edible portion of fish excludes head, scales, viscera, and inedible bones.

8. 0.2 parts per million in infant and junior foods.

9. 10 parts per million in paper food-packaging material intended for or used with human food, finished animal feed and any components intended for animal feeds. The tolerance shall not apply to paper food-packaging material separated from the food therein by a functional barrier which is impermeable to migration of PCB's.

3.4   Non-Added Substances and Adulteration

Non-added substances are subject to a less rigorous, more permissive, control scheme. This is again not a licensing framework, but a prohibitory standard the same as for "added" substances. As with add substances the FDA must first seize and sample products to assert an alleged adulteration case. Given the permissive nature of the non-added standard court challenges are sparingly few. The leading case originates from 1942 and remains good case law today.

Non-added substances are chiefly defined as inherent or naturally occurring. As the discussion above noted both the CFR and the courts addressed what constitutes non-added under Section 402. The cases below will make clear the types of substances the FDA sought to enforce under the non-added standard.

The standard the FDA must prove for non-added substances is the "ordinarily injurious" standard. It requires a lower evidentiary showing than the "may render injurious" standard. The lower evidentiary hurdle makes the non-added substance framework the preferred 402 adulteration scheme of the FDA.

3.4.1   Ordinarily Injurious Standard

Courts typically ask whether the naturally occurring substance renders the food injurious when consumed by ordinary consumers in ordinary quantities. The leading case is *United States v. 1232 Cases of American Beauty Brand Oysters.*[25] *American Beauty*, which involved fragments of shells in canned oysters. The FDA claimed the fragments were a deleterious substance and could cause injury either by damage to the mouth or esophagus.

The court looked to a number of factors to determine if the shell fragments were ordinarily injurious. The court quickly noted the shells were not "artificially created" and thus subject to the "ordinarily injurious" standard. The court also found the facility followed good manufacturing practices and modern techniques to remove the shells. It heard no evidence that the shell fragments in the American Beauty brand were any different than other brands. It also found no evidence of any complaints despite the facility distributing over fifty million cans of oysters. Thus, the court concluded the oysters were not shown to be dangerous in their ordinary use.

The court did not look for perfection. It did not expect or find any need for the facility to remove all shell fragments. In noted doing so was impossible and if that was such a requirement then the oysters would be banned entirely. The same for bones in fish it reasoned, all of which the court considered "unthinkable." Thus, not only must the FDA show a naturally occurring substance to cause injury to ordinary consumers in the ordinary use of the food, but also that complying with removal of the inherent article does

---

[25] *43* F.Supp. 749 (W.D. Mo. 1942).

not result in the effective ban of the food itself. The court introduces a cost-benefit analysis not explicitly mentioned in the statute.

United States v. 1232 Cases of American Beauty Brand Oysters

REEVES, District Judge.

This is a proceeding by the process of libel to condemn an alleged adulterated food product. Such food consists of 1232 cases of oysters, each case containing 24 cans, marked "American Beauty Brand Oysters."

As a basis for condemnation, it is alleged by the government that said article "contains shell fragments, many of them small enough to be swallowed and become lodged in the esophagus and that said shell fragments are sharp and capable of inflicting injury in the mouth."

The provision of the law invoked by the government is Section 342, Title 21 U.S.C.A., and sundry subdivisions thereof…

The claimant appeared to deny the averments of the libel and assert ownership of the product. The evidence in the case showed that in the processing of oysters for food there is a constant effort to eliminate shells and fragments thereof from the product. For this purpose many means and devices are used to reduce as nearly to a minimum as possible such shells and fragments in the product. The evidence, however, on behalf of both the government and the defense was that with present known means and devices it was impossible to free the product entirely from the presence of part shells and shell fragments. Moreover, it not only appeared, but it is a matter of common knowledge, that an oyster is a marine bivalve mollusk with a rough and an irregular shell wherein it develops and grows, and that, in the processing of the food product, it is necessary to remove this irregular, rough shell so far as that may be accomplished. The shells, therefore, are not artificially added for the purpose of growth or to aid in the processing operations.

The evidence on the part of the government was that parts of shell and shell fragments upon inspection were found in many of the cans taken from the article seized. Such parts of shell and fragments were exhibited at the trial.

There was evidence on behalf of the claimant that its processing operations were in accord with the best manufacturing practice and there was even some testimony that the means employed by it for the elimination of shell fragments were superior to the means employed by other processors engaged in similar operations. The testimony on the part of the claimant further tended to show that within the Kansas City area over a period of ten years it had sold approximately 5 million cans of its product and that no complaint had ever been made concerning the presence of shell fragments. Claimant also proved that over 50 million cans had been processed by it and distributed in its trade territory and that no complaints had ever been made of the presence of part shells or shell fragments.

It seems proper at this point to comment that in this case involving considerable testimony there was no substantial controversy as to the facts and practically no difference of opinion as to the law. There was a contention by the government that the shells as a deleterious substance were added to the product while being processed. There was no evidence to support this contention.

The excerpt from the statute heretofore quoted contemplates that there may be of necessity food products containing deleterious substances. No one who has had the experience of eating either fish or oysters is unfamiliar with the presence of bones in the fish (a deleterious substance) and fragments of shell in the oysters (also a deleterious substance).

The Congress, however, withdrew such foods from the adulterated class "if the quantity of such substance in such food does not ordinarily render it injurious to health."

The evidence on both sides was that by the greatest effort, and in the use of the most modern means and devices, shell fragments could not be entirely separated from an oyster food product. The government, in its brief, quite aptly and concisely stated its point by using the following language: "It is the character, not the quantity of this substance that controls its ability to injure."

This concession on the part of the government, properly made, upon the evidence removes the case immediately from that portion of the statute which says: "* * * such food shall not be considered adulterated under this clause if the quantity of such substance in such food does not ordinarily render it injurious to health."

Since it is the "character, not the quantity of this substance that controls its ability to injure", as stated by the government, then in the view that it is impossible to eliminate shell fragments *in toto* from the product, the use of oysters as a food must be entirely prohibited or it must be found that the presence of shell fragments is not a deleterious substance within the meaning of the law and must be tolerated to reject oyster products as a food is unthinkable.

It would be as reasonable to reject fish because of the presence of bones. Even if a greater percentage of shells and shell fragments were found in claimant's product than in that of other processors, yet this fact, under the theory of the government, would not add to the deleterious nature of claimant's product. It should be stated, however, that there was no evidence that there was an excess of shell fragments in claimant's product over that of other processors. On the contrary, a preponderance of the evidence showed that the claimant's processing methods were superior.

…Counsel for both the government and the claimant, at the trial and in their briefs, discussed the question of the right to a tolerance regulation as provided by Section 346, Title 21 U.S.C.A. This provision is for tolerance of both poisonous and deleterious substances where the presence of such substance cannot be avoided. However, that section says: "(a) Any poisonous or deleterious substance added to any food, except where such substance is required in the production thereof or cannot be avoided by good manufacturing practice shall be deemed to be unsafe for purposes of the application of clause (2) of Section 342 (a)."

Adverting to clause 2 of said Section 342 (a), it reads as follows: "* * * or (2) if it [food] bears or contains any added poisonous or added deleterious substance which is unsafe within the meaning of Section 346."

It will be seen at once that this provision does not apply where the deleterious substance inheres in the product and is not added. Further quoting from Section 346, however, note this language: "* * * but when such substance is so required or cannot be so avoided, the Administrator shall promulgate regulations limiting the quantity therein or thereon to such extent as he finds necessary for the protection of public health."

Upon the concession made by the government in this case, even if the tolerance section could be construed to apply, it is not the quantity of the substance but its character "that controls its ability to injure."

Upon the evidence in the case it must be found that the presence of shell fragments in the article sought to be condemned does not ordinarily render it injurious to health.

Under the statute and upon the evidence the government is not authorized to condemn the article seized for the reason that the processed article does not offend against the food and drug law. The claimant, therefore, should have restored to it the articles seized and the libel should be dismissed. It will be so ordered.

3.4.2   Ordinarily Injurious and the USDA

The USDA/FSIS is in a singular position when it comes to inherent or naturally occurring added substances. Raw meat and poultry naturally carries a wide range of pathogens from *Salmonella* to strains of *E. coli.* The trouble is twofold. First, the pathogens are naturally occurring and subject to the more permissive "ordinarily injurious" standard. Second, the question often becomes whether the pathogens are in fact injurious. Recall the standard asks whether ordinary consumers eating the product in the quantities and manner intended could reasonably cause injury. In the case of raw cuts of meat, the ordinary consumer will prepare the meat by cooking it to the recommended temperature. Thus, the cooking is expected to kill the bacteria and thus render the food safe. If it does not cause injury, then it fundamentally fails any test for adulteration.

The challenge FSIS finds itself in is when pathogens cause illness after cooking the meat. In some cases certain cuts of meat are traditionally cooked rare leaving an increased risk of illnesses. In other cases the pathogen persists despite proper cooking. *Salmonella* is a well-known and widely discussed example. It even carries its own name—"The Salmonella Problem" (Flippin and Eisenberg 1960).[26] Its prevalence in outbreaks is attributed to a number of factors, including proper handling and preparation of raw meat and poultry. *Salmonella* is difficult to remove from a processing facility leaving sometimes high concentrations of the bacteria on the meat.

Consumers can cross-contaminate raw meat with vegetables, cut themselves when preparing raw meat or poultry, or fail to properly sanitize counter tops and cutting boards all of which presents a risk of infection. This would seem to provide FSIS the authority to regulate *Salmonella* as a non-added substance.

FSIS and the courts continue to view *Salmonella* as outside the definition of adulteration. In *Supreme Beef* the USDA agreed with the court and Supreme Beef Inc. that *Salmonella* was not an adulterant as defined in the FMI or PPI. In the prevailing view it failed to cause injury and thus fell outside the scope of the adulteration definition. Under this paradigm prompting recalls can be difficult when they involve pathogens such as *Salmonella.*

Foster Farms Salmonella Outbreak

In March of 2013 a spike in *Salmonella* poisonings became linked to Foster Farms poultry products. It persisted through the summer with 29 states linked to the Foster Farms products by July (NBC News 2013).[27] An estimated 621 became ill, but the CDC expected the number much larger. It estimated that for every one reported case twenty-nine went unreported (CDC 2013).[28] The only measure FSIS could take was to request Foster Farms take new safety measures and urge it to initiate a voluntary recall. It was powerless to shut down the facility or mandate a recall because of the classification of *Salmonella.*

FSIS instead attempts to control *Salmonella* and other pathogens using other provisions of the FMI or PPI. For example, *E. Coli* O157:H7 is

---

[26] Flippin and Eisenberg (1960).

[27] NBC News, Foster Farms Salmonella Outbreak Expands (July 4, 2013) available at: http://www.nbcnews.com/health/health-news/foster-farms-salmonella-outbreak-expands-n148466.

[28] Department of Health and Human Services, Center for Disease Control and Prevention, 2013 Progress Report on Six Key Pathogens Compared to 2006–208, available at: http://www.cdc.gov/foodnet/data/trends/trends-2013-progress.html.

considered and controlled by FSIS as a naturally occurring added substance. This includes raw beef products, which consumers would properly cook. FSIS reasoned beef products contaminated with O157:H7 are cooked rare and pose a public health risk (FSIS Regulatory Perspective).[29] It found support in a 1993 outbreak linked to O157:H7. The outbreak was associated with hamburgers that were often undercooked by restaurants (FSIS Regulatory Perspective).[30] It also found only a few cells of the pathogen were enough to cause illness (FSIS Regulatory Perspective).[31]

FSIS also utilizes other statutory provisions to take enforcement actions against naturally occurring pathogens. The PPI and FMI contain a section in the adulteration definition that broadens the definition considerably. FSIS cites this language for raw products associated with a recall for which the ordinarily injurious standard does not apply. The language of the provision provides ample cover to take enforcement action within the facility, but may fall short of providing legal authority to require a recall.

> **Language from 21 U.S.C. 453(g)(3) or 601(m)(3)**
> Product is adulterated if it is—… unsound, unhealthful, unwholesome, or otherwise unfit for human food…

3.5   Amendments Controlling Specific Adulterations

Congress enacted a series of Amendments through the late 1950s and 1960s. Each adulterant amendment addressed a specific category of added food substances. None of the amendments concern a non-added or naturally occurring food contaminant. As the amendments were enacted the system by which the FDA deemed food adulterated grew increasingly complex. The amendments themselves provided the FDA new authority to limit or approve the use of certain added food substances. The licensing scheme was intended to clarify the prohibitory standards of Section 402, but in many ways it only made the system more unruly.

3.5.1   Pesticide Residues Amendment of 1954

The Pesticide Residues Amendment applies to a narrow category of products. The Amendment created Section 408 to address pesticides added to raw agricultural commodities, commonly referred to as RACs. Any pesticide residues that carried over into processed food products are controlled under Section 409 of the Food Additives Amendment. Additionally, any pesticide residues applied to other crops, those not defined as an RAC, remain regulated under Section 406. Prior to the 1954 Amendment all pesticides were regulated under Section 406, which sets tolerances on added poisonous or deleterious substances. Section 408 superseded Section 406 only as it pertains to RACs. Thus, the 1954 Amendment applies to a narrow subset of food regulated by the FDA.

> **Section 201(r) Definition of RAC**
> (r) The term "raw agricultural commodity" means any food in its raw or natural state, including all fruits that are washed, colored, or otherwise treated in their unpeeled natural form prior to marketing.

Section 408 is unique in that it is a proviso of the FD&C directed at the Environmental Protection Agency (EPA). The Amendment requires the EPA set tolerances for pesticide residues on RACs. All tolerances must be safe, which accounts for aggregate exposure. Under the Amendment the

---

[29] United States Department of Agriculture, Food Safety and Inspection Service, FSIS Regulatory Perspective (available at: http://www.aphl.org/conferences/proceedings/Documents/2012/2012-PulseNet-OutbreakNet/007-Edelstein.pdf).

[30] See *Id.*

[31] See *Id.*

EPA could look at a wide range of risk factors when setting tolerances. It also was authorized to grant exemptions where there was likely no risk.

Section 408 underwent dramatic changes with the passage of the Food Quality Protection Act (FQPA) of 1996. The FQPA revised Section 408 and the Federal Insecticide, Fungicide, and Rodenticide Act (FIFRA), both under the purview of the EPA. The FQPA provided a vast and complex new food safety mission to the EPA.

The FQPA set new safety standards for pesticide residues in food. It emphasized protecting the health of infants and children. It amended Section 408, meaning the new provisions will also be codified in Section 408. As with Section 408 all pesticides must be safe. The FQPA introduced a stringent new standard defining safe as "reasonable safety of no harm." The Section 408 mandate to conduct a comprehensive assessment of each pesticide's risks also continues.

Example of Risk Factors in Safety Assessment (408(b)(2)(D))

- Aggregate exposure of the public to residues from all sources including food, drinking water, and residential uses;
- Cumulative effects of pesticides and other substances with common mechanisms of toxicity;
- Special sensitivity of infants and children to pesticide; and
- Estrogen or other endocrine effects.

The FQPA mandated new reassessments. The new amendment directed the EPA to reassess all existing tolerances and exemptions issued under Section 408. The reassement must look at the existing tolerances and exemptions to again ask if they are safe as defined in the FQPA. The reassessments are a routine component of the FQPA with review expected on a 15-year cycle.

3.5.2 Food Additives Amendments of 1958

The Food Additives Amendment eased the evidentiary burden for the FDA on proving added substances were adulterated. As the Amendment related to the topic of this chapter any substance qualifying under the Amendment and not approved is adulterated. It also shifted the paradigm to reactive and enforcement based to proactive and pre market oriented. No longer would the FDA need to locate, sample, test, and seek judicial relief prior to stopping the sale of a potentially injurious substance. Under the Food Additives Amendment the FDA could evaluate many new substances before it entered the stream of commerce.

A reactive model was rigid and stifled innovation. From 1906 to 1938 the food industry grew immensely in its transition from farm to factory. The early growth of the industry was eclipsed by the advancements following World War II. The 20 years between the 1938 Act and the Food Additives Amendment witnessed a dramatic increase in technology and use of intentional added substances. The prevailing use of chemical preservatives, for example, boomed in the 1940s and 1950s. The FDA's model of enforcing after-the-fact could not keep pace with the changes and created uncertainty in the industry.

A select committee of the House of Representatives convened to investigate the use of chemicals in food. It estimated that over 700 chemicals were used at the time, but only 428 were known to be safe (House Report 1952).[32] Many estimates placed the number of chemicals added to food much higher. This raised another fault of the 1938 Act's approach to regulating added substances. In the absence of notification of an intent to use a new added substance, the FDA could not effectively monitor or learn about the substances. It was often only after a consumer became ill that the FDA would learn of the new substance. The committee concluded the 1938 system left questions unanswered, such as the risks of chronic use, and could not adequately assure the public of the additives safety.

---

[32] H.R. Rep. No. 82-2356 at 4 (1952).

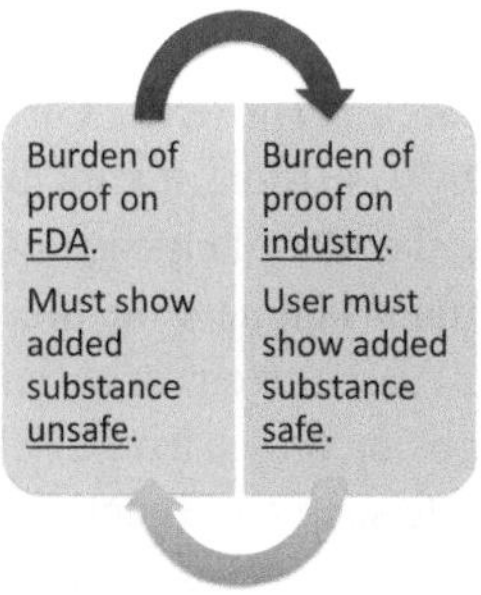

Fig. 3.7 Changes in evidentiary standard following 1958 Amendment

The Food Additives Amendment is rare because lacking a singular crisis the drafters required eight years to complete their work. Congress began hearings two years prior to the select committee conducting its investigation. It would be another six years before the Amendment would be enacted.

Although delayed result was predictable. Congress routinely shifted the burden of enforcement from the FDA to industry in drafting the food safety legislation (see Fig. 3.7). This was the result in many of the provisions of the 1938 Act. It was repeated again in the 1958 Amendment. Prior to the Amendment the burden of proof sat heavily on the FDA's shoulders. The FDA required expert evidence, studies, and data that when interpreted showed the added substance was unsafe. Under the Amendment, industry now stood in the FDA's place. Prior to marketing a new added substance the facility seeking its use must demonstrate safety to the satisfaction of the agency.

Congress believed the Amendment would not only make food safe but encourage innovation. Under the Amendment the FDA was given greater flexibility in approving food additives. It could authorize limited uses or levels of additives shown to have toxicity at higher levels. Thus, rather than banning any added substance deemed to be poisonous or deleterious, the Amendment allowed some additives below its known injurious levels.

3.5.3 Delaney Clause

The Delaney Clause is unique among the Amendments in that it is not its own statutory amendment.

Instead the Delaney Clause consists of three parallel statutory provisions. The clause is found in the Food Additives Amendment, the Color Additive Amendment, and the Animal Drug Amendment. It is a well-intentioned, but difficult to enforce provision in these three arenas.

The Delaney Clauses are prohibitory provisos aimed at carcinogens. In each case the statutory language is identical. No additive, food, color, or animal drug, will be considered safe if found to cause cancer in humans or animals. For added substances, such as pesticides, that are outside these three categories the proviso does not apply. It still may be possible for the FDA to exercise authority under the Food Additives Amendment to capture the full range of added substances found in food. Recall also the Delaney Clauses apply to a wide range of cosmetics, and drugs under the Color Additive Amendment. Thus, although the statutory language is identical in each Delaney Clause, the context of the surrounding Amendment shades its meaning.

> **Section 409**
> Provided, That no additive shall be deemed to be safe if it is found to induce cancer when ingested by man or animal, or if it is found, after tests which are appropriate for the evaluation of the safety of food additives, to induce cancer in man and animal...

The Delaney Clause requires the FDA act proactively. The plain language of the Delaney Clause requires the FDA to make a determination of safety based on two criteria. Either the added substances induces cancer when ingested or in some appropriate test or study. In short it requires evidence of cancer to some degree. In this way the Delaney Clause is not a self-executing ban on all carcinogens. It requires the development, presentation, and expert interpretation of safety data before an added substance can be banned as a carcinogen.

The Delaney Clause stops short of stipulating the means, methods, or thresholds to be used when deciding if an added substance induces cancer. The statutory provision does not go on to stipulate what types of evidence should be considered or what tests conducted. It provides an open and flexible standard to allow scientific techniques to develop. This is both helpful and harmful. It affords the FDA wide latitude in determining its obligations under the Delaney Clauses.

The discretion under the Delaney Clauses often leads to the FDA rejecting the need to ban carcinogens. Early examples emerged following the passage of the Color Additive Amendment. Prior to the passage of the Amendment the drug and cosmetic industry long used Orange No. 17 and Red No. 19. The two colors were given a provisional listing under the new Amendment and underwent further testing (Color Additives 1977).[33] The testing revealed that the color additives caused tumors in animals. This would seem to fit the plain language of the Delaney Clause and lead to a ban of the color additives.

The FDA refused to ban the colors and continued to sanction their use. The FDA concluded the colors were applied topically and in low concentrations. Thus, it determined the risk of cancer even from a lifetime of use was minimal. It placed an estimate on the risk as one in nineteen billion and one in nine million. With this conclusion in hand the FDA attempted to explain approval under the Delaney Clause. The agency cited a D.C. Circuit opinion which held executive agencies could exercise "inherent" authority to "overlook circumstances that in context may fairly be described as *de minimis*" (Color Listings 1986).[34] Thus, the FDA held it could overlook the plain language of the Delaney Clause given the low risk and developments in testing.

The FDA lost a challenge to its listing of Orange No. 17 and Red No. 19. The challenge came not from industry, but from the public. A collection of consumers sued the FDA to overturn its approval of Orange No. 17 and Red No. 19 (*Public Citizens*).[35] The D.C. Circuit court rejected the FDA's position. It concluded the FDA was bound by the plain language of the Delaney Clause. The plain language neither sets an outright ban on all carcinogens or *de minimis* exception for dyes posing only a trivial risk. The FDA's attempt to create such an exception exceeded the statutory authority and was held invalid (*Public Citizens*).[36]

The Delaney Clause means well but absent stronger language affords little protection. It relies heavily on the FDA to act proactively and preventatively. In the absence of language stating what level of carcinogen is prohibited, the FDA can approve carcinogens at some level depending on the safety data.

3.5.4 Color Additive Amendment

The Color Additive Amendment is another change to food safety laws borne from crisis. In 1950 many children fell ill after eating orange Halloween candy. The candy contained 1–2 % of FD&C Orange No. 1. It was an approved color additive, one of about seven approved shortly after the passage of the 1906 Act. The children experienced a range of symptoms, including diarrhea and rashes. A federal investigation of the incident found the color additive was a known occupational hazard for the workers handling the color additive (Goldsmith 1950).[37] Orange No. 1 was a coal-tar dye that contained benzene.

Prompted by this and other events Congressman Jim Delaney began hearings on food additives and pesticides. The process was long and deliberative. It began in 1950 and did not enact legislation until 1954. The Color Additive Amendment would not become law until ten years after the Halloween scare that prompted a new look at color additives.

---

[33] See Color Additives: Provisional Regulations; Postponement of Closing Dates, 42 Fed. Reg. 6991(1977).

[34] See Listing of D&C Orange No. 17 for Use in Externally Applied Drugs and Cosmetics, 51 Fed. Reg. 28,331 (1986); Listing of D&C Red No. 19 for Use in Externally Applied Drugs and Cosmetics, 51 Fed. Reg. 28,346, 28,539 (1986).

[35] *Public Citizen v. Young*, 831 F.2d 1108 (D.C. Cir. 1987).

[36] *Id.* at 122.

[37] Norman and Goldsmith (1950).

The Color Additive Amendment introduced a new licensing system that required FDA approval for all color additives. Only those colors listed as "suitable and safe" by the agency could be used. The approval applied to any FDA regulated product—foods, drugs, cosmetics, and medical device—it also encompassed FSIS regulated products. When enacted nearly 200 color additives were in commercial use. All were provisionally listed following passage of the Amendment, but not all would survive reassessment.

As discussed above the Color Additive Amendment also contained a Delaney Clause. The Clause continues to exercise an effect on color additive listings, with many provisionally listed colors removed due to carcinogenic effects.

Color additives will be discussed in more detail in Chapter 6 as part of the discussion on Food Additives.

3.5.5   Animal Drug Amendments of 1968

Unlike other amendments aimed at enhancing safety, one of the principal aims of the Animal Drug Amendments was clarity and simplicity. Prior to the Food Additives Amendment in 1958 animal drugs were subject approval under Section 505 and perhaps limited enforcement if drugs were considered "added" under Section 402(a)(1). The 1958 Amendment created a new layer of preclearance to animal drugs. Following the mandates of Section 409 the FDA regulated drugs administered to food producing animals under Sections 505, 409, and 507 (regulating antibiotics).

The Animal Drug Amendment created a unified approval system (see Fig. 3.8). Section 512(b) directs all drug submissions to the Center for Veterinary Medicine (CVM). Section 512 facilitates the centralization of all the information required to approve an animal drug. This includes evaluating drug residues.

Even under a unified system determination there are a numerous definitions and classification to navigate in order to determine if the Amendment applies. If a product, including medicated feed, falls under the Amendment then it must apply and gain FDA approval. Otherwise there are few hurdles to enter the market. This requires a review of 201 definitions, including subsections (v) (new animal drug), (g) (drug), (f) (food), (w) (animal feed), (s) food additive, and (p) (new drug). As the introductory chapters suggested, classification remains a crucial regulatory skill.

The Animal Drug Amendment also contains a Delaney Clause. The Delaney Clause, like those found in the Food Additives and Color Additive Amendments, prohibits drugs that induce cancer. There is a caveat in the Animal Drug Amendment's Delaney Clause. A drug found to induce cancer can be approved if the appropriate evaluation determines no residue of the drug will remain in the edible portions of the tissue. This carve out was sought by the FDA and is known as "Diethylstilbestrol (DES) Proviso," named for a growth hormone approved in 1954 (FDA Consumer Magazine 2006).[38] The FDA is again afforded latitude in determining whether or not the Delaney Clause is satisfied when approving new animal drugs.

3.6   Comparative Law

Nearly every nation regulating its food supply defines adulteration. The experience with an outbreak linked to adulterated food also enjoys universal application. It can involve simple microbiological contamination to more severe instances of intentional adulteration. The Spanish toxic oil offers an example of this type of adulteration. In 1981, 25,000 became ill and nearly 1000 died following the ingestion of denatured rapeseed oil labeled as olive oil. The denatured rapeseed oil led to what would later be called "toxic oil syndrome" (Posada de le Paz et al.).[39] Whether in

---

[38] FDA Consumer Magazine, Animal Health and Consumer Protection (2006) available at: http://www.fda.gov/AboutFDA/WhatWeDo/History/ProductRegulation/AnimalHealthandConsumerProtection/.

[39] (Posada de La Paz et al. 2001)

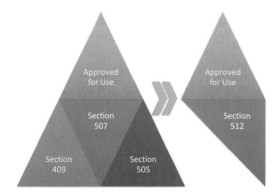

Fig. 3.8 Comparing drug approval pre- and post-1968 amendment

Spain, Germany, China, Canada, or the US adulteration is regulated to some degree.

Melamine offers another example of adulteration. It presents a unique opportunity to evaluate a global response to adulteration. Melamine caused several thousand deaths in 2007 and 2008 when it was intentionally added to pet food and baby formula to artificially increase the nutritional profiles of the products (Time 2008).[40] Depending on the regulatory framework it can be defined as both intentional and economic. The goal was fraud inspired by greed, but most countries which reacted to melamine treated it as a harmful contaminant, not a labeling offense. Nearly every country's system of surveillance missed the risk posed by the products. New Zealand led the push with persistent calls on Chinese manufacturers to investigate illnesses reported to government officials (Time 2008). It was a Chinese investigation that ultimately found evidence of the intentional fraud. Once the issue came to light subsequent testing, such as that conducted by Japan, shed light on the scope of melamine in the market.

Adulteration takes many forms but its effects remain the same no matter where found. As a potential health hazard it is the central focus of food safety agencies. Despite best efforts adulterants enter the market. No system will offer risk-free food to its people. Melamine highlights this fact and points to gaps in the safety nets of many nations. Not a single country identified melamine in the food supply prior to the New Zealand-lead call to action and Chinese investigation. The universal approach was reactive. Recent examples, such as the horse meat scandal, continue to emphasize to what extent many nations rely on post-market surveillance to control adulterants.

3.7 Chapter Summary

This chapter introduces the reader to the concept of adulteration. It covers the definition, legislative history, types, and standards for proving adulteration. It also provides a sampling of the adulteration amendments used to control various added substances. Combined the statutory provisions broadly encompass a wide range of activities with disparate standards. The singular standard in nearly every provision is the shift of proving safety off the FDA's shoulders to industry for self-enforcement.

Overview of Key Points:

- Comparison of 1906 and 1938 definition of adulteration
- Overview of the original tirade of controls in the 1938 Act—added, not-added, and tolerances
- Exploration of the types of adulteration
- Definition and regulation of economic adulteration, a pass-off misbranding type offense
- Comparison of the "may render injurious" and "ordinarily injurious" standards
- Case law surrounding the two standards
- Overview of USDA and the application of the two standards
- Discussion of the four adulteration amendments and the Delaney Clause

3.8 Discussion Questions

Research the FDA warning letter database for letters concerning economic adulteration. Then research popular news media article on economic adulteration.

1. Are the two sources equal in covering the topic?

---

[40] (Pickert 2008).

2. If there is a disparity, how can it be explained?

Identify a substance currently on the market that has been linked as a carcinogen. Has the FDA commented on its use or safety? Have the regulatory bodies of other nations? If there is a difference, e.g. banned in one country but not in the U.S., what does that say about the Delaney Clause? Explain your answer.

References

Department of Health and Human Services, Center for Disease Control and Prevention (2013) Progress report on six key pathogens compared to 2006–2008. http://www.cdc.gov/foodnet/data/trends/trends-2013-progress.html. Accessed 16 June 2014

FDA Consumer Magazine (2006) Animal health and consumer protection. http://www.fda.gov/AboutFDA/WhatWeDo/History/ProductRegulation/Animal-HealthandConsumerProtection/. Accessed 17 June 2014

Flippin HF, Eisenberg GM (1960) The *Salmonella* problem. Trans Am Clin Climatol Assoc 71:95–106

Hart LF (January 1952) A history of the adulteration of food before 1906. 7 Food Drug Cosmet Law J 5:7–8

NBC News (4 July 2013) Foster farms *Salmonella* outbreak expands. http://www.nbcnews.com/health/health-news/foster-farms-salmonella-outbreak-expands-n148466. Accessed 17 June 2014

Norman R, Goldsmith MD (1950) Dermatitis from orange I in a candy factory. AMA Arch Derm Syphilol 62(5):695–696

Pickert K (17 September 2008) Brief history of melamine. http://content.time.com/time/health/article/0,8599,1841757,00.html. Accessed 16 June 2014

Posada de La Paz M et al (2001) Toxic oil syndrome: the perspective after 20 years. Epidemiol Rev 23(2):231–247

Misbranding

4

Abstract

This chapter presents the second major prohibited act under US food law—misbranding. It fully develops the concept by examining both the regulatory requirements and defenses under the First Amendment. It begins with a discussion on the concept of "label" and "labeling" as interpreted by the courts to include any material that "accompanies" the product. It then turns to the concept of misbranding and allergen labeling. It introduces the types of claims allowed on conventional foods with a limited comparison to claims on dietary supplements. The commercial speech doctrine and its impact on misbranding is the final topic of the chapter.

4.1 Introduction

Progressing from adulteration the focus shifts away from the product to its labeling and the concept of misbranding. Since the 1906 Act the FDA has enforced a prohibition on misbranding. Unlike the previous chapter's discussion on adulteration, misbranding emphasizes what is said about a product rather than its composition. It aims primarily at fraud while touching occasionally on safety.

Labeling and marketing of food products is ubiquitous in the grocery store or in our homes. Enter a grocery store or open a cupboard and one finds dozens of brands with abundant information. Some information is helpful, such as the product name, net weight, or ingredients. Other statements stand out as marketing, such as product comparisons. Still some claims straddle the line, such as "high in fiber," "lowers cholesterol," or "all natural." Such claims are intended both for marketing purposes and to inform the consumer

about the benefits of the food. Consumers trust, often blindly, that all the information given on the label is accurate and truthful. We may confidently eat the chocolate cookies or almond candy knowing it contains chocolate and almonds only because we trust the label.

There are two chief sources of FDA authority, one more recent and detailed than the other. The 1938 Act lists misbranding as a prohibited act in Section 331. Congress also passed the Nutrition Labeling and Education Act (NLEA) in 1990. NLEA amended the FD&C providing the FDA specific authority to require nutrition labeling of many foods regulated by the agency and issue regulations about certain types of claims.

Crucial to understanding many of the regulations is familiarity with packaging terms. The FDA regulates the location of certain labeling information, such as the nutrition facts panel found on the back of all packaged foods. This chapter will introduce the terms "principal display panels," and "information panel."

M. C. Sanchez, *Food Law and Regulation for Non-Lawyers*, Food Science Text Series, DOI 10.1007/978-3-319-12472-8_4, © Springer International Publishing Switzerland 2015

This chapter will also explore the FDA's attempt to expand the labeling concept. Traditionally labeling involved product packaging. The FDA ventured into some enforcement of marketing materials, but largely steered away from most advertising activities. The FDA shares concurrent jurisdiction with the FTC who enforces the truthfulness of marketing, such as print, radio, or television advertisements. The growth of social media, websites, search engine optimization (SEO), and other uses of technology is pushing the FDA to expand its concept of labeling. As discussed below, the FDA now routinely regulates areas far removed from the product packaging.

The misbranding provisions identify areas of noncompliance over premarket requirements. The FDA does not evaluate labels prior to use in the market. The FDA will routinely identify labeling or labeling claims that constitute misbranding and notify a facility in a Form 483 or Warning Letter. Imported products pose particular labeling problems because the products often bear labels compliant with the country of origin.

4.1.1 Defining Misbranding

The Section 201 definitions do not directly define misbranding. Section 201 provides a definition for label (201(k)) and labeling (201(m)), but only an implied definition of misbranding (201(n)). In Section 201(n) a list of factors are provided to determine if a label is misleading. The implication is that misbranding under the Act means a misleading label. Broadly construed, this would be correct. There are, however, other areas of labeling deviance which do not comply with the regulations. These variants are treated as misbranding even though it would be challenging to state they are truly misleading. For example, the regulations require a name and address on the back of the label of the manufacturer or distributor. The FDA will frequently cite this as a misbranding violation.

Label, Labeling, and Misbranding Definition from Section 201

(k) The term "label" means a display of written, printed, or graphic matter upon the immediate container of any article; and a requirement made by or under authority of this Act that any word, statement, or other information appear on the label shall not be considered to be complied with unless such word, statement, or other information also appears on the outside container or wrapper, if any there be, of the retail package of such article, or is easily legible through the outside container or wrapper.

(m) The term "labeling" means all labels and other written, printed, or graphic matters (1) upon any article or any of its containers or wrappers, or (2) accompanying such article.

(n) If an article is alleged to be misbranded because the labeling or advertising is misleading, then in determining whether the labeling or advertising is misleading there shall be taken into account (among other things) not only representations made or suggested by statement, word, design, device, or any combination thereof, but also the extent to which the labeling or advertising fails to reveal facts material in the light of such representations or material with respect to consequences which may result from the use of the article to which the labeling or advertising relates under the conditions of use prescribed in the labeling or advertising thereof or under such conditions of use as are customary or usual.

The broader reach of misbranding fits with the overarching mandate in the Act. As discussed in the introductory chapters the 1906 and 1938 Acts were introduced with the twin aim of protecting the public from physical and economic harm (See Fig. 4.1). The adulteration provisions of the previous chapter tackle one of the twin aims of

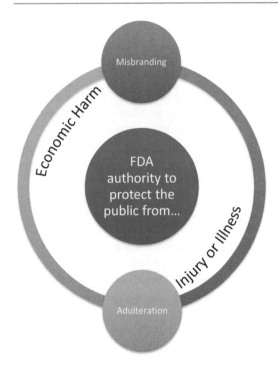

Fig. 4.1 FDA's mission

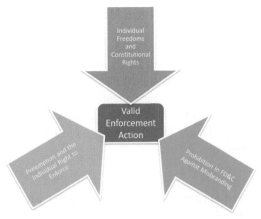

Fig. 4.2 Competing constitutional and statutory interests

product safety. Misbranding tackles the other. The individual consumer cannot evaluate the entire market place and make a choice about which labels are honest and accurate and which contain a whole host of mistruths. The FDA steps in to act as an enforcer and deterrent by verifying that labels are truthful and accurate or in the parlance of the statute not misbranded. Under this charge, the FDA, often with the courts, blessing, broadly expands the concept of misbranding and labeling.

4.1.2 Overview of Constitutional Challenges

Misbranding actions strike at some of the most closely held constitutional rights. Limitations on labeling impact both a facility's First Amendment right to free speech and the consumers freedom of choice. Constitutional rights stand in equal, if not greater, importance to the FDA's statutory mission to protect the public from fraudulent labeling.

The constitutional question often involves definitional questions. The central issue often involves what constitutes labeling. It is this reach by the FDA to regulate in new areas that consumes many of the judicial challenges. Constitutional challenges rarely feature the misbranding concept. Other questions arise, such as what type of speech is made on a label. The varying options, regulated, commercial, or ordinary, all carry case law and criteria that determine the outcome of the challenge.

Some constitutional challenges do not involve the FDA's attempt to take enforcement action. In fact it is often the FDA's inaction that prompts attempts to regulate misbranding or labeling provisions in the market. Often, this is seen where a competitor is making putatively violative claims that offer it an unfair advantage. This constitutional case involves the issue of preemption and whether a private party can take on the enforcement activities of the FDA. Recent Supreme Court case law informs much of the discussion on this topic.

Balancing these three opposing interests sets the boundaries on permissible enforcement actions. The FDA can take enforcement actions where it does not interfere with a constitutional right or freedom. Likewise, the FDA's authority proscribes some private enforcement actions. Where the three competing interests are balanced a valid misbranding grievance exists (Fig. 4.2).

4.1.3   The USDA and Misbranding

At first blush many of the definitions used under the PPI and FMI will look similar to those used by the FDA. Both the FMI and PPI define "label" and "labeling" in nearly identical terms to those used by the FD&C. FSIS does, however, employ a much longer list of acts that constitute misbranding.

FMI Sections 601(o) and (p) and PPI Section 451 (s)

A label is "a display of written, printed, or graphic matter upon the immediate container of any article."

Labeling is "all labels and other written, printed, or graphic matter (1) upon any article or any of its containers or wrappers, or (2) accompanying such article."

Misbranding Circumstances
FMI Section 601(n) and PPI Section 451 (h)

1. if its labeling is false or misleading in any particular;
2. if it is offered for sale under the name of another food;
3. if it is an imitation of another food, unless its label bears, in type of uniform size and prominence, the word "imitation" and immediately thereafter, the name of the food imitated;
4. if its container is so made, formed, or filled as to be misleading;
5. if in a package or other container unless it bears a label showing (A) the name and place of business of the manufacturer, packer, or distributor; and (B) an accurate statement of the quantity of the contents in terms of weight, measure, or numerical count: Provided, That under clause (B) of this subparagraph (5), reasonable variations may be permitted, and exemptions as to small packages may be established, by regulations prescribed by the Secretary;
6. if any word, statement, or other information required by or under authority of this chapter to appear on the label or other labeling is not prominently placed thereon with such conspicuousness (as compared with other words, statements, designs, or devices, in the labeling) and in such terms as to render it likely to be read and understood by the ordinary individual under customary conditions of purchase and use;
7. if it purports to be or is represented as a food for which a definition and standard of identity or composition has been prescribed by regulations of the Secretary under Section 607 of this title unless (A) it conforms to such definition and standard, and (B) its label bears the name of the food specified in the definition and standard and, insofar as may be required by such regulations, the common names of optional ingredients (other than spices, flavoring, and coloring) present in such food;
8. if it purports to be or is represented as a food for which a standard or standards of fill of container have been prescribed by regulations of the Secretary under Section 607 of this title, and it falls below the standard of fill of container applicable thereto, unless its label bears, in such manner and form as such regulations specify, a statement that it falls below such standard;

9. if it is not subject to the provisions of subparagraph (7), unless its label bears (A) the common or usual name of the food, if any there be, and (B) in case it is fabricated from two or more ingredients, the common or usual name of each such ingredient; except that spices, flavorings, and colorings may, when authorized by the Secretary, be designated as spices, flavorings, and colorings without naming each: Provided, That to the extent that compliance with the requirements of clause (B) of this subparagraph (9) is impracticable, or results in deception or unfair competition, exemptions shall be established by regulations promulgated by the Secretary;

10. if it purports to be or is represented for special dietary uses, unless its label bears such information concerning its vitamin, mineral, and other dietary properties as the Secretary, after consultation with the Secretary of Health and Human Services, determines to be, and by regulations prescribes as, necessary in order fully to inform purchasers as to its value for such uses;

11. if it bears or contains any artificial flavoring, artificial coloring, or chemical preservative, unless it bears labeling stating that fact: Provided, That, to the extent that compliance with the requirements of this subparagraph (11) is impracticable, exemptions shall be established by regulations promulgated by the Secretary; or

12. if it fails to bear, directly thereon or on its container, as the Secretary may by regulations prescribe, the inspection legend and, unrestricted

by any of the foregoing, such other information as the Secretary may require in such regulations to assure that it will not have false or misleading labeling and that the public will be informed of the manner of handling required to maintain the article in a wholesome condition.

The USDA follows an entirely different model of policing misbranding. Whereas the FDA allows products to enter the market before taking enforcement action, FSIS preapproves every label used prior to a product entering the market. The difference is in part due to the statutory language in the FMI and PPI. In Sections 607(d) and 457(c) the statute states products may only be sold that are not misleading and which are "approved by the Secretary" of the USDA. This "approved by" language is absent from the FD&C. It is also the textual hook the USDA interprets as mandating preapproval of all food labels before the products are offered for sale.

FMI Sections 607(d) and PPI Section 457(c)
(d) Sales under false or misleading name, other marking or labeling or in containers of misleading form or size; trade names, and other marking, labeling, and containers approved by Secretary

No article subject to this subchapter shall be sold or offered for sale by any person, firm, or corporation, in commerce, under any name or other marking or labeling which is false or misleading, or in any container of a misleading form or size, but established trade names and other marking and labeling and containers which are not false or misleading and which are approved by the Secretary are permitted.

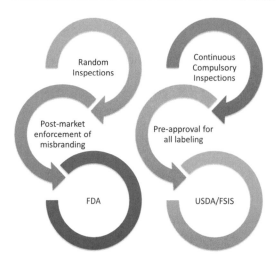

Fig. 4.3 Comparing FDA and FSIS enforcement models

The full ramifications of FSIS's authority can now be seen. In Chapter 2 the text outlined FSIS's unique continuous compulsory inspection model. In labeling, FSIS exercises similar control with preapproval for all labels. FSIS is given much stricter controls to ensure meat and meat products that enter the market are both safe and genuine (See Fig. 4.3).

FSIS attempts to ease the potential preapproval burden using two tools. Prior approval is granted in one of two ways. FSIS will either provide a sketch approval or generic approval. Sketch approval involves a review by the Labeling and Program Delivery Staff whereas generic approval merely requires compliance with applicable regulations. FSIS regulations approve some labels without submitting a sketch or concept label. Although it appears FSIS is abandoning its prior approval labeling mandate by using its generic approval system it technically continues to approve all labels. Akin to continuous inspection taking on one meaning for slaughter facilities and another for processing plants, prior approval can be achieved in a variety of ways. FSIS holds that prior approval is maintained either by FSIS staff reviewing draft labels or FSIS develops regulations that a facility must comply before entering the market. Thus, it can be said all FSIS labels are preapproved.

4.2 Defining the Label

4.2.1 When Written Material Becomes Labeling

Independent Books and Literature
One of the key threshold questions when discussing FDA enforcement of misbranding is distinguishing labeling from advertising. This will inform the gamut of statutory and constitutional issues raised by the FDA, industry, or the courts. In the absence of a clear sense of whether the written material in question is labeling or advertising, there is no indication what statutory criteria is to be applied in evaluating misleading or misbranding. This is of particular importance given the concurrent jurisdiction exercised by the FTC over advertising.

The statutory definition itself is intentionally broad and flexible. Nearly any written materially associated with a food product can be captured under Section 201(k). In addition, 201(m)(2) carries a textual hook, "accompanying such article," that would allow for easy expansion of the concept. Rather than freezing the label concept to the period of the Act, Congress sought to provide the FDA flexibility to keep pace with labeling and marketing practices. The definition gains new elasticity when courts abandon the strict construction standard for interpreting the Act.

Courts interpret the Act broadly. As discussed in Chapter 1 courts routinely cite the drafter's intent to interpret the Act broadly. Courts also find a liberal construction consistent with the Act's intent to protect the public from harm. Whether that harm is physical or economic is immaterial. Thus, an expansive definition meets little resistance from courts reviewing the FDA's efforts to stretch the concept.

One of the first cases to consider what constituted labeling found literature written by an independent author as labeling. The case, *United States v. 250 Jars 'Cal's Tupelo Blossom U.S. Fancy Pure Honey'*, was decided by the Sixth

Circuit in 1965.[1] In *Cal's Honey* the court reviewed a misbranding of the retail store of Cal's Tupelo Honey displayed with copies of a booklet titled "About Honey" (*Cal's Tupelo Honey*).[2] The booklets were placed on the shelf above the honey. An FDA inspector entered the store acting as customer and requested information about honey. The clerk handed the undercover agent a newsletter containing an article "Eat Honey and Increase Your Vitality" (*Cal's Tupelo Honey*).[3] This was apparently the common practice of the store. The FDA seized 71 copies of the newsletter from the store. The newsletter contained a number of statements about the benefits of eating honey. This included a claim that the honey was a "panacea for various diseases and ailments that have plagued man from time immemorial" (*Cal's Tupelo Honey*).[4] The store also mailed the newsletter to prospective clients as a type of direct-mail marketing practice.

The court found the book and newsletter as labeling based on the connection to the retailer's products. The court began by reasoning the misbranding provisions of the Act were to be liberally construed because it was "passed to protect unwary customers in vital matters of health..." (*Cal's Tupelo Honey*).[5] From this lens the court concluded it must uphold the action against the retailer or risk opening a "...loophole through which those who prey upon the weakness, gullibility, and superstition of human nature..." (*Cal's Tupelo Honey*).[6] This combined with the "immediate connection" of the literature and the sale of the products resulted in a finding that the independent materials were part of the food label or labeling. Thus any written material used to effectuate the sale constitute labeling under the Act.

United States v. 250 Jars 'Cal's Tupelo Blossom U.S. Fancy Pure Honey'
McALLISTER, Senior Circuit Judge.

The United States brought a libel proceeding for the condemnation, under Section 304(a) of the Federal Food, Drug and Cosmetic Act, as amended, Title 21, U.S.C.A. § 334(a) of a quantity of allegedly misbranded honey, sold by claimant-appellant at one of its stores located in Detroit, Michigan.

Upon stipulated facts, and after argument, the District Court entered an order of condemnation for misbranding of the honey, from which claimant appeals.

It appears that the honey which was condemned was displayed on shelves in appellant's retail store, on top of which were placed copies of a booklet, "About Honey," which were sold to any customer desiring to purchase them. On the store premises, when the honey was seized, were 71 copies of a newspaper-type mailing piece, containing an article "Eat Honey and Increase Your Vitality." The booklet and the newspaper leaflet were shown to a drug inspector, acting as a prospective customer, in response to his request for information about honey. The information contained in the foregoing publicity material was also mailed to prospective customers in order to promote the sale of the honey.

Appellant contends that the booklet, "About Honey," by an independent author, which was on sale in the Book Department of its retail store, and the newspaper leaflets located in a back room of the store, did not constitute misbranding; and that the inferences from the evidence relied upon by the District Court were negated by uncontroverted facts.

Judge Freeman found that the booklet and the newspaper leaflet constituted labeling and misbranding of the honey, and that a reading of the booklets and mailing leaflets resulted in the inescapable conclusion

---

[1]   344 F.2d 288 (6th Cir. 1965).

[2]   *Id*. at 289.

[3]   *Id*.

[4]   *Id*.

[5]   *Id*.

[6]   *Id*.

that such honey was intended to be used as a drug, since the literature made the rather remarkable claim that honey is a panacea for various diseases and ailments that have plagued man from time immemorial. Furthermore, Judge Freeman held that the Act was passed to protect unwary customers in vital matters of health and, consequently, must be given a liberal construction to effectuate this high purpose, and that the Court would not open a loophole through which those who prey upon the weakness, gullibility, and superstition of human nature can escape the consequences of their actions. Upon a review of the record and the briefs of the parties, we are in accord with the trial judge's reasoning and his conclusion.

It should here be observed that, since the determination of the District Court in this case, there has been decided by the Court of Appeals for the Second Circuit, a case involving the sale of food products in a store in which misleading books, relating to food, were also sold, and a judgment of condemnation by the District Court in that case, was reversed. However, that case differs from the one before us in that, although misleading books were sold in the same store that sold the product which was seized and condemned, there was no immediate connection of the misleading books with the sale of the product, and, as the court said, the books were not put to such use by the seller of the product.

As we have said in prior cases, it is not the policy or practice of this court, in reviewing cases on appeal where a District Court has rendered a comprehensive opinion with which we find ourselves in full agreement, to rewrite such an opinion and, in a sense, deprive the trial court of the credit of its careful consideration of the issues and arguments, and complete determination of the cause.

In accordance with the foregoing, the judgment of the District Court is affirmed for the reasons set forth in the opinion of Judge Freeman…

Comparing the decision in *Cal's Honey* to two other cases provides a clearer sense of when independent literature constitutes labeling. In district court case preceding *Cal's Honey* a maker of blackstrap molasses closely coordinated the promotion of its product through a book written by an independent author (*Plantation Molasses*).[7] The maker of the blackstrap molasses placed copies of the book along with jars of the molasses in the store window. With the display was a sign stating customers could order all the products required by the "Hauser diet" including blackstrap molasses (*Plantation Molasses*).[8] The sign invited customers in with the sign stating "Come in for full information" (*Plantation Molasses*).[9] A customer intrigued by the sign and inquiring about the diet would be handed a copy of the book and directed to the pages discussing the use and benefits of blackstrap molasses (*Plantation Molasses*).[10] In *Cal's Honey* the book was placed near the honey or mailed to customers, but was never overtly marketed like the book on the Hauser diet.

Yet simply having books in the same store is not enough to constitute labeling. The Second Circuit reviewed a case around the time the Sixth Circuit reviewed *Cal's Honey*. The case focused on a condemnation ruling against Balanced Foods, Inc., a maker of sterling cider vinegar and honey (*Sterling Vinegar Honey*).[11] Balanced foods carried its products and two

---

[7] *United States v. 8 Cartons 'Plantation 'The Original' etc. Molasses'* 103 F. Supp. 626 (W.D.N.Y. 1951).

[8] *Id.* at 627.

[9] *Id.*

[10] *Id.* at 628.

[11] *U.S. v. 24 Bottles "Sterling Vinegar Honey,"* 338 F.2d 157 (2d Cir. 1964).

books, "Folk Medicine" and "Arthritis and Folk Medicine" along with other products (*Sterling Vinegar Honey*).[12] The books, vinegar and honey were seized by the FDA from its warehouse in New York. The district court found the books were labeling and condemned the "labeling" as misbranded (*Sterling Vinegar Honey*).[13] The books promoted the use of sterling cider vinegar as suitable for medicinal use (*Sterling Vinegar Honey*).[14] The products itself made no claims other than the contents of the bottle. The Second Circuit provides the first boundary on what constitutes labeling. It reasoned:

> On the other hand, labeling does not include every writing which bears some relation to the product. There is a line to be drawn, and, if the statutory purpose is to be served, it must be drawn in terms of the function served by the writing (*Sterling Vinegar Honey*).[15]

The court held that there must be some "immediate connection" for written material to constitute labeling, such as joint promotion (*Sterling Vinegar Honey*).[16] In assessing joint promotion, the court looked at evidence of shelving together, proximity of the products, and displays featuring the books and the products (*Sterling Vinegar Honey*).[17] Merely carrying two related products was insufficient to function as a connection between the products (Fig. 4.4).

Images, Social Media, and Other "Labeling" Materials

The general framework that emerges from judicial precedent is that any material that bears a strong relationship to the sale of the product is labeling. Further, case law clarifies the FDA's policy on when images, website material, social media and metadata constitute labeling. In both *Cal's Honey* and *Plantation Molasses* the court

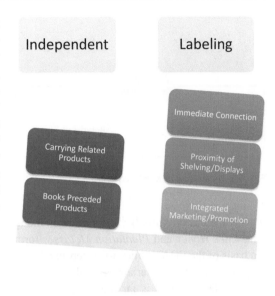

Fig. 4.4 Illustration of factors weighed when assessing whether independent literature is labeling

focused on whether the written material "accompanies" the food product. This lead the court to conclude that any immediate connection would mean the written material accompanied the food product. An earlier case provides another definition to "accompanies."

The Supreme Court reviewed the question of when labeling accompanies a product in the drug context. In 1948 the Supreme Court decided *Kordel v. United States*.[18] *Kordel* involved a drug company mailing drugs and explanatory pamphlets to its retailers in separate packages often at different times. The pamphlets explained the benefits and effectiveness of the drugs, which Kordel wrote based on private and public research. The court found it was not necessary for the written material to physically accompany the products (*Kordel*).[19] Instead "accompanies" refers to a textual relationship between the written material and the product. When written materials "supplements or explains" the product, then

---

[12] *Id.* at 158.

[13] *Id.*

[14] *Id.*

[15] *Id.* at 158–159.

[16] *Id.* at 159.

[17] *Id.*

---

[18] 335 U.S. 345, 69 S.Ct. 106, 93 L Ed. 52 (1948).

[19] *Id.* at 349–350.

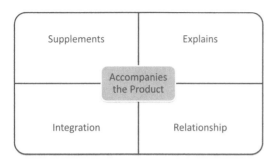

Fig. 4.5 Illustration of *Kordel* criteria

it accompanies it (*Kordel*).[20] This decision along with *Cal's Honey* and *Plantation Molasses* form the basis of the FDA's policy to sweep-in a broad range of marketing materials as labeling.

The FDA's policy is far reaching. In policy statements for every regulated category the FDA cites to *Kordel*. Under *Kordel* it interprets "labeling" as defined in the Act to include brochures, booklets, video, sound, images, websites, social media, and metadata. Courts continue to agree. The central question to determine whether material constitutes label is not its form, but whether it "accompanies" the regulated product. To accompany the regulated product the material must "supplement or explain" it (see Fig. 4.5).

The FDA does not always meet the supplements-or-explains standard. It may seem defeating to read the case law giving the FDA copious coverage to deem virtually all material labeling. Still there are instances where the courts limit the labeling concept. For example, a district court in 2013 dismissed an attempt to bring a false advertising case against Frito-Lay (*Frito-Lay North America*).[21] The plaintiffs sought to include Frito-Lay's website as part of the company's label. It cited the language, "www.fritolay.com," found on the packaging as evidence of integrated marketing (*Frito-Lay North America*).[22] The court

disagreed that the website was enough under *Kordel*:

> The Court does not find that the language on the www.fritolay.com website constitutes labeling under the FDCA, because as cited by Plaintiffs, none of the website language **explains or supplements** the individual Named Products such that the website could generally be found to **"accompany"** the Named Products. Even though the Named Products' labels ask consumers to visit the website, they do not state that the website will inform consumers of the details of the Named Products' nutritional facts, and none of the language Plaintiffs cite is drawn closely enough to the Named Products themselves to merit the website's being found to constitute "labeling" (emphasis added).[23]

Listing a website is commonplace in the market today. The district court found how the website is listed matters as much as what is on the website. Frito-Lay simply listed the website and did not add language like "learn more" or "get more nutrition information" to suggest the website supplemented or explained the product. Also, the website was fairly generic. It contained photos and company information, but no specific statements that could be construed as explaining or supplementing the information on the package. Together the evidence could not support the supplements-or-explains standard provided in *Kordel*.

4.2.2 Regulatory Components of a Label

In addition to the global concept of what constitutes labeling is the more focused regulatory question on the regulatory required components of labeling. The components discussed here fit the traditional notion of labeling relating specifically to product packaging. Under the regulations of food labeling two chief packaging areas are defined, each carrying certain information and disclosures.

---

[20] *Id.* at 350.

[21] 2013 U.S. Dist. LEXIS 47126, at *4–5 (N.D. Cal. Apr. 1, 2013).

[22] *Id.* at *18.

[23] *Id.* at *18–19.

Fig. 4.6 FDA labeling guidance identifying PDP and net quantity statement

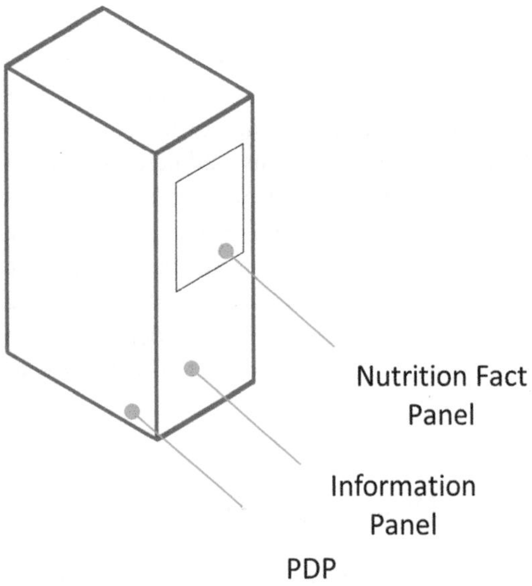

Fig. 4.7 Illustration identifying location of information panel

The principal display panel is a regulatory name for the well-known face of the package. It is the main label panel either facing out from the shelf or alternatively at the top of packaging. The CFR defines the principal display panel or PDP as "that portion of the package label that is most likely to be seen by the consumer at the time of purchase" (21 CFR Part 101).[24] The PDP must contain the name of the food, known as the statement of identity, and the net quantity statement (see Fig. 4.6). This sounds technical, but is seen every day when we shop in a market or examine the cupboard.

The information panel is another regulatory term for an intimately familiar part of the packaging. The information panel is the panel that contains the nutrition facts and ingredients listing (see Fig. 4.7). The regulations require the information panel "immediately to the right of the PDP, as displayed to the consumer" (21 CFR Part 101).[25] If the packaging does not accommodate useable space immediately to the right of the PDP then the next label immediately to the right suffices (21 CFR Part 101).[26] The in-

formation panel cannot contain any "intervening material" with solely the required name and address of the manufacturer, packer or distributor, the nutrition labeling, any required allergy labeling, and the ingredient list (21 CFR Part 101).[27]

The regulations contain a wealth of detail covering every facet of the packaging. This includes the types of information required, the location, and in some cases even the size of font used. Understanding this element of labeling remains a critical task for daily operations and compliance.

4.3   What Constitutes Misbranding

Equipped with an understanding of what is a "label" or "labeling" the focus shifts to when that label violates the Act. This will involve a review of the full range of written material from social media to product packaging including images and video material.

---

[24] 21 CFR 101.1.

[25] 21 CFR 101.2(a).

[26] *Id.* at 101(2)(a)(1).

[27] 21 CFR 101.2(b) and (d).

There are several grounds for finding an FDA-regulated product misbranded. This section will discuss the most common violations as well as provide coverage of unique areas of labeling, such as allergens, "all natural" claims, and organic. Irrespective of the category or type of misbranding the overall prohibition is against labeling material that is misleading. As will be seen there is a grand gradient of mislabeling from the missing regulatory components to the egregious claims about product efficacy or benefits.

4.3.1 Misleading Labels

The central statutory prohibition for all labeling is against misleading labels. To protect the public the overarching aim of labeling enforcement is against misleading labeling. Labeling is one aspect of food law consumers are all familiar with. We all exhibit our own product preferences and interact with the label to assess its attributes, in particular products new to our pantry. Some will hone in on the ingredient list others immediately look for hallmark claims such as "all natural" or "high in fiber." This practice builds in a misleading-radar. As consumers we can all assess the credibility of labeling and marketing claims to varying degrees. This section will fine-tune that misleading-radar to pick up the nuances of statutory provisions and case law interpretations.

Starting with the FD&C a broad prohibition is laid down. Section 201 broadly proclaims food is deemed misbranded if misleading. It goes on to provide some criteria for finding misleading, including omitting material facts about the consequences of using the product. Otherwise the statute provides no indication of when a label is misleading and thus misbranded. This is known as the "materiality analysis" whereby omitted facts are evaluated to determine if they were material facts. The plain meaning of misleading encompasses more than factual omissions.

Misleading nutritional information qualifies for misbranding under Section 201(n). This was seen in *United States v. An Article of Food...*

Nuclomin (Nuclomin).[28] The Eighth Circuit reviewed on appeal a vitamin/mineral tablet known as "Nuclomin." The FDA deemed the supplement misbranded and seized it (*Nuclomin*).[29] As will be discussed in Chapter 6 dietary supplements are regulated as food products. The district court concluded many of the ingredients listed on the label were of no nutritional value or at such "minute" levels as not to offer a nutritional value to the supplement (*Nuclomin*).[30] On appeal the Eight Circuit found "substantial evidence" that the disputed ingredients in Nuclomin were either "not needed in the human diet" or the levels were so small as to offer no value to the consumer (*Nuclomin*).[31] Although technically compliant by listing all of the dietary ingredients the ingredients could "persuade a purchaser that the product possessed greater nutritional value" than it or its competitors contained (*Nuclomin*).[32]

The FDA is not required to present evidence that consumers were misled. The maker of Nuclomin argued on appeal that the FDA failed to prove any of its customers were actually misled. The Eight Circuit cited to prior precedent holding that the "fact that no purchasers have actually been misled is not a defense under the Act" (*Nuclomin*).[33]

United States v. An Article of Food... Nuclomin
LAY, Circuit Judge.
This is an appeal from an in rem proceeding brought under the Federal Food, Drug, and Cosmetic Act against a special dietary product "Nuclomin" claiming it is misbranded in violation of Section 403(a) of the Act, 21 U.S.C. § 343(a).1 Hunt Investment, Inc., owner of Nuclomin, intervened. Jurisdiction rests under 21 U.S.C. § 334. The district court upheld the government seizure and condemnation on the basis that

---

[28] 482 F.2d 581 (8th Cir. 1973).

[29] *Id.* at 583.

[30] *Id.*

[31] *Id.* at 584.

[32] *Id.* at 583.

[33] *Id.*

several ingredients listed on the label were "either of no nutritional value per se or the quantities are so minute as not to enhance the nutritional value of the tablets." The district court, the Honorable John K. Regan presiding, found that such label was false and misleading in that it could persuade a purchaser that the product possessed greater nutritional value than it actually did.

The basic issues on appeal include (1) whether the Food and Drug Administration (FDA) possessed the authority to prohibit the sale of a product that lists, as required by the regulations, completely safe ingredients that may be unnecessary or insignificant; (2) whether sufficient proof was presented to establish that the questioned ingredients were not needed or were included in inadequate amounts; and (3) whether the product label was in fact misleading. We affirm the trial court's ruling.

THE FDA'S AUTHORITY

The government does not challenge the factual accuracy of the Nuclomin label; rather it claims that the label is misleading to the public because some of the ingredients are either not needed in human nutrition or are included in such insignificant amounts as to be valueless. Specifically, the government attacks the vitamin constituents choline, inositol and p-aminobenzoic acid, the mineral elements potassium, magnesium and calcium succinate, and the amino acids found in the yeast extract. It is undisputed that these ingredients are consumed daily by the public and are completely safe.

2. The Nuclomin label reads:

"Amino Acid Complexed Trace Mineral With Multi-Vitamins "A concentrated source of vitamins, minerals and other nutritional factors, plus the micronutrients associated with the amino acids and polypeptides as found in a special yeast extract." Dosage "Adults, orally as a dietary supplement, two tablets per day. Each two tablets contain MDR" Vitamins "Vitamin A............ 5000 Units 125% Vitamin D............. 1000 Units 250% Vitamin E............ 5 Units * Vitamin B1............ 2.5 mg. 250% Vitamin B2 (Riboflavin)............ 5.0 mg. 400% Vitamin B6............ 1.5 mg. * Niacinamide............ 25.0 mg. 250% Panthothenic Acid (as Cal. Pan....)...... 25.0 mg. * Choline (as Bitartate)........... 50.0 mg. * Inositol............ 50.0 mg. ** Vitamin C............ 50.0 mg. 166% Vitamin B-12............ 1.0 meg. * p-Aminobenzoic Acid............ 10.0 mg. ** "Minerals "Copper............ 1.0 mg. * Iodine............ 0.075 mg. 75% Manganese............ 2.0 mg. * Iron............ 10.0 mg. 100% Potassium............ 20.0 mg. * Zinc............ 2.0 mg. * Magnesium............ 20.0 mg. * "In associate with: "Calcium Succinate............ 50 mg. "MDR: Minimum Daily Requirement. "* Need in human nutrition, established Requirement not determined. "** Need in human nutrition not established."

Most of claimant's arguments relate to Section 403(j), 21 U.S.C. § 343(j), relating to the misbranding of special dietary articles. This, however, overlooks the direct authority of the government to bring a condemnation suit for violation of Section 403(a) pertaining to misbranding because of the use of a misleading label... Section 403(a), 21 U.S.C. § 343(a), clearly states that a food is misbranded if its labeling is false or misleading in any particular. Therefore, Section 403(j) is not applicable, and the broad proscription of Section 403(a) is.

The claimant asserts that it is in compliance with the applicable regulations, 21 C.F.R. §§ 125.3(a)(2) and 125.4(a)(2), in that the product label contains a statement of the quantity of such vitamin or mineral in a specified quantity of the product and also bears a statement concerning whether the need or requirement in human nutrition has been established. However, Sections 125.3(a)(4) and 125.4(a)(4) point out that:

"Compliance with the provisions of subparagraphs (2) and (3) of this paragraph shall not be construed as relieving any food which purports to be or is represented for special dietary use by reason of its [vitamin or mineral] property from the application of Section 403(a) and 201(n) of the act, as in the case where the need for such [vitamin or mineral] in human nutrition is not substantially supported by the opinion of experts qualified by scientific training and experience to determine such needs."

Thus even though the Nuclomin label is technically accurate and further meets the regulations' disclosure requirements, it must also comply with Section 403(a) and not be misleading. Realizing that "the Food, Drug, and Cosmetic Act is to be given a liberal construction consistent with the Act's overriding purpose to protect the public health…" we hold that the FDA had the authority under Section 403(a) to seize and condemn the special dietary supplement Nuclomin.

SUFFICIENCY OF PROOF

The testimony of the government's witnesses, Dr. Thomas D. Luckey (a professor of biochemistry at the Missouri School of Medicine and chairman of the graduate nutrition program), and Dr. Harold L. Rosenthal (a professor of physiological chemistry at Washington University Dental School who has done considerable research in the field of nutrition, primarily the metabolism of Vitamin B-12 and amino acids) and the appellant's witness, Dr. Edward Doisy, Jr., (a nutritionist, biochemist, and nonpracticing physician) was admittedly conflicting. Nevertheless, there was substantial evidence for the trier of fact to believe that the disputed ingredients in Nuclomin were either not needed in the human diet or that the amount of the ingredient was so small that it would have no value. For example, Dr. Luckey testified that choline is not a vitamin and is not needed in the human diet because it is produced in human tissues from food. He further stated that the amount of choline in Nuclomin (50 mg) was of no value whatsoever. The government's expert further opined that inositol is also not a vitamin, and all three experts generally agreed that any requirement for inositol is not known. Dr. Luckey testified that p-aminobenzoic acid (PABA) is not required in nutrition, is not a vitamin, and is not needed by man. He further clarified that PABA is one of three constituent parts of the vitamin folic acid; however, the body is unable to take PABA and use it to make folic acid. All witnesses confirmed that potassium is a necessary nutrient, present in all cells, and contained in most all foods. The doctors disagreed on the daily requirement of potassium (from 500 mg to 4000 mg), however, they all felt that the 20–32 mg of potassium in Nuclomin would be of little or no value except in the most dire circumstances. Dr. Luckey and Dr. Rosenthal were in agreement that the 15–20 mg of calcium that Nuclomin would provide per day was meaningless with respect to the recommended daily allowance of 750–800 mg. Finally, the amino acids, polypeptides and other nutritional factors in the yeast extract were concluded to be insignificant for any dietary value by Dr. Luckey, Dr. Rosenthal and Dr. Wixom (a professor of biochemistry at the School of Medicine, University of Missouri in Columbia, who is a specialist in amino acid and protein metabolism and nutrition).

In general, Dr. Luckey testified that he had never heard of anyone being deficient in choline or inositol, and if a person were in need of potassium or magnesium, a therapeutic treatment would consist of greatly higher doses than are present in Nuclomin. Dr. Rosenthal believed Nuclomin to be beneficial for its content of vitamins A, C, D, B-12, riboflavin and niacinamide, but the elements challenged by the government provided no special dietary value. On cross-examination, Dr. Rosenthal stated that even he had been confused by the Nuclomin label. On the basis of the overall record this court cannot hold the district court's finding that the questioned ingredients were not needed or were included in inadequate amounts clearly erroneous.

WHETHER MISLEADING

The claimant urges that the FDA has failed to prove that any of Nuclomin's customers were actually misled by the product label, relying on the *United States v. 119 Cases*... "New Dextra Brand Fortified Cane Sugar..." This case simply held that the government failed to carry its burden of proof through its nutritional experts that the label used was misleading. Other courts have held that although admissible on the issue of whether a label is false or misleading, the fact that no purchasers have actually been misled is not a defense under the Act...

In several contexts the claimant has portrayed Nuclomin as a safe or harmless product and, on the basis of its other nutritionally accepted ingredients, a beneficial dietary supplement. The safety of a product, however, cannot be a basis for substantiating the legality of a seized article. Numerous authorities have held that it is immaterial to a question of misbranding whether the condemned article is inherently dangerous or harmful or whether it may in some way be beneficial...

In *United States v. 95 Barrels*... Apple Cider Vinegar... the Supreme Court found a vinegar label describing its contents as "apple cider vinegar made from selected apples" to be misleading to the public because the seized product was made from dehydrated apples rather than from fresh apples—even though the contested vinegar was similar in color, taste and consistency to vinegar processed from fresh apple cider and equally wholesome. The Court explained:

> The statute is plain and direct. Its comprehensive terms condemn every statement, design, and device which may mislead or deceive. Deception may result from the use of statements not technically false or which may be literally true. The aim of the statute is to prevent that resulting from indirection and ambiguity, as well as from statements which are false. It is not difficult to choose statements, designs, and devices which will not deceive. Those which are ambiguous and liable to mislead should be read favorably to the accomplishment of the purpose of the act. The statute applies to food, and the ingredients and substances contained therein.

We accept the well reasoned opinion of the district court. The Nuclomin label defines itself as a dietary supplement and lists the challenged ingredients among known nutritional vitamins and minerals. As the district court found, this ambiguity could represent by indirection that these elements contributed some additional benefit when in fact they do not.

The claimant urges that an added consideration in judging whether the Nuclomin label is misleading is that Nuclomin is distributed only to doctors for their patients, not to the general public. However, Nuclomin is not a prescription drug, therefore, nothing prevents it from being sold to the public. Moreover, licensed physicians are not exempt from the Act, and the fact that a seized article may only be sold to or used by physicians does not stay the full trust of the Act.

Judgment affirmed.

Fig. 4.8 Illustration of FDA misbranding requirements

Misleading largely preserves its plain meaning through judicial interpretation. Misleading then means some element of the label is or could be confusing, deceptive, or false. The test for the FDA which is demonstrating whether consumers are misled further secures the agency's success. In *U.S. v. Strauss*, for example, the court noted it is not the label's effect on a "reasonable consumer" but the "ignorant, the unthinking, and the credulous" consumer (*Strauss*).[34] A bit of an insult it would seem, but recall the intent of the 1938 Act discussed in Chapter 1. The drafters of the Act determined consumers could not protect themselves. Thus, it would seem the FDA is given a low-bar to clear in order to provide maximum protection.

Safety is not a central concern only the accuracy of the text in question. The FDA must present some evidence to demonstrate how or why the label could be deceptive, which may include actual customer testimony. Demonstrating that actual users of the product were misled, however, is neither a requirement nor a defense. Finally, technical compliance with the regulatory aspects of the label does not shield a product from misbranding actions (see Fig. 4.8).

Nuclomin also teaches the industry about the best basis for challenging a misbranding action brought by the FDA under Section 201(n). Rather than throw the kitchen-sink with every argument from technical compliance, safety, and actual customer defense, the best defense focuses on the evidence. A challenge to a misbranding action should, to the extent possible, build counter evidence to demonstrate truthfulness or accuracy.

Fig. 4.9 Example from FDA guidance on use and placement of statement of identity

The alternative is to undermine the agency's experts or the overall sufficiency of its evidence.

4.3.2 Common or Usual Name of Food

There are certain areas where misbranding requires the FDA to make little evidentiary showing. The FDA issues regulations for certain labeling claims that are deemed necessary to accurately inform the consumer about the product. Deviating from the regulations or omitting the regulated text is considered misleading without any additional evidence.

The common or usual name of a food is one such area of regulation. The FDA both in the regulations and its guidance documents requires that food contain the common or usual name (see Fig. 4.9). If a food with an established name uses a new one or if a food subject to a standard of identity does not "bear the name specified in the standard" it is misleading. (21 CFR Part 101).[35] The

---

[34] 999 F.2d. 692, 697 (1993).

[35] 21 CFR 101.3(b)(2).

issue is rather straightforward for standardized foods and less so for nonstandardized products.

Each standardized food regulation stipulates the name or names of food products. The FDA currently sets identity standards for roughly 300 products in 20 categories. For instance, the standard of identity for sherbet requires that the labeling stipulates how fruit and nonfruit sherbets are named on the label (21 CFR Part 101).[36] Care must be taken to ensure the details in the identity match the product being labeled. There are shades of differences with some products, such as with fruit butter (21 CFR 150.110), fruit jelly (21 CFR 150.140), and fruit jam (21 CFR 150.160). To carry the name fruit butter but the ingredients of fruit jam would be misleading and thus a misbranding offense.

How to Name Sherbet 21 CFR 135.140(f)(1) (i)–(iii)

(f) Nomenclature. (1) The name of each sherbet is as follows:

(i) The name of each fruit sherbet is "___ sherbet", the blank being filled in with the common name of the fruit or fruits from which the fruit ingredients used are obtained. When the names of two or more fruits are included, such names shall be arranged in order of predominance, if any, by weight of the respective fruit ingredients used.

(ii) The name of each nonfruit sherbet is "___ sherbet", the blank being filled in with the common or usual name or names of the characterizing flavor or flavors; for example, "peppermint", except that if the characterizing flavor used is vanilla, the name of the food is "___ sherbet", the blank being filled in as specified by § 135.110(e)(2) and (5)(i).

(2) When the optional ingredients, artificial flavoring, or artificial coloring are used in sherbet, they shall be named on the label as follows:

(i) If the flavoring ingredient or ingredients consists exclusively of artificial

flavoring, the label designation shall be "artificially flavored".

(ii) If the flavoring ingredients are a combination of natural and artificial flavors, the label designation shall be "artificial and natural flavoring added".

(iii) The label shall designate artificial coloring by the statement "artificially colored", "artificial coloring added", "with added artificial coloring", or "___, an artificial color added", the blank being filled in with the name of the artificial coloring used.

Nonstandardized foods are named following general principles provided in the regulations. The regulations identify the criteria on how to name a non-standardized food, but the regulations are simply general guidelines. The primary section that will guide naming a nonstandardized food is 21 CFR 102.5(a). The principles first require the common or usual name if it already exists, then a coined term may be used if the common name is not known. The name must be simple, direct, uniform and not confusingly similar to related foods. The FDA Warning Letter database provides insight into how the agency interprets and enforces this provision.

Naming Non-Standardized Food 21 CFR 102.5(a)

The common or usual name of a food, which may be a coined term, shall accurately identify or describe, in as simple and direct terms as possible, the basic nature of the food or its characterizing properties or ingredients. The name shall be uniform among all identical or similar products and may not be confusingly similar to the name of any other food that is not reasonably encompassed within the same name. Each class or subclass of food shall be given its own common or usual name that states, in clear terms, what it is in a way that distinguishes it from different foods.

---

[36] 21 CFR 135.140(f)(1)(i)–(iii).

FDA warning letters focus on full range of criteria provided in the regulations. In a 2009 Warning Letter the FDA deemed the name "Raw Parmesan Cheeze" as misbranded on two grounds (Awesome Foods Inc.).[37] First it found that the product did not "accurately describe the basic nature of the food" because it lacked milk which all cheeses have as their starting ingredient (Awesome Foods Inc.).[38] Second the name "Raw Parmesan Cheeze" was confusingly similar to "parmesan cheese" (Awesome Foods Inc.)[39] On this and other bases the FDA determined the food was misbranded. Other examples abound in warning letters from "tub cheese" (Paul Meserve 2005)[40] to "Kimchi" (Yon's Foods 2009).[41] When naming nonstandardized products creativity must be tethered to clarity to ensure consumers and the FDA cannot claim that the statement of identity is misleading.

4.3.3   Country of Origin Labeling

Country of origin labeling is an increasing interest for consumers and industry. Congress passed a Country of Origin Labeling (COOL) requirement as part of a farm bill in 2002 (Notice of Country of Origin).[42] The initial COOL requirement was the country of origin labeling for fresh beef, pork, and lamb, but not processed meat products. An expansion to the law was passed in 2009 to include fresh fruits, nuts, and vegetables.

COOL introduces a new element of compliance and a healthy dose of controversy.

COOL is administered by the USDA Agricultural Marketing Services (AMS). Despite having implications for the FDA and FSIS as a farm bill, Congress designated the AMS as the primary branch of the USDA to administer the law. As a mandatory rule it will be enforced by FSIS and the FDA as a form of misbranding. The AMS will be responsible for promulgating regulations and providing the FDA, FSIS, and industry information about how to comply with the rule.

The rule only covers fresh meat and produce. Exempt from the COOL requirements are processed foods. As interpreted by the USDA this term exempts a wide swath of the market place. FSIS explains processed foods as a COOL immune commodity that:

1. Has undergone specific processing resulting in a change of character (for example, cooking, curing, smoking, restructuring); or
2. Has been combined with another food component (FSIS COOL).[43]

This exemption will leave out both FDA and USDA products. From smoked hams to roasted almonds COOL will not apply. The focus is squarely on produce and raw meat.

Meats will be labeled depending on where the animal is born, raised, and slaughtered. A meat product can only be labeled "Product of the USA" if the animal's entire production cycle occurs in the USA (FSIS COOL).[44] If meat is from an animal born in another country and slaughtered in the USA, then it must be labeled "Product of the USA, Country A" (FSIS COOL).[45] Animals imported to the USA for immediate slaughter are labeled in the same manner. All imported fresh meat must bear a label indicating the country of origin. This applies to fresh cuts of meat and ground products. Thus, it could be that a ground

---

[37] FDA warning letter, Awesome Foods, Inc. (December 2009) available at: http://www.fda.gov/iceci/enforcementactions/warningletters/2009/ucm195206.htm.

[38] *Id.*

[39] *Id.*

[40] FDA warning letter, Paul Meserve Distributor (May 2005) available at http://www.fda.gov/iceci/enforcementactions/warningletters/2005/ucm075419.htm.

[41] FDA Warning Letter, Yon's Foods LLC (August 2009) available at: http://www.fda.gov/iceci/enforcementactions/warningletters/2009/ucm180665.htm (finding the name Kimichi absent a descriptive phrase such as "Pickled Spicy Korean Cabbage" did not comply with 21 CFR 102.5(a)).

[42] Codified at 7 U.S.C. § 1638a.

[43] USDA, FSIS, Country of Origin Labeling for Meat and Chicken available at: http://www.fsis.usda.gov/wps/portal/fsis/topics/food-safety-education/get-answers/food-safety-fact-sheets/food-labeling/country-of-origin-labeling-for-meat-and-chicken/country-of-origin-labeling-for-meat-and-chicken.

[44] *Id.*

[45] *Id.*

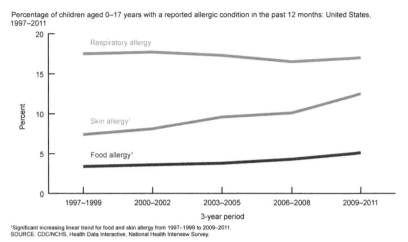

Percentage of children aged 0–17 years with a reported allergic condition in the past 12 months: United States, 1997–2011

¹Significant increasing linear trend for food and skin allergy from 1997–1999 to 2009–2011.
SOURCE: CDC/NCHS, Health Data Interactive, National Health Interview Survey.

Fig. 4.10 CDC trends in allergic conditions in children report

meat product contains three or more countries on the labeling depending on how the meat is mixed. FSIS will continue to enforce COOL for beef, pork, and poultry with the FDA enforcing the provisions for all meat or poultry not covered by the FMI or PPI.

The COOL controversy focuses on two disputes. One dispute involves a lawsuit by a meat industry group known as the American Meat Institute. The other is a complaint filed by Canada and Mexico to the World Trade Organization (WTO). As is common with USDA labeling rules the challenge is not to individual enforcement actions but to new labeling requirements. The preapproval process removes the likelihood of misbranding actions once a product is in the market. The AMI lost its lawsuit which sought to challenge the rule on the basis of freedom of speech. The case will be discussed in more detail in Sect. 5 below.

4.3.4   Allergen Labeling

Labeling to this point in the discussion largely avoided the issue of safety. As was seen in Section 3.1 the FDA is under no obligation to demonstrate safety concerns to prevail under the misleading standard for misbranding. Omitting food allergens from the label are a form of misbranding. Food allergen declaration also gathers around the concept of safety. Allergens appear

not on the label for consumer preference, but in many cases as a matter of life and death.

Food allergens only recently moved to the forefront of food safety. Prior to 2004 allergic reactions were not a legislative or regulatory priority. A decade later food allergens topped the list for annual number of recalls. This reflects a surge in consumers with food allergies, in particular, children. The CDC statistics point to significant growth in children with food allergies over the past 20 years (see Fig. 4.9 and 4.10)

The 1938 Act proved inadequate to address labeling allergens. Under the provisions of the Act a food containing two or more ingredients is required to list the entire ingredient by their common or usual name in order of predominance (FD&C).[46] Compliant labels still may not be clear to consumers making an assessment on whether the food contained or was derived from a food allergen. Thus, Congress passed the Food Allergen Labeling and Consumer Protection Act of 2004 (FALCPA).

Congress designated eight food or food groups as "major allergens" requiring prominent disclosure on the packaging. The "major allergens" comprise 90% of known allergens (FALCPA).[47] Consumers sensitive to other foods or food groups remain protected under the 1938 Act requiring listing of ingredients by common or usual name.

---

[46] 21 USC 403(i).

[47] Section 202(2).

FD&C Section 201(qq) Major Allergens

1. Milk
2. Egg
3. Fish
4. Crustacean shellfish
5. Tree nuts
6. Wheat
7. Peanuts
8. Soybeans

Compliance with FALCPA requires one mandatory activity and one voluntary activity. All labels must now bear the common name of the allergen in the ingredient list. For example, rather than listing "lethcin" or "whey" a label must label the ingredient as "letchin (soy)" or "whey (milk)." In addition the familiar "contains" disclaimer may also appear on the label. There is flexibility in how the disclaimer appears, but at a minimum it must fulfill three criteria.

Guidance for Industry: Questions and Answers Regarding Food Allergens, including the Food Allergen Labeling and Consumer Protection Act of 2004 (Edition 4)

1. The word "Contains" with a capital "C" must be the first word used to begin a "Contains" statement. (The use of bolded text and punctuation within a "Contains" statement is optional.¬)
2. The names of the food sources of the major food allergens declared on the food label must be the same as those specified in the FALCPA, except that the names of food sources may be expressed using singular terms versus plural terms (e.g., walnut versus walnuts) and the synonyms "soy" and "soya" may be substituted for the food source name "soybeans."
3. If included on a food label, the "Contains" statement must identify the names of the food sources for all major food allergens that either are in the food or are contained in ingredients of the food.

Food allergen labeling is more akin to economic adulteration than other labeling provisions. Recalling from Chapter 3 economic adulteration was a form of misbranding, but the designation as an adulterant allowed the FDA to seize products with little evidentiary showing. Likewise, the FALCPA enables the FDA to seize products not conforming to the labeling requirements or mandate a recall for undeclared allergens. This is unique among the misbranding violations discussed in this chapter.

Noticeably absent from the list is gluten. In 2013 the FDA issued a draft rule on "gluten free" labeling, which became effective the autumn of 2014. Prior to the rule "gluten free" was not a uniform term that would inform a consumer of the level of wheat in a product. As the healthy eating trend grew the term "gluten free" appeared on an increasing number of products. The trend was troubling for the estimated 1.5–3 million Americans experiencing symptoms of celiac disease (Gluten Free Rule).[48] Due to the absence of clarity in the market on when a product was gluten free as declared consumers with celiac struggled to find suitable products.

The new rule sets a limit on the amount of gluten in products making the claim "gluten free." Under the new rule a packaged product making the claim "gluten free," "free of gluten," "without gluten," or "no gluten" must contain less than 20 parts per million of gluten (Gluten Free Rule).[49] The new requirements were enacted as part of the FALCPA. Now a product making a "gluten free" claim that exceeds the threshold amount of gluten can be deemed misleading and a misbranding action can be taken against the facility.

---

[48] 78 FR 47154.

[49] *Id.*

4.3.5   Organic and Natural Food Labeling

Perhaps the most appealing and marketable statements on labels are "organic" and "all natural." The words "all natural" are estimated to sell $ 40 billion in the USA alone (Washington Post 2014).[50] The term "organic" generates an estimated $ 9.5 billion in sales (Washington Post 2014).[51] The terms are attractive to brands sold in the market. For consumers there is a bond of trust about what the claims mean. The question becomes how the terms are regulated and to what extent they are misleading.

In framing this discussion it is important to distinguish organic from natural. The term natural implies many qualities to consumers ranging from minimally processed, free of synthetic preservatives, sweeteners, colors, or flavorings to antibiotic- and growth hormone-free. Meat and poultry are the only food subject to controls on the meaning of "natural." FSIS requires "natural" products to be free of artificial colors, flavors, sweeteners, preservatives, and ingredients (FSIS Policy Book).[52] FSIS also requires minimum processing and the labeling to explain the term natural, such as "no artificial ingredients" (FSIS Policy Book).[53] The FDA does not define or control the term natural or all natural. It made an attempt in 1989 but abandoned the effort in 1993 due to resource limitations (Proposed Rule 1993).[54]

Organic refers to food and food production. Prior to 2002 the term organic was also unregulated and subject to various voluntary certifications by private organizations or States. In 2002 the USDA National Organic Standards were promulgated (see Fig. 4.11). The standard applies to all agricultural commodities or products both raw

Fig. 4.11 USDA organic emblem

and processed. Thus, even FDA-regulated products are subject to the USDA program.

In order to be claimed "organic" a product must meet the USDA standard. The USDA sets standards for livestock, crops, and multi-ingredient foods. The standards are codified in 7 CFR Section 205. In general, each organic standard describes specific requirements that must be verified by a USDA-accredited certifying agent prior to a product being labeled USDA organic.

The confusion over the term "all natural" is leading to a wave of litigation. The "all natural" statement is the fools-gold of labeling. It offers tantalizing rewards, but also attracts needless litigation. In Chapter 7 the concept of labeling lawsuits will be discussed. There this type of lawsuit, where consumers claim fraud, will be discussed. The lawsuits do not focus on safety issues; the granola bar for example, poses no risk of illness or injury, but its use of artificial sweeteners or preservatives may not support the "all natural" claim. Thus, a consumer complains they paid a higher price for the product labeled "all natural" than the product containing similar ingredients but not bearing the statement.

Organic and "all natural" labeling provide two lessons. First, the regulatory process can be seen progressing from a disparate decentralized system defining the term "organic" to the controlled system used today. The process was long in the making, but provided clarity, consistency, and confidence in the market place. Second, the term "all natural" provides insight into the risks of voluntary statements. There is no requirement to declare a product "all natural" merely for the market drive to do so. How that statement can

---

[50] Washington Post 2014.

[51] *Id.*

[52] USDA, FSIS, Food Standard and Labeling Policy Book.

[53] *Id.*

[54] 58 Fed. Reg. 2302 (January 6, 1993).

Fig. 4.12 Types of regulated claims

open a company to unintended risks will be fully explored in Chapter 7.

4.4   Labeling Claims and Misbranding

Misbranding encompasses more than the required components of a label. In addition to the name of the food, listing of ingredients, or the net weight as examples, marketing requires something more said about why a consumer should select one product over another. These statements are generally referred to as claims.

There are three types of claims regulated by the Act (see Fig. 4.12). Each type of claim allows a specific description of the product to be made. The three types of claims are health, nutrient, and structure/function claims. Claims that do not follow the regulations risk enforcement as misbranding. Some claims also risk reclassification as a non food product, namely a drug. When this occurs the offending claim is called a "disease" or "new-drug" claim. Nutrient content claims are the most closely regulated type of claims. Structure/function claims follow general guidelines and are often used to make the most expansive type of claims.

4.4.1   Types of Claims

Nutrient Content Claims

The first type of claim is known as a nutrient claim. It can also be referred to as a nutrient content claim, which more accurately describes the claim's purpose. NLEA permits the use of

these claims on labels so long as they are authorized by the FDA. Nutrient claims describe the nutritional level of the food product or its ingredients. There are several claims that qualify as nutrient content claims. Familiar terms like "free," "high," "low," and "lite" are claims that describe the level of a nutrient in the product. Other claims, such as "more," "reduced," and "lite" compare the nutrient level by comparison to another food. Other claims that characterize the level of a nutrient also fall under the nutrient content claims regulations. For instance, a quantitative statement would declare 200 mg of sugar, but the statement "only 200 mg of sugar" characterizes the level of the nutrient. Thus, simple changes in language can trigger the regulations.

Most nutrient content claim regulations apply to nutrients with established daily values. If the daily value is not set then there is no way to quantify the level of nutrient present or how it compares to other products. The daily value will be reflected in the nutrition facts panel. The regulations also require that the nutrient content claim be based on the amount of food consumers ordinarily eat. This is formally known as the Reference Amount Customarily Consumed per Eating Occasion (RACC) or the reference amount. Occasionally, but not always, the reference amount is the same as the serving size.

Health Claims

Health claims describe a relationship between the food or its ingredient and a reduced risk of disease or health condition. The FDA exercises control of health claims either using NLEA, the 1997 Food and Drug Administration Modernization Act (FDAMA), or the Agency's Interim Guidance on qualified health claims. A health claim must always relate the food or its ingredient to a disease or health-related condition. If this basic structure is absent then the claim falls outside the regulations and is misbranded. An example is the claim that fruits and vegetables are part of a healthy diet. This is a general dietary guidance or pattern and not a health claim.

NLEA sets criteria for the FDA to review and approve of health claims. Only those health claims approved by the FDA may be used on labeling. NLEA authorizes the FDA to conduct an exhaustive review based on a petition citing scientific literature that establishes a relationship between the food substance and the disease. The evidence must show "significant scientific agreement" (SSA) to be approved (21 CFR Part 101).[55] This is a high bar of proof to clear.

A lower threshold of proof is used for health claims based on authoritative statements. Under the FDAMA a health claim will be authorized by submitting notification to the FDA of the claim based on an authoritative statement from certain scientific bodies of the US Government or the National Academy of Sciences. Examples of suitable agencies include the CDC and NIH. Although a lower evidentiary hurdle, an authoritative statement still must meet certain criteria.

FDA Guidance on Notification and Authoritative Statements[56]

FDA also believes it is necessary to clarify what constitutes an authoritative statement under FDAMA. FDAMA itself states that an authoritative statement: (1) is "about the relationship between a nutrient and a disease or health-related condition" for a health claim, or "identifies the nutrient level to which the claim refers" for a nutrient content claim, (2) is "published by the scientific body" (as identified above), (3) is "currently in effect," and (4) "shall not include a statement of an employee of the scientific body made in the individual capacity of the employee."

In addition, given the legislative history of Sections 303 and 304 of FDAMA, FDA currently believes authoritative statements also should: (5) reflect a consensus within

Table 4.1 Standardized qualifying language for qualified health claims

Scientific ranking[a]	FDA category	Appropriate qualifying language[b]
Second level	B	…"although there is scientific evidence supporting the claim, the evidence is not conclusive"
Third level	C	"Some scientific evidence suggests… however, FDA has determined that this evidence is limited and not conclusive"
Fourth level	D	"Very limited and preliminary scientific research suggests… FDA concludes that there is little scientific evidence supporting this claim"

[a] From guidance for industry and FDA: interim evidence-based ranking system for scientific data

[b] The language reflects wording used in qualified health claims as to which the agency has previously exercised enforcement discretion for certain dietary supplements. During this interim period, the precise language as to which the agency considers exercising enforcement discretion may vary depending on the specific circumstances of each case

the identified scientific body if published by a subdivision of one of the Federal scientific bodies, and (6) be based on a deliberative review by the scientific body of the scientific evidence.

Qualified health claims are authorized under an interim guidance document. The qualified health claims process is intended to provide approval to claims that cannot meet the significant scientific agreement standard for health claims. Such is the case when there is only emerging evidence of a relationship between food and a reduced risk of a disease or health-related condition. The standard is reduced to require a petition only meet a "credible evidence" standard based on the quality and strength of the scientific evidence. The claim is approved through issuance of an enforcement discretion letter in which the agency asserts a right to take enforcement action but discretion to refrain from doing so. The claim is subject to qualifying language depending on the level of evidence supporting the claim (Table 4.1).

---

[55] 21 CFR 101.14(c).

[56] FDA, Notification of a Health Claim or Nutrient Content Claim Based on an Authoritative Statement of a Scientific Body, available at: http://www.fda.gov/Food/GuidanceRegulation/GuidanceDocumentsRegulatoryInformation/ucm056975.htm.

Structure/Function Claims

Structure/function claims are made on a range of products from dietary supplements, to conventional foods and drugs. There are only a few regulations surrounding the claims making their use both appealing and risky. The Dietary Supplement Health and Education Act of 1994 (DSHEA) establishes a structure/function rule for dietary supplements making such claims. This rule will be discussed in more detail in Chapter 6. Conventional foods seeking to make a more direct link between the product and a disease or health-related condition also use structure/function claims. Here the DSHEA rule is only persuasive.

Conventional foods making structure/function claims are still subject to the general prohibition against making misleading claims. A prominent example is from Kellogg's. During the swine flu scare of 2009 it released boxes of its Rice Krispies that bore the claim, "Now Helps Support Your Childs Immunity." This is a classic structure/function claim. Given the timing of its release it implied protection against the swine flu. Before the FDA could take action and in the face of public outcry the company revised its packaging.

4.4.2 Restrictions on Claims

The greatest risk in making claims is reclassification. This is most common with structure/function claims, but the risk remains for health claims. When a product begins to promote benefits that fall outside the *Nutrilab* paradigm of intended use for food then there is a risk of reclassification as a drug. When this occurs the FDA typically states the claim is a disease or new drug claim because it expressly or implicitly suggests treatment of a disease and lacks new drug approval.

There is also the risk of either the FDA or private individuals taking legal action on the basis that the claim is misleading. The FDA will take up a misbranding action when the technical components of making claims are not followed or when the claim ventures in to areas outside its desired food classification. There is also a fine line between marketing and fraud. When that line is crossed consumers are eager to seek judicial relief, especially if there is demonstrable economic harm from paying a higher price for the product.

Claims should be used with caution and special attention paid to the regulations. Caution over creativity is always advised.

4.5 Constitutional Challenges

Speech is a particular sensitive constitutional right. It is one that is closely held and fiercely defended. It seems that no matter what action the FDA or FSIS takes on labeling, from new regulations to enforcement actions, the outcry revolves around freedom of speech. There is a difficult balancing act in reviewing freedom of speech issues. A court hearing such a challenge will be confronted by broad statutory language, a mandate to liberally construe the statute to protect the public and the immutable constitutional right to speech of the challenger. It also touches on the issue of a consumer's freedom of choice.

Other constitutional questions involve the ability for private individuals to challenge false or misleading claims in the place of the FDA. This is known as preemption. The FDA is the sole enforcer of the FD&C, but where it fails to act consumers assert a right to pursue judicial relief using State law causes of action.

4.5.1 Freedom of Speech and Choice

The First Amendment of the US Constitution provides the protection of free speech. It applies to the States through the Fourteenth Amendment. The First Amendment provides two protections. It protects from censorship or the impediment of speech and also from the compulsion of speech. In the case of food law this means a protection from speaking freely about a product and compelled disclosure requirements. The FDA attempts both through enforcement and regulation.

Commercial Speech Doctrine

To understand the constitutionality of a food facility's right to free speech it must be determined what type of speech the facility is making. The Supreme Court through judicial precedent constructed a three-tiered system of speech deeming some speech more protected than others. The three categories are regulated speech, commercial speech, and pure speech.

The FDA naturally seeks the category of speech that would impose the fewest limitations on its powers to regulate labeling. In this case that would be regulated speech. Areas of extensive regulatory authority, such as securities and antitrust, are subject to regulated speech controls. Commercial speech receives an intermediate level of review while pure speech receives the greatest protection. Courts, however, continually apply the commercial speech label to FDA activities. There has been little expansion of the regulated speech category beyond economic interests.

In a complex area of law two cases provided the need insight on the commercial speech doctrine. One could spend an entire course discussing freedom of speech in all its nuances. For our purposes two historical cases provide the precedent needed to understand labeling laws and free speech. Those cases are *In the Matter of R.M.J.*[57] and *Central Hudson Gas & Electric Co. v. Public Service Commission.*[58] In *Central Hudson Gas* the Supreme Court held invalid an order by the State utility commission prohibiting the utility

including in the monthly bill inserts discussing controversial issues of public policy. In reaching its holding, the court provided the first articulation of the four-part analysis for commercial speech. The court stated:

> In commercial speech cases, then, a four-part analysis has developed. At the outset, we must determine whether the expression is protected by the First Amendment. For commercial speech to come within that provision, it at least must concern lawful activity and not be misleading. Next, we ask whether the asserted governmental interest is substantial. If both inquiries yield positive answers, we must determine whether the regulation directly advances the governmental interest asserted, and whether it is not more extensive than is necessary to serve that interest (*Central Hudson Gas*).[59]

In *R.M.J.* the Supreme Court reversed the reprimand of an attorney for unprofessional conduct. The attorney used the words "real estate" in a newspaper advertisement instead of the approved term "property" (*R.M.J.*).[60] The court held the use of the unapproved term had not been shown to misleading to the public and was thus afforded First Amendment Protection. The court provided a succinct summary of the commercial speech doctrine:

> …Truthful advertising related to lawful activities is entitled to the protections of the First Amendment. But when the particular content or method of the advertising suggests that it is inherently misleading or when experience has proved that in fact such advertising is subject to abuse, the States may impose appropriate restrictions. Misleading advertising may be prohibited entirely. But the States may not place an absolute prohibition on certain types of potentially misleading information…if the information also may be presented in a way that is not deceptive…

> Even when a communication is not misleading, the State retains some authority to regulate. But the State must assert a substantial interest and the interference with speech must be in proportion to the interest served…Restrictions must be narrowly drawn, and the State lawfully may regulate only to the extent regulation furthers the State's substantial interest…(*R.M.J.*).[61]

---

[57] 445 U.S. 191 (1982).

[58] 447 U.S. 557 (1980).

[59] *Id.* at 566.

[60] *R.M.J. Id.* at 203.

[61] *Id.*

At its simplest we can say commercial speech is only protected if it is not misleading. If it is not misleading, then it is afforded protection under the four-part test in *Central Hudson Gas*, which, through its analysis, asks whether the government's interest in the regulation is substantial and if the regulation is narrowly drawn.

From this vantage point two general observations can be made. When the FDA finds labeling misleading or requires basic labeling information, such as the common or usual name of the food and ingredients, it will likely encounter little resistance under the First Amendment. If however, the FDA attempts to revise or restrict alleged new drug or disease claims, it will likely struggle under the First Amendment. This particularly so when the statements in question are presented truthfully can cannot easily be ascribed the misleading designation. This concept will be explored more with dietary supplements in Chapter 5.

Compelled Speech

Regulatory mandated disclosures are often challenged on First Amendment grounds. The court reviewing the challenges will utilize the commercial speech doctrine to determine whether there is a substantial government interest in requiring the disclosure and whether the regulation is narrowly tailored. All of the components of the *Central Hudson Gas* test must be met.

The results vary depending on the regulation in question. For example, the Sixth Circuit held an Ohio food labeling regulation violated the First Amendment (*Int'l Dairy Foods Assoc.*).[62] The Ohio Department of Agriculture prohibited the claim "this milk is from cows not supplemented with rbSt." The court found the claim sought by the dairy association, namely "rbST free," factual and not misleading (*Int'l Dairy Foods Assoc.*).[63] The court required that the government provide evidence to satisfy each prong of the *Central-Hudson* test, namely that its regulation was the

least restrictive means to achieve its regulatory goal (*Int'l Dairy Foods Assoc.*).[64]

By comparison, the D.C. Circuit ruled against a challenge to invalidate the COOL rule. The American Meat Institute challenged the rule under *Central Hudson Gas* arguing there was no substantial interest in requiring disclosure of the country of origin other than satisfying customer curiosity (*American Meat Institute*).[65] The court disagreed and applied *Central-Hudson* to the issue. It held that in the case of mandating a disclosure the least restrictive alternative element of *Central Hudson Gas* is unnecessary because commercial actors only enjoy a minimal First Amendment interest in not providing purely factual information (*American Meat Institute*).[66] This type of disclosure does not infringe a commercial actor's First Amendment rights as long as it "reasonably relate[s]" to preventing deception of consumers among other purposes. On this basis the court upheld COOL.

4.5.2 Preemption and Private Litigation

Private litigation known as labeling lawsuits continues to gain traction. The topic will be discussed in more detail in Chapter 7. For our purposes here the question becomes to what extent can a private individual or a class of individuals pursue litigation alleging food labeling is misleading. The question is raised often by the companies seeking to defend against such suits. The concept is known as preemption. This concept essentially means where the FDA exercises its valid authority to act or could potentially act in that area of regulation, then it bars States from enacting competing or conflicting legislation. Since many private actions rely on State law if the law is preempted by federal law, then the claim is barred.

---

[62] *International Dairy Foods Association, et. al. v. Boggs*, 622 F.3d 628 (6th Cir. 2010).

[63] *Id.* at 637.

[64] *Id.* at 640.

[65] 746 F.3d 1065, 1074 (D.C. Cir. 2014).

[66] *Id.* at 1072 citing *Zauderer v. Office of Disciplinary Counsel*, 471 U.S. 626, 651–52 n.14 (1985).

Preemption can be either express or implied. Express preemption or field preemption is when the Statute makes clear what preemption it intended. For example, NLEA in Section 6(c) states it "shall not be construed to preempt any provision of state law, unless such provision is expressly preempted." In some instances preemption is implied, such as with the FD&C's general exclusivity clause. The exclusivity clause declares all violations of the FD&C must be prosecuted "by an in the name" of the federal government (FD&C).[67] Thus, it is implied any misbranding actions are preempted from private or State enforcement.

Preemption remains a complex issue highly dependent on the type of claim. It also can be circumvented depending on the use of other applicable federal law. For example, the Supreme Court recently allowed Pom Wonderful's suit against Coca Cola under the Lehman Act to proceed. Pom claimed in its suit that Coca Cola's juice products were deceptive. Coca Cola raised the defense that the suit was barred on the basis of preemption. The district court agreed with Coca Cola holding, held that Pom Wonderful's claims regarding the name and label of the juice were preempted by the FD&C. The Supreme Court overturned the ruling noting the Lehman Act and FD&C coexisted since 1946. Thus, a claim on the basis of misleading can be brought under a separate federal law in some instances, such as the Lehman Act.

4.6   Comparative Law

Labeling accuracy like adulteration enjoys worldwide regulation. Every nation exercises some authority to ensure food is accurately labeled. This raises issues in two areas suitable for a comparative law discussion. The first is the impact of one nation's labeling requirements on exporting nations. An issue was raised with the COOL rule and the complaint filed by Canada and Mexico with the WTO. The second is allergen labeling.

The USA is not alone in experiencing a compelling demand for accurate allergen labeling.

The WTO functions as a neutral entity to hear disputes between member nations. It is primarily regulated by the Technical Barriers to Trade Agreement (TBT Agreement) and General Agreement on Tariffs and Trade (GATT). At its most basic the WTO hears complaints about unfair trade practices. That is those practices which unfairly discriminate against foreign nations exports in favor of domestic products. This is essentially the dispute with the COOL rule used by the USDA. To comply with the rule foreign producers would need to segregate cattle intended for the US market in order to know which label to use. Canada and Mexico in its complaint allege this would drive up the costs and unfairly discriminate against foreign meat. The WTO hears a wide range of food-related complaints from labeling requirements to new standards on GMOs or hormones in meat production. It highlights the real impact of conflicting approaches to labeling and food safety.

Sensitivity to food allergens among both children and adults is widely prevalent across the globe. This has resulted in a wide variety of labeling regulation for food allergens. Each nation takes its own approach to determining how allergens will be controlled and which allergens are priorities for regulation. Estimates place the number of allergens subject to regulation by various nations between 5 and 13 (Gendal 2012).[68] Five foods, wheat/cereals, eggs, milk, peanuts, and crustacean are among the most commonly listed. Some nations, such as Canada, develop a list of criteria for determining which allergens to list (Gendal 2012; see Fig. 4.13). Other nations, such as Japan, rely on a review of scientific literature (Gendal 2012). Japan's model results in the largest list of allergens, currently at twenty-five-give foods including banana and gelatin (Japan Allergen FAQ; see Fig. 4.14).[69] Despite the

---

[67] 21 USC § 337(a).

[68] Gendel 2012.

[69] Standards and Evaluation Division, Department of Food Sanitation, *FAQ on Labeling System for Foods Containing Allergens*, available at: http://www.caa.go.jp/foods/pdf/syokuhin13.pdf.

Table 1
Food allergen regulatory frameworks.

Jurisdiction	Source
Codex Alimentarius	Codex General Standard for the Labeling of Prepackaged Foods (CODEX STAN 1-1985, Rev. 2010)
Argentina	Argentine Food Code Section 235 Seventh (Information from WTO Notification G/TBT/N/ARG/252)
Australia/New Zealand	Australia New Zealand Standard 1.2.3 – Mandatory Warning and Advisory Statements and Declarations
Barbados	Barbados National Standard Specification for Labeling of Prepackaged Foods BNS 5: Part 2: 2004 (Second Revision)
Canada	Regulations Amending the Food and Drug Regulations (1220 – Enhanced Labeling for Food Allergen and Gluten Sources and Added Sulfites) – Canada Gazette 145 (4) February 16, 2011
Chile	Amendment to Article 107 of the Food Health Regulations, Decree No. 977/96 (Information from WHO Notification G/TBT/N/CHL/95)
China	General Rules for the Labeling of Prepackaged Foods GB7718-2011 (Information from USDA GAIN Report CH110030)
European Union	Regulation No. 1169/2011 and Annex II; European Commission Directive 2007/68/EC and Annex IIIa; European Commission Directive 2003/89/EC and Annex IIIa, Regulation No. 1169/2011 and Annex II
Hong Kong	Food and Drugs (Composition and Labeling) (Amendment) Regulation 2004 Labeling Guidelines on Food Allergens, Food Additives and Date Format.
Japan	Standards and Evaluation Division, Department of Food Sanitation, Pharmaceutical and Medical Safety Bureau, Ministry of Health, Labour and Welfare; FAQs on Labeling System for Foods Containing Allergens
Korea	Korean Food and Drug Administration (Information from USDA ARS GAIN Report KS1102)
Mexico	Official Standard for General Labeling Specification of Pre-Packaged Foods and Non-Alcoholic Beverages (Information from USDA GAIN Report MX0505)
Mongolia	Technical Regulation on Trading and Supplying of Specified Import Goods and Products Only with the Labels Which Appear in Mongolian Language in the Domestic Market
Papua New Guinea	Food Sanitation Regulation 2007
Philippines	Rules and Regulations Governing the Labeling of Prepackaged Foods Products Distributed in the Philippines (Amending Administrative Order No. 88-B s. 1984) (Information from WTO Notification G/TBT/N/PHL/128)
St. Vincent and The Grenadines	Standard for Labeling of Pre-Packaged Foods, SVGNS 1 Part 3: 2000 (Rev. 2009)
Switzerland	Ordinance of the Federal Department of Home Affairs on Food Labeling and Advertising of Foods (Information from WTO Notification G/TBT/N/CHE/106)
Ukraine	Technical Regulations on Rules for Marking of Food Products (Information from WTO Notification G/TBT/N/UKR/52)
United States	Food Allergen Labeling and Consumer Protection Act of 2004[a]

[a] FALCPA applies to products regulated by the Food and Drug Administration. The labeling of most meat and poultry products is regulated by the Food Safety and Inspection Service in the Department of Agriculture and the labeling of most alcoholic beverages is regulated by the Alcohol and Tobacco Tax and Trade Bureau in the Department of the Treasury. Both of these agencies have developed policies to encourage allergen labeling consistent with the requirements of FALCPA.

Fig. 4.13 Tables from Gendel on food allergen regulatory framework

Table 2
Allergenic foods.[a]

	Codex[b]	European Union[c]	Australia/New Zealand	Canada	China	Hong Kong	Japan	Korea	Mexico	United States
Wheat/Cereals	X	X	X	X	X	X	X[d]	X[e]	X	X
Eggs	X	X	X	X	X	X	X	X	X	X
Milk	X	X	X	X	X	X	X	X	X	X
Peanut	X	X	X	X	X	X	X	X	X	X
Fish	X	X	X	X	X	X		X[e]	X	X
Crustaceans	X	X	X	X	X	X	X[d]	X[e]	X	X
Soy	X	X	X	X	X	X		X	X	X
Tree Nuts	X	X	X	X	X	X			X	X
Sesame		X	X	X						
Shellfish/Mollusks		X		X						
Mustard		X		X						
Celery		X								
Lupine		X								
Other							X[d]	X[e]		

[a] As of January 1, 2012 based on the sources listed in Table 1.
[b] The following countries use CODEX wording in their regulatory frameworks: Barbados, Chile, Papua New Guinea, Philippines, St. Vincent and The Grenadines. The wording for the Papua New Guinea uses the term "shellfish" instead of "Crustacean." It is not clear if this is intended to include Molluscan shellfish. Mongolia cites the CODEX standard by reference.
[c] The following countries use the EU Annex IIIa wording in their regulatory frameworks: Argentina, Switzerland, Ukraine.
[d] Shrimp and crab are the only Crustaceans listed. Grains includes wheat and buckwheat. "Other" includes foods for which labeling is recommended but not required: abalone, squid, salmon roe, oranges, kiwifruit, beef, walnuts, salmon, mackerel, soybeans, chicken, bananas, pork, matsutake mushrooms, peaches, yams, apples, gelatin.
[e] Mackerel is the only fish, crab and shrimp are the only Crustacean listed. Grains includes wheat and buckwheat. "Other" includes pork, peaches, and tomatoes.

Fig. 4.14 Tables from Gendel on allergens listed by various nations

variance, allergen labeling is one aspect of label regulation that enjoys near universal agreement.

4.7 Chapter Summary

This chapter covered the breadth of labeling and misbranding. It outlined the nimble concept of labeling, which encompasses nearly every form of material closely linked to the product. The chapter defined the terms misbranding and the case law interpreting what constitutes "misleading" statements. It looked at the types of various claims and regulatory requirements for labeling and the associated risks of noncompliance. It concluded with a look at the constitutional issues raised when regulating commercial speech. There it was seen the FDA's authority is limited by the doctrine in *Central Hudson Gas*, which among other questions asks whether the speech is inherently misleading and if the regulations narrowly tailored to regulate the speech. Constitutional questions of labeling will be explored again in the next chapter with a look at labeling on dietary supplements.

Overview of Key Points
- Define misbranding and "labeling"
- Review case law defining when material "accompanies" a label
- Compare the FDA and USDA approaches to labeling

- Define misbranding and the various acts that lead to the prohibited act
- Explore the concept of allergen labeling and its unique connection to safety
- Touch on "all natural" labeling and the potential for litigation
- Review the types of claims and regulatory components of a label
- Constitutional questions raised by the regulation of speech
- Explanation of commercial speech doctrine and the case law applying it to food labeling
- Introduction to the concept of preemption

4.8 Discussion Questions

Search your cupboard or local market and find examples of each type of labeling claim: nutrition, health, and structure/function. Provide the legal authority that either supports the claim's legality or potential illegality.

Why would FSIS receive greater authority to regulate meat and meat products, both in terms of labeling and adulteration, than the FDA?

References

Gendel SM (2012) Comparison of international food allergen labeling regulations. Regul Toxicol Pharmacol 63:279–285

Standards and Evaluation Division, Department of Food Sanitation, *FAQ on Labeling System for Foods Containing Allergens*. http://www.caa.go.jp/foods/pdf/syokuhin13.pdf. Accessed 11 July 2014

Washington Post (2014) Wonkblog, The word "natural" helps sell $ 40 billion worth of food in the U.S. every year—and the label means nothing, (24 June 2014) http://www.washingtonpost.com/blogs/wonkblog/wp/2014/06/24/the-word-natural-helps-sell-40-billion-worth-of-food-in-the-u-s-every-year-and-the-label-means-nothing/. Accessed 10 July 2014

Regulation of Dietary Supplements and Other Specialized Categories

5

Abstract

This chapter covers areas of special regulation. The primary topic will be dietary supplements, followed by the seven rules promulgated under the Food Safety Modernization Act. Other topics will include the regulation of seafood, juice, milk, water, and ice. Each of these areas deserves discussion apart from the coverage of general topics under the FD&C because of specific legislation or rules promulgated to address unique enforcement considerations raised by these products. FSMA fits into this discussion both because many of the rules are product specific, but also because as the first substantive change in 75 years coverage of the rules is ill suited for other chapters.

5.1 Introduction

There are certain categories of food products subject to additional regulation or particular scrutiny during enforcement. These food products must comply with both the general provisions of the FD&C and additional requirements unique to their classification. The additional regulations typically came in the form of amendments to the Act, which augmented the FDA's authority over ingredients and labeling in an effort to address public and enforcement concerns. Other changes came in the form of new regulations issued by the FDA. The use of additional amendments or regulations not only serves the public, but also industry. Absence of a product specific amendment compliance requires applying broad ambiguous categories.

The area of greatest growth in consumer use and market share is dietary supplements. Estimates place the overall sales of dietary supplements well over 20 billion dollars annually. Not only has the market share changed since the Act was passed in 1906 and 1938, but the range of dietary supplement products and marketing practices are radically different. A dietary supplement for long periods of the past century simply referred to a vitamin containing one or more nutrients, like vitamins A or K. In today's market dietary supplements, or nutraceiticals as they are often called, refer to a wide range of products from natural and traditional remedies to new synthetic ingredients. The products dance along the fine edge of food and drug, with claims extolling benefits such as arthritis and flu relief.

As the dietary supplement market developed the regulations needed to grow as well. Under the 1906 and 1938 Act dietary supplements were regulated as foods. At least this was so long as the labeling did not indicate an intended use for therapeutic purposes in which case it would be regulated as a drug. Under an Amendment known as the Dietary Supplement Health and Education Act (DSHEA), Congress provided the FDA greater control and flexibility over ingredients and labeling. This allowed industry to gain approval of new dietary ingredients and make more claims that spoke the benefits and functions of ingredients in their products without risking reclassification.

Seafood and juice are subject to additional regulations due to safety concerns. The FDA convened several committees and consulted with Congress about the risks posed by seafood and juice products. The FDA found a history of repeated instances of serious public health issues related to juice and seafood warranted additional preventative controls (GAO 2008). In the case of juice the FDA found a number of health risks from unpasteurized juice. In both cases the FDA promulgated new regulations to require Hazard Analysis and Critical Control Point (HACCP) as a preventative control to minimize health hazards associated with juice and seafood. The FDA released the seafood HACCP regulation in 1995 and proposed a rule requiring HACCP for juice products in 1998, which was finalized in 2001.

The challenge in working with continual Amendments is piecing the entire Act together. Often when the Amendments are enacted the provisions do no expressly repeal incongruent language in the Act. This can leave the impression that an Amendment is fundamentally revising the 1938 Act. Instead the intent is not only to maintain the status quo, but add a new layer for the products identified in the Amendments. The sections below will explore how the Amendments interact and add to the existing provisions of the Act.

5.2 Dietary Supplements

5.2.1 Regulation 1906–1994

Dietary supplements enjoy a long history in the USA beginning in 1920. Early supplements consisted of a small number of ingredients usually common food constituents. Chapter 6 in its discussion of food additives provides black currant oil as an early example of a dietary supplement. Cod liver oil containing vitamins A and D is

often considered the first dietary supplement in the USA (Hutt 1986).[1] As scientific methods developed an increasing number of nutrients were identified. Fortification of foods began using these isolated nutrients in order to guard against common nutrient deficiencies (Hutt 1984).[2] This provides insight into the early idea of dietary supplements. The products were intended to supplement the diet by providing nutrients missing from daily meals or from a poor diet.

Regulation Under the 1906 Act
Dietary supplements emerged only in the last decade of the 1906 Act. The FDA's predecessor the Bureau of Chemistry did not begin to conduct investigations on dietary ingredients and vitamins until 1916. It only issued its first consumer guide to the public in 1925. Products were on the market claiming fortification or the presence of some nutrient in the food product that would act as a remedy against some disease, but in general dietary supplements were not a distinct category.

Under the 1906 Act the distinction between food and drug mattered little. The 1906 Act was a mere five pages in length and consisted chiefly of prohibitions against adulteration and misbranding. It defined "drug" and "food" in Section 6, but left regulation of both products nearly identical. It would be nearly 80 years later that *Nutrilab* would introduce the "intended use" concept and not until the 1938 Act would regulation of the two products change. Thus, under the 1906 Act products whose packaging contained enough claims to qualify as a drug had legal actions pursued under the drug provisions. Otherwise, the legal actions were taken under the food provisos of the Act. Little else separated early regulation.

Typical claims regulated as drug claims under the 1906 Act simply listed diseases or ailments. A court case from 1911 provides a sample of the type of claims that lead to product seizures by the FDA. Claims that could not be easily reclassified and regulated under the drug provisions of the 1906 Act were regulated as misleading claims.

---

[1] Hutt (1986).

[2] Hutt (1984).

> Grant's Hygienic Crackers. No predigested stuff are they, but solid food for work or play.
>
> Just read what leading doctors say of Grant's Hygienic Crackers for Constipation, Indigestion,
>
> Dyspepsia and Sour Stomach. Ideal food for general family use. A daily regulator.
>
> Aweek's trial will convince you. Eaten daily in place of bread will keep the system in perfect
>
> order, Recommended & prescribed by leading physicians & dentists.

As public interest in dietary supplements swelled the need for greater regulation grew. Following increasing concerns about manufacturers taking advantage of the public's interest in vitamins, the FDA brought six legal actions against vitamin claims in 1935. That year it also established a Vitamin Division to address what the agency deemed was exploitation of the vitamin market. The focus was not only on claims, but new ingredients or compounds with claimed health benefits.

The FDA secured one criminal prosecution in the last year of the 1906 Act. In 1939 as the 1938 Act was being passed and enacted, the FDA brought a criminal case against a vitamin product containing milk, sugar, wheat starch along with vitamins A-G (Hutt 1986).[3] It claimed the product would cure high blood pressure, low blood pressure, dropsy, toxic goiter, and heart disease (Hutt 1986).[4]

The 1906 Act, often criticized for lacking either the temerity or the legal authority to regulate dietary supplements, did attempt to take action. The early enforcement actions under the 1906 Act would influence and shape enforcement in the years to come under a new version of the Act.

Regulation Under the 1938 Act

The 1938 Act changed the regulation of dietary supplements in several ways. It provided new authority for the FDA to regulate the labeling of food for "specialty dietary uses" under the misbranding provisions. This provision remains valid today and is currently codified as Section 343(j). The 1938 Act also expanded the FDA's authority to include "labeling" and not just the "label" as it was in the 1906 Act. In Chapter 4 this concept was explored in detail and in particular the FDA's ability to broadly construe labeling to include any material that "accompanies" the label. The FDA also gained the ability to strictly enforce labeling under a reduced burden of proof for misbranding. No longer did the FDA need to show "false and fraudulent" but only "false or misleading in any particular" (2906 Act; FD&C).

21USC 343(j)

(j) Representation for special dietary use

If it purports to be or is represented for special dietary uses, unless its label bears such information concerning its vitamin, mineral, and other dietary properties as the Secretary determines to be, and by regulations prescribes as, necessary in order fully to inform purchasers as to its value for such uses.

Regulation under the 1938 Act initially focused on three groups of foods. First, the FDA considered and developed regulations around staple foods fortified with vitamins and minerals. This category was strictly regulated with standards of identities, limits on the amount of added supplemental nutritional ingredients, and barring any health claims. This included a prohibition on marketing conventional fortified foods as containing fewer calories or as high in fiber (Continental Baking Company 1976).[5] Such claims are commonplace on today's market. A second

---

[3] Hutt 1986.

[4] *Id.*

[5] FDA(October 1976).

category involved those areas identified by the FDA through regulation as conventional foods represented for special dietary uses. Early areas of regulation included infant foods and foods for diabetics (CFR 105).[6] Left unregulated on the market were dietary supplements or vitamins. Neither the Act nor any initial regulations identified or classified dietary supplements as a unique group in the market. Thus, dietary supplements were regulated either as conventional food or food for special dietary purposes.

Early regulation placed some dietary supplements exclusively in the realm of drug regulation. Despite any labeling that may misconstrue a product's use as a drug all parenteral supplements were regulated as drugs (FDA 1945).[7] If it was not consumed orally, then beginning in 1945 the FDA regulated the product as a drug. As will be seen this mirrors the current definition of a dietary supplement under the Act. Thus, if it was labeled with drug properties or benefits or not taken orally the FDA deemed it a drug. This left only products with an intended use to supplement the daily diet regulated as foods.

After ten years of attempted regulation of dietary supplements as conventional foods for special dietary purposes a new informal framework emerged. The regulations for special dietary food were not designed, and therefore did not work well, to regulate claims on vitamins, minerals, and other dietary supplements. The claims ranged from the harmless to what the FDA Commissioner in 1951 would call quackery (Commissioner Crawford).[8] A criminal case examined in Chapter 4, *United States v. Kordel*, provides an example of the more outlandish claims. Kordel's pamphlets included claims the products work to cure every ailment from failing eyesight, arthritis, paralysis, sterility to general weakness (*Kordel*). The year after the FDA secured its victory in the Supreme Court in *Kordel* it instituted multiple seizures of a popular brand.

_____

[6] 21 CFR 105.65;67.

[7] FDA TC2-A (1945).

[8] Crawford (June 1951).

> **Excerpt from Circuit Court opinion in *United States v. Kordel* (164 F.2d 913, 916 (7th Cir. 1947)**
>
> A study of the three pamphlets reveals that the products therein described are recommended for relieving stomach agonies, general weakness, anemia, premature old age, high blood pressure, liver troubles, failing eyesight, sore feet; maintaining blood energy, muscular activity, sound teeth and gums, healthy skin, hair and eyes, normal functioning of the pituitary and thyroid glands, stomach, intestines, colon, liver and kidneys; and preventing arthritis and stiff joints, excess weight, catarrh, nervous breakdown, sterility, and paralysis.

In 1949, the FDA began a case against Nutrilite Food Supplement that would ultimately end in informal guidelines governing the entire supplement industry. The FDA's attempt to seize Nutrilite's products lead to protracted litigation focused on constitutional issues. After two years of legal battles the FDA and Nutrilite entered into a consent decree and injunction. The agreement carried far-reaching repercussions. The Nutrilite Consent Decree, as it became known, served as informal FDA policy and industry guidelines for the next ten years.

It would only be additional litigation in the 1960s that would expand and revise the FDA's enforcement policy. The cases rested on one of two theories. One line of cases asserted enforcement based on claims made in promotional literature, which was deemed part of the "labeling." A second series of cases focused on claims made for a product based on its formulation.

The FDA first pursued its enforcement of dietary supplements on the basis of labeling in promotional materials in *Abbott*. *Abbott* involved the use of a booklet titled "Vitamins for Your Family" which exclaimed a number of benefits about Abbott's vitamin and mineral products. *Abbott* extended the provisions of the Nutrilite Consent Decree because the FDA determined some claims permitted under the decree could remain misleading absent qualifying explanations. For Abbott to come into compliance it simply needed to add qualifying explanations to its labeling.

Excerpt from FDA Foods Notice of Judgment in re Abbott

1. "It is difficult, if not impossible, to obtain adequate nutrition from a diet of ordinary foods, due to nonuniform distribution of vitamins and minerals in the various articles of food, and due to the adverse effects on the nutritive quality of foods ordinarily consumed due to poor quality soil; weather; agricultural, processing, and marketing practices; and home preparation; which will result in practically everyone suffering from, or being in danger of suffering from, inadequate vitamin and mineral nutrition unless a vitamin or vitamin and mineral supplement is added to the diet;"

2. "The regular consumption of 'a vitamin tablet' is a suitable corrective for all aspects of inadequate nutrition due to poor eating habits; fad diets; eating only the foods one likes and ignoring other needed foods; limiting the diet to foods easily prepared, attractive, and pleasing in appearance and taste;"

3. The addition to the diet of the "insignificant quantity" of five milligrams per day of potassium is of importance "when compared with either the large amount of potassium present in the body, needed by the body, or supplied to the body by the ordinary diet;"

4. "Additional quantities of vitamins, far in excess of the amount recommended for adequate nutrition, will provide additional benefits to persons in good health; and will assist in returning a sick or injured person, or one convalescing from an operation, to good health;"

5. "The body has a greatly increased need for vitamins when the individual is under mental stress, tension, or strain, is physically fatigued, or is suffering from injury, infection, or illness, or is undergoing surgery;"

6. The products are adequate "for the treatment and prevention" of a number of specific diseases mentioned in the leaflet, and "these diseases are quite likely to occur unless the ordinary diet of the usual person in this country is supplemented by a vitamin or vitamin and mineral supplement;" and

7. Anyone suffering from one or more of a number of specific symptoms listed in the labeling "is suffering from a dietary deficiency and can eliminate the symptoms and conditions by adding a vitamin supplement to their diet."
(Hutt 1986)[9]

The FDA also required Abbott to change its formulation for one product. The product contained only 5 mg of potassium, but claimed on its labeling to provide "extra insurance" "extra potency," and similar statements (Hutt 1986)).[10] The FDA determined the claims were misleading when referring to the "insignificant quantity" of potassium in the product. Abbott could either change its labeling or reformulate in order to keep its current labeling claims (Hutt 1986).[11]

It would be a later series of cases that would become associated with the FDA's attempts to regulate a product's formulation under the misbranding provisions of the Act. The first case came in 1964 with FDA taking action against Dextra Brand Sugar. Dextra fortified its sugar

---

[9] Hutt 1986.

[10] *Id.*

[11] *Id.* at 56–57.

with 19 added vitamins and minerals (*Dextra* 1963).[12] The FDA argued the labeling was false and misleading under the Act on three grounds. First, Dextra claimed or implied that the American diet was deficient in vitamins and minerals. Second, it informed consumers the added vitamins and minerals were present in the sugar in nutritionally significant amounts. Third, it stated sugar was the preferred carrier for fortification (*Dextra* 1963). The district court rejected the FDA's arguments on the basis there was no "persuasive evidence" consumers could misconstrue the labels in the way the FDA contended. The court noted the FDA's testimony pointed merely to a preference in fortification, which the Act did not allow (*Dextra* 1963). On appeal the FDA dropped all but one claim that stated the Dextra Sugar could nutritionally improve any diet (*Dextra* 1964).[13] In a unanimous *per curium* opinion the circuit court quickly dispensed of the FDA's remaining argument. The circuit court upheld the district court's conclusion that the consumer was allowed to choose from among products that were fortified and those that were not (*Dextra* 1964). Thus, under *Dextra* a consumer must evaluate the label to assess the level of nutrients added and whether or not to buy a fortified or non fortified product. The FDA could not expand the definition of misleading to make the choice for the consumer.

Dextra District Court Opinion

CHOATE, District Judge.

1. This is a civil action in rem arising under Section 304 of the Federal Food, Drug, and Cosmetic Act, 21 U.S.C. § 334.

2. The United States, Libelant herein, instituted this consolidated action by the filing of a Libel of Information at Jacksonville, Florida alleging that

449 cases, more or less, of an article of food labeled in part "New Dextra Brand Fortified Cane Sugar," had been shipped in interstate commerce from Ottawa, Ohio to Jacksonville, Florida, on or about July 21, 1961, and was misbranded when introduced into and while in interstate commerce within the meaning of the Federal Food, Drug, and Cosmetic Act, 21 U.S.C. § 331 et seq., in a number of ways.
   A similar libel was filed in Tampa, Florida alleging that a shipment of "New Dextra Brand Fortified Cane Sugar" had been shipped in interstate commerce from Ottawa, Ohio to Tampa, Florida in interstate commerce, on or about July 21, 1961.

3. Pursuant to Monitions in both of these actions, the United States Marshal at Jacksonville, Florida, seized 585 cases of the libeled sugar on December 20, 1961, and the United States Marshal, at Tampa, Florida, seized 106 cases of the libeled sugar, on December 18, 1961.

4. Upon the stipulation of both parties, the Jacksonville and the Tampa cases were consolidated and removed to this Court for disposition, and the issues of fact and law in each case are identical.

5. A claim for the seized article was duly filed by the Sugarlogics Southern Corporation. This company is a subsidiary of the Dextra Corporation. As hereafter noted, at the time the seized article was manufactured and shipped, claimant's principal offices were in Delray Beach, Florida. They are now located in Miami, Florida.

6. It was established by stipulation that the bags of the res, manufactured at Ottawa, Ohio and shipped in interstate commerce as described above,

---

[12] *United States v. 119 Cases…"New Dextra Brand Fortified Cane Sugar,"* 231 F. Supp. 551 (S.D. Fla. 1963).

[13] *United States v. 119 Cases…"New Dextra Brand Fortified Cane Sugar,"* 334 F.2d 238 (5th Cir. 1964).

consist of approximately 96% sugar produced from beets and 5% sugar produced from cane.

7. Pursuant to stipulation of the parties, the shipment of the res in interstate commerce is admitted. Claimant and its parent, the Dextra Corporation, have been and are now selling Dextra Brand Fortified Sugar in interstate commerce. However, it has limited its distribution efforts to test marketing because the Food and Drug Administration has opposed the sale of the product. The serious risk of seizure proceedings resulting from this opposition has precluded the company from making substantial investments necessary for major marketing of the product until the company's rights to sell the product have been clarified.

8. The first assertion of the libelant is to the effect that the res is mislabeled because the name "DEXTRA" implies that the product is comprised of dextrose rather than sucrose. The Court notes that the root of all words found in an unabridged dictionary bearing the "dext" prefix is from the Latin meaning pertaining to the right or right hand, or dextrous, or fortunate. The Government has not sustained the charge that the registered trademark "Dextra" as used on the labels of the article in issue represents and suggests to consumers that the article is composed of dextrose. No evidence of consumer reaction was introduced; the only evidence presented by the Government was the conjectural opinions of several of its expert nutritional witnesses. On the other hand, the record affirmatively establishes that dextrose is physically different in appearance from granu-

lated sugar and is sold through drug channels; that Dextra Brand Fortified Sugar was not labeled, sold or promoted in any manner to imply or suggest to consumers that the product contains dextrose, which is an inferior sweetening agent; and that, in fact, consumers have not regarded the product as being comprised of dextrose.

9. Secondly, the Government alleges that the label of the seized article of food contains statements which represent, suggest, and imply:

a. That the American diet is deficient in vitamins and minerals and that Dextra Sugar will correct this implied deficiency;

b. That the nutritional content of diets generally is significantly improved by the use of the seized article;

c. That Dextra Sugar when used in the ordinary diet is significantly more nutritious than any other sugar;

d. That the article under seizure is of significant value because it restores vitamins and minerals lost in the refinement of cane juice;

e. That all of the vitamins and minerals in the article are present in nutritionally significant amounts for special dietary use.

The label complained of has the following statements:

(on the front panel of the label)

"New!"

"Dextra Brand Fortified Sugar"

"Fortified with Vitamins and minerals"

(on the backside panel of the label)

"Now, at long last, many of the vitamins and Minerals lost in the refinement of cane juice have been restored to DEXTRA Fortified Cane Sugar."

"Almost any diet can be nutritionally improved by the use of DEXTRA Fortified

Cane Sugar in place of sweetening agents containing only "empty" calories—calories unaccompanied by nutrients."

"MORE NUTRITIOUS THAN ANY OTHER SUGAR!"

The representations above referred to are also made by listing 19 ingredients of the seized sugar and comparing the amounts of each of these ingredients in the seized sugar with the amounts present in ordinary sugar.

...OPINION OF THE COURT

This proceeding involves the question whether claimant's product, consisting of sugar fortified with vitamins and minerals, is misbranded and in violation of Section 403 of the Federal Food, Drug, and Cosmetic Act. While a number of charges are asserted in the libels of information filed herein, the Government's principal challenge is on a novel basis—that the offering of a fortified sugar, truthfully labeled to disclose such fortification, is misleading "per se" to consumers. At the outset it is important to note that despite the sweeping nature of the consumer deception which this product is charged to create, the Government at the trial presented no actual evidence that consumers were misled by the product. The Government has chosen to rest its case on opinion evidence of several nutritionists despite the fact that in a seizure proceeding, the "burden is upon the Government to prove the ground for forfeiture alleged in the libel ...by a fair preponderance of the evidence." ...It is clear that the Government failed to meet its burden in this case.

The Government's witnesses' testimony was largely directed to their views regarding the most preferable means of supplying vitamins and minerals to consumers, and whether the fortification of sugar complied with a Statement of General Policy on fortification issued by the Food and Nutrition Board of the National Research Council. Such testimony plainly is not pertinent here. Section 403 of the Federal Food, Drug, and Cosmetic Act permits the seizure and condemnation of goods only if they are misbranded, and that plainly means only if the labeling of the product is false or misleading.

Section 301 of the Act merely empowers the Food and Drug Administration to issue "a reasonable definition and standard of identity" so that consumers who purchase it can obtain "assurance that they will get what they reasonably expect to receive". Such standards have no bearing on the sale of a single, unique food product such as Dextra Brand Fortified Sugar.

The Government charges that "mere mention" on the labels of Dextra Brand Fortified Sugar of the fact that the product is fortified and the listing of the vitamins and minerals contained therein could be construed by consumers to suggest or imply that vague generality known as the "American diet" is deficient in the supply of vitamins and minerals, and that use of this product would overcome this deficiency.

The Government also challenges the product as inherently deceptive on the ground that the disclosures regarding fortification misrepresent the product's nutritional significance in comparison with ordinary sugar. However, the Government's witnesses did not dispute that this product is an effective carrier of the vitamins and minerals added to respondent product, and that ordinary sugar contains none of these nutrients, and is commonly referred to in nutritional literature by the derogatory term, "empty calories". Indeed, the Government's own witnesses appeared to concede that in comparison with ordinary sugar, the product in fact was significantly more nutritious.

The sole basis of the Government's charges is that the added nutrients are of no value because they are already in adequate supply in the American diet. This is clearly an untenable basis for holding the product misbranded.

It is clear that the true basis for the objection to the fortification of sugar is not that the vitamins and minerals added to the sugar are of no nutritional value, but rather, that the Food and Drug Administration does not regard sugar as a preferable vehicle for fortification, or for addition of vitamins where a deficiency exists. In short they quarrel over the vehicle.

The basic flaw in the Government's case against the product is that it is seeking, under the guise of misbranding charges, to prohibit the sale of a food in the marketplace simply because it is not in sympathy with its use. But the Government's position is clearly untenable. The provisions of the Federal Food, Drug, and Cosmetic Act did not vest in the Food and Drug Administration or any other federal agency the power to determine what foods should be included in the American diet; this is the function of the marketplace. Under Section 403 of the Act, Congress expressly limited the Government's powers of seizure to those products which are falsely or deceptively labeled. As the Supreme Court aptly stated in rejecting a similar attempt to overreach the authority granted by the Federal Food, Drug, and Cosmetic Act:

"In our anxiety to effectuate the congressional purpose of protecting the public, we must take care not to extend the scope of the statute beyond the point where Congress indicated it would stop."

…The Court does not undertake to constitute itself an arbiter of nutritional problems involved in determining more or less desirable agents for vending vitamin and mineral supplements to the consumer. The Congress did not provide the necessity of such determination. Neither will the Court permit a federal agency to appoint itself such an arbiter under the guise of prosecuting an action under the Act in question. Plainly only Congress can or should regulate the use of vitamins and then only to prevent public injury.

A second case struck at the concept of unneeded nutrients in a decision that directly contradicted *Dextra*. The FDA brought a charge of misbranding against Vitasafe Corporation claiming the long list of ingredients in its supplements were of "no nutritional significance for dietary supplementation" (Vitasafe 1964).[14] The long list of ingredients included:

…vitamin K (menadione), rutin, lemon bioflavonoid complex, monopotassium glutamate, l-lysine monohydrochloride, dessicated liver, sodium caseinate, leucine, lysine, caline, histidine, isoleucine, phenylalanine, threonine, tryptophane, manganese, potassium, zinc, magnesium, sulfur, calcium, and phosphorous.

The district court agreed with the FDA's view. It held that a listing of a nutrient on the label implied it enhanced the nutritional value of the supplements (*Vitasafe* 1964). Based on expert testimony the court concluded some of the nutrients in the Vitasafe products, such as lemon bioflavonoid, monopotassium glutamate, were of no nutritional value at all and others were not present in sufficient quantity to be of any nutritional value (*Vitasafe* 1964). On appeal the trial court's holding was affirmed with no additional explanation (*Vitasafe* 1965).[15]

---

[14] *United State v. An Undetermined Number of Shipping Pacakges…Labeled in Part "Vitasafe Formula M,"* 225 F. Supp. 266 (D.N.J. 1964).

[15] *United States v. Vitasafe Corp.*, 345 F.2d 864 (3d Cir. 1965).

Vitasafe District Court Opinion

LANE, District Judge.

This action, which arises under the Federal Food, Drug, and Cosmetic Act, 21 U.S.C. § 301 et sEq. , was initiated by the government's filing of a libel of information in this court wherein the United States, having seized a quantity of vitamin and mineral capsules and labeling for the articles located at Middlesex, New Jersey, under powers granted the United States by 21 U.S.C. § 334, sought to have these articles condemned. The items were in the possession of the Vitasafe Corporation, Division of Consolidated Sun Ray, Inc., and also in the possession of the United States Post Office after delivery for shipment in interstate commerce. The asserted ground for seizure was that the designated articles were misbranded within the meaning of 21 U.S.C. §§ 343(a), 352(a), and 352 (f) (1). The items were concededly introduced into and traveling in interstate commerce within the meaning of the act.

Pursuant to monition, there were seized at the Vitasafe Corporation's premises 906 packages containing Vitasafe capsules, and approximately 3,730,000 pieces of written, printed, and graphic material designed to promote the sale of the Vitasafe capsules.

The libel was amended to specifically include the products "Vitasafe CF" and "Vitasafe Queen Formula with Royal Jelly Supplement for Women," by name, which were found at the premises of Claimant when the monition was executed. A further amendment was allowed pursuant to Rule 15(a) of the Federal Rules of Civil Procedure, so as to include the allegation that the seized articles were misbranded while held for sale after shipment in interstate commerce, as well as when introduced into and while in interstate commerce.

The United States alleges that the Vitasafe capsule, as an article of food within the meaning of 21 U.S.C. § 321 (f), is misbranded under 21 U.S.C. 343 (a) in that:

1. Its labeling, when viewed as a whole, represents, suggests and implies that the nutritional needs for men and women differ, and that the "Formula M" capsules are designed to satisfy the special needs of men as contrasted to the "Formula W" capsules which are designed to satisfy the special needs of women, which representations, suggestions, and implications are contrary to fact;

2. The listing on the label of, and references in the labeling to, certain ingredients implies and suggests that the nutritional value of Vitasafe capsules is enhanced by the presence of such ingredients, when in fact such implications and suggestions are false and misleading in that the presence of these ingredients is of no nutritional significance for dietary supplementation. The ingredients so listed are: vitamin K (menadione), rutin, lemon bioflavonoid complex, monopotassium glutamate, l-lysine monohydrochloride, dessicated liver, sodium caseinate, leucine, lysine, caline, histidine, isoleucine, phenylalanine, threonine, tryptophane, manganese, potassium, zinc, magnesium, sulfur, calcium, and phosphorous.

3. The statements in the labeling to the effect that the quoted "Minimum Adult Daily Requirements" (MDR) are a recommendation of the Food and Nutrition Board, National Academy of Science—National Research Council, are false and misleading because the Food and Nutrition Board has not recommended any "Minimum Daily Requirements" but has established "Recommended Dietary Allowances" (RDA) which

differ from the MDR and are not the allowances designated in the labeling as the "Minimum Daily Requirements."

4. The overall impression suggested and implied by the statements in the labeling concerning the large amounts of common foods that must be consumed in order to furnish quantities of nutrients equal to the quantities of such nutrients present in one Vitasafe capsule is false and misleading since such large quantities of food would not be needed to supply the necessary dietary requirements for these nutrients and since the labeling does not list all the various nutrients furnished by the stated quantities of food designated in the labeling.

It is further alleged that the Vitasafe capsule, as a drug within the meaning of 21 U.S.C. § 321(g) is misbranded under 21 U.S.C. § 352(a) in that:

5. Its labeling contains false and misleading representations, suggestions, and implications that the article is an adequate and effective treatment for depression, tension, weakness, nervous disorders, lethargy, lack of energy, lassitude, impotence, aches and pains, aging, impaired digestion, loss of appetite, skin infections, lesions and scaliness, night blindness, photophobia, fatigue, headache, insomnia, diarrhea, edema of the legs, hypersensitivity to noise, swelling, redness, soreness and burning of the tongue, impairment of memory, inability to concentrate, dermatitis, cracking of the lips, lesions at the corners of the mouth, growth failure in children, sore, swollen and bleeding gums, defective calcification of the bones, lowered resistance to disease and lowered vitality, which

representations, suggestions, and implications are false and misleading since the article is not an adequate and effective treatment for the disease conditions and symptoms as stated and implied.

6. Its labeling contains false and misleading representations, suggestions, and implications that practically everyone in this country is suffering from or is in danger of suffering from a dietary deficiency of vitamins, minerals and proteins which is likely to result in specific deficiency diseases, such as scurvy, as well as a great number of non-specific symptoms and conditions, which threatened deficiency is represented as being due to loss of nutritive value of food by reason of the soil on which the food is grown, and the storage, processing, and cooking of the foods, which representations, suggestions, and implications are false and misleading since they are contrary to fact.

FINDINGS OF FACT

A study of all the exhibits and the expert testimony outlined above leads this court to the following conclusions:

1. The labeling of the seized article, when viewed as a whole, does represent, suggest and imply that a woman, because of sex alone, has different nutritional needs than a man, and "Vitasafe Formula W" will satisfy these special needs of women as contrasted to "Vitasafe Formula M" which will satisfy the special needs of men.

These representations are false and misleading since there is no difference in the nutritional requirements of non-pregnant, non-lactating women as compared to men, except for iron in women of childbearing age, which need is adequately satisfied by the normal diet readily available and normally consumed. With the exception of iron, nutritional need is the same for men and women.

The quantities required for some nutrients differ depending on size, weight or activity, but this holds substantially true for both men and women. No reference is made in any of the labeling to conditions of pregnancy and lactation and any special need resulting.

With respect to iron content, both the men's formula and women's formula contain the same quantity of iron so that the "Vitasafe Formula W" and "Formula M" may be used interchangeably by either men or women with no difference in effect. All this is contrary to the implications of the labeling at issue.

2. The labeling of the seized res, when viewed in its entirety, does represent, suggest and imply that the nutritional value of Vitasafe capsules is enhanced by the presence in these capsules of the following ingredients in the stated amounts:

Vitamin K (Menadione) -.05 mg. Monopotassium Glutamate - 20 mg. L-lysine Monohydrochloride - 7 mg. Sodium Caseinate (protein)—100 mg. (Formula "M") 50 mg. (Formula "W") Leucin - 8 mg. (Formula "M") 4 mg. (Formula "W") Lysine - 7 mg. (Formula "M") 3.5 mg. (Formula "W") Valine -6 mg. (Formula "M") 3 mg. (Formula "W") Histidine -2.8 mg. (Formula "M") 1.4 mg. (Formula "W") Isoleucine -5 mg. (Formula "M") 2.5 mg. (Formula "W") Phenylalamine -4 mg. (Formula "M") 2 mg. (Formula "W") Threonine -4 mg. (Formula "M") 2 mg. (Formula "W") Tryptophan - 1 mg. (Formula

"M").15 mg. (Formula "W") Manganese −5 mg. Potassium −2 mg. Zinc -.5 mg. Magnesium - 3 mg. Calcium - 75 mg. (Formula "M") 50 mg. (Formula "W") Phosphorous -58 mg. (Formula "M") 39 mg. (Formula "W") Choline Bitartrate −31.4 mg. (Formula "M") 30. mg. (Formula "W") Inositol - 15 mg. (Formula "M") 10 mg. (Formula "W") Lemon bioflavonoid -5 mg. Sulfur - 22 mg. Royal Jelly - 550 mcg.

The evidence produced at trial proves that the normal or ordinary diet supplies amounts of the above-listed ingredients greatly in excess of those necessary for good nutrition. Furthermore, with the exception of the fat soluble vitamins A and D, vitamins ingested in excess of those required are excreted and make no nutritional contribution. The evidence further proves that the following ingredients represented to be present in the Vitasafe product have no nutritional value whatever, namely rutin, lemon bioflavonoid, monopotassium glutamate, sulfur, choline bitartrate, inositol and royal jelly. Other ingredients, such as all the amino acids are in such small quantities as to be of no significant value when compared to the quantities of such ingredients which are required and which are present in the average diet. Because the ingredients here designated are either of no nutritional significance per se or are contained in the Vitasafe product in such minute quantities, these ingredients do not enhance the nutritional value of the Vitasafe capsules. Consequently, the representation and suggestion referred to in this finding is false and misleading.

3. The labeling of the seized res states and represents that the "Minimum Adult Daily Requirements" are a recommendation of the Food and Nutrition Board of the National Academy of Science National Research Council.

   This representation is false. "Minimum Adult Daily Requirements" is a list of essential nutrients with designated amounts established by the United States Food and Drug Administration and set forth in their Regulations at 21 C.F.R. 125.3 (vitamins) and 125.4 (minerals).

   The Food and Nutrition Board of the National Academy of Sciences has published its book, "Recommended Dietary Allowances." This is a list of nutrients with specified amounts, which differ in the number of nutrients listed and the amounts of such nutrients from those set forth as "Minimum Daily Requirements."

4. Although correct if read literally and carefully, the labeling of the seized res represents and suggests to the ordinary reader that it is necessary to eat enormous quantities and varieties of foods in order to obtain the variety of vitamins and minerals in the amounts provided by one Vitasafe capsule.

   This representation is false and misleading because the variety and quantity of foods referred to in the labeling of the seized res provide many times the amounts of nutrients as well as additional nutrients than are supplied by one Vitasafe capsule.

5. The labeling of the seized res represents, suggests, and implies that Vitasafe capsules are an adequate and effective treatment of or preventive for the following symptoms and conditions which are referred to in this labeling: depression, tension, weakness, nervous disorders, lethargy, lack of energy, lassitude, impotence, aches and pains, aging, impaired digestion, loss of appetite, lesions and scaliness, night blindness, photophobia, fatigue, headaches, insomnia, diarrhea, edema of the legs, hypersensitivity to noise, swelling, redness, soreness and burning of the tongue, impairment of memory, inability to concentrate, dermatitis, cracking of the lips, lesions at the corners of the mouth, growth failure in children, sore, swollen and bleeding gums, defective calcification of the bones, lowered resistance to disease, and lowered vitality.

   This representation is false and misleading. The evidence produced at trial conclusively proves that the above designated symptoms or conditions are caused by and associated with a great number of serious pathological diseases. Further, although some of these symptoms may be associated with vitamin and mineral deficiencies, the likelihood of their being caused by or associated with vitamin or mineral deficiencies in the United States today is very small.

   There is a danger involved in the use of this type of labeling insofar as a person having one or more of the above-listed symptoms may resort to a Vitasafe product as a cure. Such a person may continue taking this vitamin product for a long continued period, as he is urged to do in the labeling of this product, and thereby fail to obtain competent medical help to correct his physical illness.

6. The labeling of the seized article represents it to be of value because it contains "lipotropic factors." The

evidence shows that such factors are substances which affect the mobilization of fat, particularly in the liver. Consequently, these factors are drugs within the meaning of 21 U.S.C. § 321(g) in that their only known use is to affect the structure and function of part of the body. Further, the labeling of the seized article fails to bear adequate directions for use of Vitasafe as a drug within the meaning of 21 U.S.C. § 352(f) (1). Nowhere in the labeling does there appear the conditions for which the Vitasafe capsule is to be used as a lipotropic factor or the collateral measures necessary for the safe use of Vitasafe by a layman as a lipotropic factor.

7. The article under seizure is both a food within the meaning of 21 U.S.C. § 321(f) because its labeling recommends its use as and represents it to be of value as a dietary and nutritional supplement, and also a drug within the meaning of 21 U.S.C. § 321(g) because its labeling recommends its use as and represents it to be of value as a curative or preventive of disease conditions in man affecting the structure and function of the body of man.

Vitasafe and *Dextra* added conflicting views to the FDA's *ad hoc* policy. Rather than promulgate rules or seek Congressional amendments the FDA pursued a series of enforcement actions in courts. In *Dextra* the Fifth Circuit rejected the FDA's enforcement on the basis of formulation as misbranding under the Act. The Third Circuit, however, agreed and upheld the policy. The resulting conflict provided the framework for the next three decades, augmented occasionally by a new decision. Confusion replaced predictability on the regulatory landscape and a patchwork of opinions displaced any chance of uniformity.

Following *Nutrilab* "intended use" proved the greatest tool to reclassify dietary supplements with alleged disease claims as new drugs. Chapter 1 introduced the concept of intended use and the definition of food. Chapter 4 also indicated that when new drug or disease claims are made the product is enforced both as misbranded and as lacking the appropriate premarket clearance for its new classification as a drug. Intended use provided a convenient vehicle for regulating dietary supplement. In this environment, as it is today, definitions mattered then more than ever. If labeling indicated any medicinal use it was regulated a drug.

Vitamin Mineral Amendments (1976) and NLEA (1990)

Attempts by the FDA to issue regulations on dietary supplements failed. The FDA attempted self-regulation in 1962 with a proposal to set dosage levels of supplements. Under its proposal it would have permitted the sale of single nutrient dietary supplements with levels close to the US Recommended Daily Allowance (RDA) and limited the number of multi vitamin and mineral products (RDA Proposal).[16] The proposed regulations also classified "high potency" dietary supplements as drugs (RDA Proposal). The proposal sparked an outcry that would overwhelm the rule-making process. Eventually, the proposed rule was abandoned.

Congress began a slow intervention into dietary supplement regulation beginning in 1976. The Vitamin Mineral Amendment of 1976 was the first attempt by Congress to define and regulate vitamins and minerals making therapeutic claims. The Amendment struck at the FDA's earlier attempts to set maximum levels of vitamins and mineral or deem vitamins or minerals drug based on threshold of the added vitamin or mineral. The Amendment was the first

---

16   27 Fed. Reg. 5815 (1962).

to specifically address dietary supplements, but its reactionary posture narrowed the scope of the legislation. It reacted to the rallying cry of industry against the FDA's attempt to subject increasing numbers of supplements to the drug regulations.

---

21 USC Sec. 350(1)

(1) Except as provided in paragraph (2)—

(A) the Secretary may not establish, under Section 321 (n), 341, or 343 of this title, maximum limits on the potency of any synthetic or natural vitamin or mineral within a food to which this section applies;

(B) the Secretary may not classify any natural or synthetic vitamin or mineral (or combination thereof) as a drug solely because it exceeds the level of potency which the Secretary determines is nutritionally rational or useful;

(C) the Secretary may not limit, under Section 321 (n), 341, or 343 of this title, the combination or number of any synthetic or natural—

(i) vitamin,

(ii) mineral, or

(iii) other ingredient of food, within a food to which this section applies.

The Nutritional Labeling and Education Act (NELA) of 1990 established nutrient content and claims for food products. NLEA required food manufacturers to include nutritional labeling and directed the FDA to establish procedural regulations for making health and nutrient content claims. Nearly 15 years after the FDA brought enforcement action against Continental Baking for making fiber claims, the agency was now charged with setting requirements for making such claims. NELA also required labels declare any vitamin or mineral supplementation. Health claims opened the door for dietary supplements to make certain claims about the benefits of the nutrients it contained. The FDA determined the criteria for approving health claims would apply to conventional foods and dietary supplements. Later actions by the agency would belie this assertion.

Health claims describe a relationship between the food or its ingredient and a reduced risk of disease or health condition. Dietary supplements must make health claims under NLEA. The provisions of the 1997 Food and Drug Administration Modernization Act (FDAMA) did not include dietary supplements. A health claim must always relate the food or its ingredient to a disease or health-related condition. If this basic structure is absent then the claim falls outside the regulations and is misbranded. An example is the claim that fruits and vegetables are part of a healthy diet. This is a general dietary guidance or pattern and not a health claim.

NLEA sets criteria for the FDA to review and approve of health claims. Only those health claims approved by the FDA may be used on labeling. NLEA authorizes the FDA to conduct an exhaustive review based on a petition citing scientific literature that establishes a relationship between the food substance and the disease. The evidence must show "significant scientific agreement" to be approved.[17] This is a high bar of proof to clear. The lower bar of using authoritative statements under FDAMA does not apply to dietary supplements.

5.2.2   Regulation Under DSHEA

Introduction to DSHEA and New Definitions

Confusion best described the era of enforcement from 1906 until 1994. The FDA confronted a rapidly growing market ill-equipped with the statutory controls needed to protect the public from the unscrupulous and hyperbolic claims of dietary supplement manufactures. It pursued an *ad hoc*

---

[17]   21 CFR 101.14(c).

enforcement policy that morphed depending on judicial outcomes. Even the judiciary struggled to develop a unified consistent interpretation of the Act as it related to dietary supplements. This period came to a much needed close with the passage of the DSHEA in 1994.

The passage of DSHEA was prompted by the FDA's aggressive enforcement of NELA as it applied to dietary supplements. Shortly before the enactment of NELA the FDA grew concerned over the adverse health effects associated with L-tryptophan. L-tryptophan was an ingredient in a supplement associated with an outbreak of eosinophilla myalgia syndrome, which lead to the FDA issuing a consumer advisory warning the public about the supplement. In the wake of the L-tryptophan scare the FDA established a task force to examine dietary supplements. The task force recommended the agency regulate supplements as drug when medicinal claims are made or food additives in the absence of such claims. The expanded ability to make health claims under NELA was not considered by the task force signaling an aggressive enforcement stance by the agency.

Congress intervened and passed DSHEA to restrict the FDA's ability to impose strict controls on dietary supplements. DSHEA defined dietary supplement and set a statutory framework for making claims on products. This framework included new Good Manufacturing Practices (GMPs) for the dietary supplement industry. Dietary supplements now must comply with Part 111 rather than Part 110 that outlines food GMPs.

The definition of dietary supplement under DSHEA embraces a wide range of products. Its primary limitation is not in what a dietary supplement may contain, but on its application. Dietary supplement are only those products which are consumed orally. Any product claiming dietary supplement status which is topical, injectable, or inhalable is not a dietary supplement.

The term dietary ingredient also broadly encompasses functional ingredients that supplement the diet, but excludes added substances like fillers, preservatives or emulsifiers. Dietary supplements are also required to be labeled as "dietary supplements" and not be misrepresented as conventional foods. The limitations though few continue to provide ample ground for enforcement.

21 USC 201(ff)

(ff) The term "dietary supplement"—

(1) means a product (other than tobacco) intended to supplement the diet that bears or contains one or more of the following dietary ingredients:

(A) a vitamin;

(B) a mineral;

(C) an herb or other botanical;

(D) an amino acid;

(E) a dietary substance for use by man to supplement the diet by increasing the total dietary intake; or

(F) a concentrate, metabolite, constituent, extract, or combination of any ingredient described in clause (A), (B), (C), (D), or (E);

(2) means a product that—

(A)(i) is intended for ingestion in a form described in Section 411(c)(1)(B)(i); or

(ii) complies with Section 411(c)(1)(B)(ii);

(B) is not represented for use as a conventional food or as a sole item of a meal or the diet; and

(C) is labeled as a dietary supplement; and

(3) does—

(A) include an article that is approved as a new drug under Section 505 or licensed as a biologic under Section 351 of the Public Health Service Act (42 U.S.C. 262) and was, prior to such approval, certification, or license, marketed as a dietary supplement or as a food unless the Secretary has issued a regulation, after notice and comment, finding that the article, when used as or in a dietary supplement under the conditions of use and dosages set forth in the labeling for such dietary supplement, is unlawful under Section 402(f); and

(B) not include—

(i) an article that is approved as a new drug under Section 505, certified as an antibiotic under Section 507 7, or licensed as a biologic under Section 351 of the Public Health Service Act (42 U.S.C. 262), or

(ii) an article authorized for investigation as a new drug, antibiotic, or biological for which substantial clinical investigations have been instituted and for which the existence of such investigations has been made public, which was not before such approval, certification, licensing, or authorization marketed as a dietary supplement or as a food unless the Secretary, in the Secretary's discretion, has issued a regulation, after notice and comment, finding that the article would be lawful under this Act.

Misrepresentation as conventional food is a particular concern for liquid dietary supplements, such as energy drinks. In the Guidance Document, "Distinguishing Liquid Dietary Supplements from Beverages" the FDA lists a number of factors which will be weighed in assessing intended use. The FDA provides a range of familiar criteria in evaluating intended use, which include for example, "product or brand name, packaging, serving size and total recommended daily intake (i.e., the volume in which they are intended to be consumed), composition, recommendations and directions for use, statements or graphic representations in labeling or advertising, and other marketing practices" (Guidance Document on Liquid Dietary Supplements).[18] Simple mistakes such as adding a nutrition facts panel or using a conventional beverage serving size can override the clear statement "dietary supplement" on the label. The result is typically a determination the product is adulterated.

The low barriers of entry combined with potential profits from a large market lead to a wide range of "dietary supplements" entering the market. Unlike drugs or medical devices where nearly every device or drug is assessed by the FDA prior to entering the market, foods can freely and quickly enter the market. This often leads to products labeled as dietary supplements entering the market with a multitude of applications. This includes an FDA recall and consumer notice about injectable vitamin C and the infamous case of AeroShot which sold inhalable caffeine (FDA Notice on Injectable; FDA AeroShot Warning Letter). If vitamin C is taken orally it is unquestionably a dietary supplement, but offer it as injections and there is no plausible coverage for it under the provisions of DSHEA. AeroShot clouded its opportunity to expand the definition of ingestible by including labeling that stated the product was both inhalable and ingestible. The FDA not only raised safety questions, but also found the labeling misleading.

---

[18] FDA (Jan 2014 2014).

Breathable Foods, Inc. 3/5/12

Department of Health and Human Services

Public Health Service
Food and Drug Administration

5100 Paint Branch Parkway
College Park, MD 20740

March 5, 2012

WARNING LETTER

Mr. Thomas Hadfield, CEO
Breathable Foods, Inc.
300 Tech Square, Suite 301
Cambridge, MA 02139

Re: 279597

Dear Mr. Hadfield:

This to advise you that the Food and Drug Administration reviewed your website at www.aeroshots.com in February 2012 and has determined that the product AeroShot is misbranded within the meaning of section 403 of the Federal Food, Drug, and Cosmetic Act (the Act) [21 U.S.C. § 343]. We also have safety questions about the product, as described below. You can find the Act and FDA regulations through links on FDA's home page at www.fda.gov.

Labeling

Your AeroShot product is misbranded within the meaning of section 403(a)(1) of the Act [21 U.S.C. § 343(a)(1)] in that your labeling is false or misleading.On the one hand, you indicate that AeroShot is intended for inhalation. For instance, the product label prominently features the claim "BREATHABLE ENERGY Anytime, Anyplace," as well as the instruction, "Puff in." In addition, your website includes headlines that describe your product as "inhalable caffeine" and tout the invention of a "caffeine inhaler." Your website also describes the product as "airborne energy" that "delivers a unique blend of 100 mg of caffeine and B vitamins in about 46 puffs."

Despite these suggestions that your product is intended for inhalation, you indicate in other statements that the product is intended for ingestion. For instance, your label characterizes AeroShot as a dietary supplement, and your website describes the product as "ingestible food "and instructs users to swallow the product. Your website further states that "Breathable Foods is revolutionizing the delivery of nutrients to the mouth for ingestion . . ." and that "AeroShot provides a safe shot of caffeine and B vitamins for ingestion."

By definition, dietary supplements must be intended for ingestion. See sections 201(ff)(2)(A) and 411(c)(1)(B) of the Act [21 U.S.C. §§ 321(ff)(2)(A) and 350(c)(1)(B)]. Your labeling is false and misleading because your product cannot be intended for both inhalation and ingestion.The functioning of the epiglottis in the throat keeps the processes of inhalation and ingestion mutually exclusive. The epiglottis is a cartilaginous structure that prevents choking or coughing during ingestion. The act of ingestion enables the tongue to push down on the larynx, which in turn elevates the hyoid bone, drawing the larynx upwards. This latter action forces the epiglottis to fold back, covering the entrance to the larynx and the airways, preventing food, drink and particulates from entering the airways and respiratory tract. When a person inhales, however, the epiglottis maintains its upright position, enabling air and particulate matter to enter the airways and ultimately the lungs. A product intended for inhalation is not a dietary supplement.

Your AeroShot product is also misbranded within the meaning of section 403(y) of the Act [21 U.S.C. § 343(y)] in that the label fails to bear a domestic address or domestic phone number through which the responsible person, as described in section 761(b) of the Act [21 U.S.C. § 379aa-1(b)], may receive a report of a serious adverse event associated with the product.

We also note that the Supplement Facts panel on the label of your AeroShot product does not comply with 21 CFR 101.36(e)(6) in that the dietary ingredients declared under 21 CFR 101.36(b)(2)(i) (ending with vitamin B12) are not separated from the dietary ingredients described in 21 CFR 101.36(b)(3) (starting with caffeine) by a heavy bar.

Safety

In addition to the misbranding violations described above, we have safety questions about possible effects of your product. As summarized above, your labeling suggests in several places that AeroShot should be inhaled. Because of those suggestions, consumers may attempt to inhale your product, causing it to enter the lungs. FDA is concerned about the safety of any such use because caffeine is not typically inhaled through the lungs, and the safety of such use has not been well studied. Your website addresses this issue in part by stating that "[d]ecades of research have shown that particles above 10 microns in size, if inhaled, fall out in the mouth and do not penetrate the respiratory tract," and also that "[o]ur powders are of a median size much larger than 10 microns." Please provide us with references to the research you cite so that we can evaluate that research. Please also submit the evidence you relied on in stating that the median size of your powders is "much larger than 10 microns."

Furthermore, although you have issued a statement in which you assert that "AeroShot is not recommended for those under 18 years of age," some of your labeling indicates otherwise. Your label states that AeroShot is "not intended for people under 12 . . . ." This suggests that the product is suitable for children 12 and over. Please provide us with any safety evidence you have relied upon related to the use of your product by children and adolescents so that we can evaluate that evidence. In addition, in light of your statements that AeroShot is "not intended for people under 12" and is "not recommended for those under 18 years of age," we question why your website states that your product is designed to be used when "[h]itting the books" and "study[ing] in the library." These activities are commonly performed by children and adolescents. Indeed, your reference to these activities seems to target this population.

Finally, we note that your labeling conveys contradictory messages about the use of your product in combination with alcohol. On the one hand, your website includes a posting of a news interview in which the inventor of your product, David Edwards, states that he is not encouraging the mixing of AeroShot with alcohol. On the other hand, your website includes clips of news videos related to AeroShot, as well as links to news articles related to the product. Several of these news items refer to the use of your product in combination with alcohol or as a "party drug." Even though these news items express health concerns about taking AeroShot while drinking alcohol, your posting of the news items on the website where you promote and sell AeroShot publicizes such use. Any such publicity may have the effect of encouraging the combination of your product with alcohol---a scenario that raises safety concerns, as peer-reviewed studies show that ingesting these two substances together is associated with risky behaviors, such as riding with a driver who is under the influence of alcohol, which can lead to hazardous and life-threatening situations. Data and expert opinion also indicate that caffeine decreases the perception of intoxication, meaning that individuals who consume caffeine along with alcohol may consume more alcohol than they otherwise would and become more intoxicated than they realize. At the same time, caffeine does not change blood alcohol content levels, and thus does not reduce the risk of harm associated with drinking alcohol.

This letter is not intended to be an all-inclusive review of your product and its labeling. It is your responsibility to ensure that all of your products comply with the Act and its implementing regulations. You should take prompt action to correct the violations cited above. Failure to do so may result in regulatory action without further notice. Such action may include, but is not limited to, seizure or injunction.

Please respond in writing within 15 working days from your receipt of this letter. Your response should outline the specific actions you are taking to correct the violations cited above and to prevent similar violations in the future. Your response should include documentation, such as revised labels or other useful information, which would assist us in evaluating your corrections. If you cannot complete all corrections before you respond, we expect that you will explain the reason for the delay and state when you will correct any remaining violations.

Please send your reply to Quyen Tien, Compliance Officer, Food and Drug Administration, Center for Food Safety and Applied Nutrition, Office of Compliance (HFS-608), 5100 Paint Branch Parkway, College Park, MD 20740.

Sincerely,

/s/

Michael W. Roosevelt
Acting Director
Office of Compliance
Center for Food Safety
and Applied Nutrition

Basis for Finding Adulteration

The one premarket control under DSHEA is also a basis for determining whether a dietary supplement is adulterated. As with most foods the FDA chiefly regulates the safety of dietary supplements through post-market enforcement. For a small group of dietary supplements the FDA requires a 75-day premarket notification. This applies for any "new dietary ingredient," a term defined in Section 350b(d). This notification places the burden of proving that the new dietary ingredient is not safe on the manufacturer.

---

21 USC 350b(d)

(d) "New dietary ingredient" defined
For purposes of this section, the term "new dietary ingredient" means a dietary ingredient that was not marketed in the United States before October 15, 1994 and does not include any dietary ingredient which was marketed in the United States before October 15, 1994.

Section 350b requires premarket notification for any supplement containing a new dietary ingredient. There is no exemption for a supplement containing multiple ingredients where only one is deemed "new." The supplement in its entirety must be withheld from the market until the notification process is complete. In the notification the manufacturer must establish that the new dietary ingredient is safe on one of two grounds. Section 350(b) states that a new dietary ingredient will be deemed adulterated unless all new dietary ingredients in the product have been present without chemical alteration in the food supply as articles used for food or other evidence of safety establishing the dietary ingredient when used under the conditions recommended or suggested in the labeling of the dietary. This is similar to what will be seen in Chapter 6 on food additive exemptions but is distinct.

---

21 USC 350b(a)

(a) In general
A dietary supplement which contains a new dietary ingredient shall be deemed adulterated under Section 342 (f) of this title unless it meets one of the following requirements:

1. The dietary supplement contains only dietary ingredients which have been present in the food supply as an article used for food in a form in which the food has not been chemically altered.
2. There is a history of use or other evidence of safety establishing that the dietary ingredient when used under the conditions recommended or suggested in the labeling of the dietary supplement will reasonably be expected to be safe and, at least 75 days before being introduced or delivered for introduction into interstate commerce, the manufacturer or distributor of the dietary ingredient or dietary supplement provides the Secretary with information, including any citation to published articles, which is the basis on which the manufacturer or distributor has concluded that a dietary supplement containing such dietary ingredient will reasonably be expected to be safe.

The notification will be a combination of food additive petition and GRAS affirmation (see, Chapter 6). The criteria for claiming a new dietary ingredient complies with Section 350b mirrors the GRAS criteria used in assessing food additives. Similar to the GRAS exemption, new dietary ingredients under Section 350b(a) depend on levels of usage and intended use. Like a food additive petition the notification requires the manufacturer to submit adequate supporting information, including published articles or safety studies. If the criteria in 350b(a) do not provide coverage for the new dietary ingredient, then it will be deemed adulterated under the Act.

The general adulteration provisions also apply. Dietary supplements typically are enforced under the "may render injurious" standard because they are manufactured products. Under the "may render injurious standard" the supplement will be deemed adulterated if it contains an added substance that is injurious to health under the conditions recommended on the labeling. As herbal products continue to gain in market share the nonadded standard may also apply to dietary supplements due to the risk of residues from pesticides. It is important to consider the role of pesticides in growing these products.

Labeling Requirements

DSHEA clarified how dietary supplements could make claims beyond the narrow confines of NELA and past enforcement actions. It specifically allows structure/function claims, general well-being claims, and claims related to classical nutrient deficiency disease. These claims must bear the now all-too familiar disclaimer stating the FDA has not evaluated the claim. This change is codified in Section 343 as an amendment to the misbranding provisions. It also required that the manufacturer notify the FDA within 30 days of marketing the product with one of the claims. The statute left undefined when a statement would be construed as "treating, mitigating, or curing a disease." This would come in later regulations promulgated by the FDA.

21 USC 350b(a)

(6) For purposes of paragraph (r)(1)(B), a statement for a dietary supplement may be made if—

(A) the statement claims a benefit related to a classical nutrient deficiency disease and discloses the prevalence of such disease in the United States, describes the role of a nutrient or dietary ingredient intended to affect the structure or function in humans, characterizes the documented

mechanism by which a nutrient or dietary ingredient acts to maintain such structure or function, or describes general well-being from consumption of a nutrient or dietary ingredient,

(B) the manufacturer of the dietary supplement has substantiation that such statement is truthful and not misleading, and

(C) the statement contains, prominently displayed and in boldface type, the following: "This statement has not been evaluated by the Food and Drug Administration. This product is not intended to diagnose, treat, cure, or prevent any disease."

Four years after the enactment of DSHEA the FDA issued a rule on structure function claims. The sole purpose of the structure/function rule was to identify the types of statements that may be made without prior FDA review about "the effects of dietary supplements on the structure or function of the body ("structure/function claims") and to distinguish these claims from claims that a product diagnoses, treats, prevents, cures, or mitigates disease (disease claims)" (Structure Function Rule).[19] The rule defined direct and implied disease claims.

The FDA retained the existing definition of disease under the Structure Function Rule. The FDA developed a definition for disease in 1993 when issuing regulations under NELA. It debated whether to update the definition given the context of DSHEA. It ultimately retained that definition and added ten criteria to assist manufacturers and re-labelers in determining whether a claim qualifies as a disease claim.

21 CFR 101.93(g)(1) Definition of Disease Claim

…damage to an organ, part, structure, or system of the body such that it does not

---

[19]   FDA, Structure Function Rule, 65 FR 4 (2000).

function properly (e.g., cardiovascular disease), or a state of health leading to such dysfunctioning (e.g., hypertension); except that diseases resulting from essential nutrient deficiencies (e.g., scurvy, pellagra) are not included in this definition.

The criteria in 21 CFR 101.93(g)(2) also assist the FDA in enforcement actions by supporting a conclusion that a statement is an implied disease claim. The criteria stretch to cover even symptoms strongly associated with a disease or health condition. For example, joint pain or inflammation, is deemed so strongly linked to arthritis that using the terms would be the same as using the name of the disease. In other words it is a prohibited disease claim to state "mitigates arthritis" and using symptoms, such as "mitigates inflammation," is the same as using the disease name. The criteria provide the FDA *carte blanche* to find a disease claim on the full "context" of the label which includes images.

21 CFR 101.93(g)(2) Criteria for Finding a Disease Claim

(2) FDA will find that a statement about a product claims to diagnose, mitigate, treat, cure, or prevent disease (other than a classical nutrient deficiency disease) under 21 U.S.C. 343(r)(6) if it meets one or more of the criteria listed below. These criteria are not intended to classify as disease claims statements that refer to the ability of a product to maintain healthy structure or function, unless the statement implies disease prevention or treatment. In determining whether a statement is a disease claim under these criteria, FDA will consider the context in which the claim is presented.

A statement claims to diagnose, mitigate, treat, cure, or prevent disease if it claims, explicitly or implicitly, that the product:

(i) Has an effect on a specific disease or class of diseases;

(ii) Has an effect on the characteristic signs or symptoms of a specific disease or class of diseases, using scientific or lay terminology;

(iii) Has an effect on an abnormal condition associated with a natural state or process, if the abnormal condition is uncommon or can cause significant or permanent harm;

(iv) Has an effect on a disease or diseases through one or more of the following factors:

(A) The name of the product;

(B) A statement about the formulation of the product, including a claim that the product contains an ingredient (other than an ingredient that is an article included in the definition of "dietary supplement" under 21 U.S.C. 321(ff)(3)) that has been regulated by FDA as a drug and is well known to consumers for its use or claimed use in preventing or treating a disease;

(C) Citation of a publication or reference, if the citation refers to a disease use, and if, in the context of the labeling as a whole, the citation implies treatment or prevention of a disease, e.g., through placement on the immediate product label or packaging, inappropriate prominence, or lack of relationship to the product's express claims;

(D) Use of the term "disease" or "diseased," except in general statements about disease prevention that do not refer explicitly or implicitly to a specific disease or class of diseases or to a specific product or ingredient; or

(E) Use of pictures, vignettes, symbols, or other means;

(v) Belongs to a class of products that is intended to diagnose, mitigate, treat, cure, or prevent a disease;

(vi) Is a substitute for a product that is a therapy for a disease;

(vii) Augments a particular therapy or drug action that is intended to diagnose, mitigate, treat, cure, or prevent a disease or class of diseases;

(viii) Has a role in the body's response to a disease or to a vector of disease;

(ix) Treats, prevents, or mitigates adverse events associated with a therapy for a disease, if the adverse events constitute diseases; or

(x) Otherwise suggests an effect on a disease or diseases.

Concurrent Jurisdiction with the FTC

The FTC exercises joint jurisdiction with the FDA in enforcing dietary supplement advertising. As discussed in previous chapters the FDA exercises primary control over food and dietary supplement labeling. The agencies operate under a working agreement, which outlines the boundaries of authority. Under the working agreement the two agencies will either act concurrently or through case referrals. The FTC claims jurisdiction over advertising and marketing and the FDA is charged with regulating the labeling. Labeling leaves the FDA in control of the standards on what constitutes misbranding, in particular when a claim becomes a new drug claim.

The FTC maintains a guidance document that outlines how it enforces dietary supplement claims. It requires advertising to be truthful and not misleading and to substantiate all product claims (FTC Guidance).[20] The guidance also provides greater clarity than the FDA in certain areas such as claims based on traditional uses (FTC Guidance). In such instances it raises the possibility of a conflict where promotional language would comply with FTC regulations but violate FDA regulations and enforcement policy.

In the past decade the FTC and FDA are shifting to collaborate rather than compete. Not only will the agencies refer cases but also pool resources in reviewing websites. There are now numerous examples where the FDA names a joint effort with the FTC in warning letters. These joint letters, signed by both the FDA and FTC, cite both sets of regulations. The continued cooperation suggests a greater number of websites will be reviewed and enforced than if one agency acted alone.

---

[20] Federal Trade Commission, Dietary Supplements: An Advertising Guide for Industry, available at: http://www. business.ftc.gov/documents/bus09-dietary-supplements-advertising-guide-industry.

Vitalmax Vitamins 2/11/13

UNITED STATES OF AMERICA
FEDERAL TRADE COMMISSION
BUREAU OF CONSUMER PROTECTION
WASHINGTON, D.C. 20580

DEPARTMENT OF HEALTH
AND HUMAN SERVICES
FOOD AND DRUG ADMINISTRATION
SILVER SPRING, MD 20993

TO: Alan Serinsky
info@vitalmaxvitamins.com
www.vitalmaxvitamins.com

FROM: The Food and Drug Administration and the Federal Trade Commission

RE: Unapproved Products Related to the 2012/2013 Flu Season; and Notice of Potential Illegal Marketing of
Products to Prevent, Treat or Cure Flu Virus

DATE: February 11, 2013

WARNING LETTER

This is to advise you that the United States Food and Drug Administration ("FDA") and the United States
Federal Trade Commission ("FTC") reviewed your websites at the Internet addresses
www.vitalmaxvitamins.com and www.healthyanswers.com in January 2013. The FDA has determined that
your website www.vitalmaxvitamins.com offers a product for sale that is intended to diagnose, mitigate,
prevent, treat or cure the Flu Virus in people. This product has not been approved or cleared by FDA for use
in the diagnosis, mitigation, prevention, treatment, or cure of the Flu Virus. This product is called BodyGuard
and is labeled as a dietary supplement. The marketing of this product violates the Federal Food, Drug, and
Cosmetic Act (FFDC Act). 21 U.S.C. §§ 331, 352. We request that you immediately cease marketing
unapproved and uncleared products for the diagnosis, mitigation, prevention, treatment, or cure of the Flu
Virus.

Some examples of the claims on your website www.vitalmaxvitamins.com that evidence that the product you
offer is intended for the use(s) described above include:

On the webpage titled, "Body Guard":

- "Boost Your Immune System and Fight Cold and Flu - Naturally!"

On the webpage titled, "Why In the World Are You STILL Catching Cold and Flu
Bugs?" (http://www.vitalmaxvitamins.com/lp/2012/bodyguard/?utm_source=prweb&utm_medium=BGLP1PR
#order2), which offers the Body Guard product for sale, we note the following claims:

- "Why in the World Are You STILL Catching Cold and Flu Bugs?

Just recently I was … astounded by the amount of wheezing…sniffling…and coughing that I had to endure
during my airplane flight! One year ago I would have walked off that plane certain that within days I would
succumb to some sort of virus or flu. But today, it's a whole new story.

Now available to the general public is a revolutionary immune booster … This all-natural remedy contains a combination of nutrients so potent, it can help shield you from a variety of bacteria, germs and viruses that would normally put you out of commission.

This miraculous remedy is called BodyGuard™ and it happens to be the single most powerful formula to help guard your health! … This unique formula sends germs packing in a way that antibiotics and drugs never could.

In fact, the healing nutrients in these tiny tablets [BodyGuard™] can help you:
o   Fight colds, infections and respiratory problems.
o   Safeguard you from deadly flu viruses.
o   Relieve you from stuffy noses, chills, hoarseness and other cold symptoms!"

● "BodyGuard™ also contains nutrients such as … Pau d'Arco extract. This South American rainforest herb has superior antiviral properties that virtually wipe out cold and flu bugs! What's more, BodyGuard™ even contains Elderberry extract—a nutrient proven to help clobber most flu and viruses.

And scientific studies prove it's true:

One study published in the winter 1995 issue of the *Journal of Alternative and Complementary Medicine* showed that 93% of influenza-B flu patients given elderberry extract in supplement form were almost completely symptom-free within two days!

In a second study published in the International Journal of Medical Research, 90% of the study participants had influenza A. Patients who took this extract recovered in an average of 3 days, versus 7 days for the placebo group!"

● "All you need to do is simply take a couple capsules of BodyGuard™ everyday and you can freely travel anywhere without thinking twice about what germ…bacteria…virus…or flu that might invade your body! I'm so sure you'll be spending fewer days in bed and more days playing golf or cards that I'm committed to offering you this 100% money-back risk-free guarantee!"
● " [F]or just pennies per day, you'll protect yourself from nasty germs, bacteria and viruses that can … leave you unprotected against deadly viruses and flu or worse!"

Your products are not generally recognized as safe and effective for the above referenced uses and, therefore, these products are "new drugs" under section 201(p) of the Act [21 U.S.C. § 321(p)]. New drugs may not be legally marketed in the U.S. without prior approval from FDA as described in section 505(a) of the Act [21 U.S.C. § 355(a)]. FDA approves a new drug on the basis of scientific data submitted by a drug sponsor to demonstrate that the drug is safe and effective.

Furthermore, your product is offered for conditions that are not amenable to self-diagnosis and treatment by individuals who are not medical practitioners; therefore, adequate directions for use cannot be written so that a layperson can use these drugs safely for their intended purposes. Thus, your product is misbranded within the meaning of section 502(f)(1) of the Act [21 U.S.C. § 352(f)(1)], in that the labeling fails to bear adequate directions for use. The introduction of a misbranded drug into interstate commerce is a violation of section 301(a) of the Act [21 U.S.C. § 331(a)].

The marketing and sale of unapproved or uncleared Flu Virus-related products is a potentially significant threat to the public health. Therefore, FDA is taking urgent measures to protect consumers from products that, without approval or clearance by FDA, claim to diagnose, mitigate, prevent, treat or cure Flu Virus in people.

You should take immediate action to ensure that your firm is not distributing, and does not distribute in the future, products intended to diagnose, mitigate, prevent, treat or cure the Flu Virus that have not been approved or cleared by the FDA. The above is not meant to be an all-inclusive list of violations. It is your responsibility to ensure that the products you market are in compliance with the FFDC Act and FDA's implementing regulations. We advise you to review the products you distribute, including the claims made for those products in websites, product labels, and other labeling and promotional materials, to ensure that the products you distribute are not intended for uses that render them misbranded in violation of the FFDC Act. 21 U.S.C. §§ 331, 352. Within 15 days, please send an email to FDAFLUTASKFORCECFSAN@fda.hhs.gov, describing the actions that you have taken or plan to take to address your firm's violations. **If your firm fails to take corrective action immediately, FDA may take enforcement action, such as seizure or injunction for violations of the FFDC Act, without further notice.** Firms that fail to take corrective action may also be referred to FDA's

Office of Criminal Investigations for possible criminal prosecution for violations of the FFDC Act and other federal laws.

If you are not located in the United States, please note that unapproved and uncleared products intended to diagnose, mitigate, prevent, treat, or cure the Flu Virus offered for importation into the United States are subject to detention and refusal of admission. We will advise the appropriate regulatory or law enforcement officials in the country from which you operate that FDA considers your product listed above to be an unapproved or uncleared product that cannot be legally sold to consumers in the United States.

Please direct any inquiries concerning this letter to FDA at FDAFLUTASKFORCECFSAN@fda.hhs.gov or by contacting Katrina L. Dobbs at 240-402-5163.

In addition, it is unlawful under the FTC Act, 15 U.S.C. § 41 et seq., to advertise that a product can prevent, treat, or cure human disease unless you possess competent and reliable scientific evidence, including, when appropriate, well-controlled human clinical studies, substantiating that the claims are true at the time they are made. See *FTC v. Direct Mktg. Concepts*, 569 F. Supp. 2d 285, 300, 303 (D. Mass. 2008), *aff'd*, 624 F.3d 1 (1st Cir. 2010); *FTC v. Nat'l Urological Group, Inc.*, 645 F. Supp. 2d 1167, 1190, 1202 (N.D. Ga. 2008), *aff'd*, 356 Fed. Appx. 358 (11th Cir. 2009); *FTC v. Natural Solution, Inc.*, No. CV 06-6112-JFW, 2007-2 Trade Cas. (CCH) P75,866, 2007 U.S. Dist. LEXIS 60783, at * 11-12 (C.D. Cal. Aug. 7, 2007). More generally, to make or exaggerate such claims, whether directly or indirectly, through the use of a product name, website name, metatags, consumer testimonials, or other means, without rigorous scientific evidence sufficient to substantiate the claims, violates the FTC Act. See *In re Daniel Chapter One*, No. 9239, slip op. 18-20, 2009 WL 516000 (F.T.C.), 17-19 (Dec. 24, 2009) (http://www.ftc.gov/os/adjpro/d9329/091224commissionopinion.pdf1), *pet. for review den.*, 2010 WL 5108600 (D.C. Cir. Dec. 10, 2010).

The FTC strongly urges you to review all claims for your products and ensure that those claims are supported by competent and reliable scientific evidence. Violations of the FTC Act may result in legal action seeking a Federal District Court injunction or Administrative Cease and Desist Order. An order also may require that you pay back money to consumers. Please notify FTC via electronic mail at healthproducts@ftc.gov, within fifteen (15) working days of receipt of this letter, of the specific actions you have taken to address FTC's concerns. **If you have any questions regarding compliance with the FTC Act, please contact Richard Cleland at 202-326-3088.**

Very truly yours,

/S/
Mary K. Engle
Associate Director
Division of Advertising Practices
Federal Trade Commission

/S/
Michael W. Roosevelt
Acting Director
Office of Compliance
Center for Food Safety
   and Applied Nutrition

5.2.3   Free Speech and Labeling Claims

As the FDA became increasingly involved in approving claims constitutional questions arose. Challengers asserted First Amendment protection when the FDA banned claims, revised disclaimers to health claims, or deemed the claims unsubstantiated. The claims in question were often health claims or qualified as health claims. As discussed in Chapter 4 commercial speech is only protected if it is not misleading. If a claim

is not misleading, then it is afforded protection under the four-part test in *Central Hudson Gas*, which in its analysis asks whether the government interest in the regulation is substantial and if the regulation is narrowly drawn.

Curing Potentially Misleading Speech with Disclaimers

A series of cases from the D.C. Circuit developed a new doctrine for determining whether claims were protected by the First Amendment. In a two-part case extending over two years the D.C. Circuit and district courts reviewed claim submissions by dietary supplement marketers Duke Pearson and Sandy Shaw (*Pearson I*).[21] Pearson sought FDA approval for four health claims, but none were approved. Person sought approval for claims stating:

1. "Consumption of antioxidant vitamins may reduce the risk of certain kinds of cancers."
2. "Consumption of fiber may reduce the risk of colorectal cancer."
3. "Consumption of omega-3 fatty acids may reduce the risk of coronary heart disease."
4. "8 mg of folic acid in a dietary supplement is more effective in reducing the risk of neural tube defects than a lower amount in foods in common form" *(Person I)*.[22]

In order to make the claims Pearson needed to pass the "significant scientific agreement standard." The FDA concluded there was adequate supporting evidence but that it was "inconclusive" for one reason or another and thus failed to give rise to "significant scientific agreement" (*Pearson I*).[23] The FDA refused to consider adding a disclaimer stating, "The FDA has determined that the evidence supporting this claim is inconclusive" (*Pearson I*).[24] In response to the First Amendment challenge the FDA argued in court that claims failing to meet the SSA standard were inherently misleading and not afforded constitutional protection (*Pearson I*).[25] It also asserted it was not obligated under *Central Hudson Gas* to consider disclaimers over a ban for potentially misleading claims (*Pearson I*).[26]

The court in *Pearson I* articulated a preference for disclosure over suppression of speech. Using the *Central Hudson Gas* test the court found there was no "reasonable fit" between the FDA's goal of protecting the public health and preventing consumer fraud by banning a claim

without consideration of adding a disclaimer (*Pearson I*).[27] The court reasoned the commercial speech doctrine expressed a "preference for disclosure over outright suppression" of speech, including disclaimers (*Pearson I*).[28]

The court did not rule out bans but considered suppression appropriate in limited circumstances. It stated the FDA could only ban a claim "where evidence in support of a claim is outweighed by evidence against the claim" making it incurable by disclaimer (*Pearson I*).[29] The evidence supporting the claim must be qualitatively superior to the evidence against it to justify a ban. The FDA could not provide such an explanation for banning the four claims and the court ordered the FDA to reconsider Pearson's claims.

The FDA unsuccessfully attempted to correct the basis for denying the claims. Following the decision in *Pearson I* the FDA continued to refuse to authorize Pearson's claims about folic acid. It dubbed the claims inherently misleading and issued a final rule prohibiting any claims about the benefits of folic acid only later to revise the rule to allow four model claims (*Pearson II*).[30] The district court applied the guidelines from *Pearson I* and found folic acid claims were not inherently misleading. It reasoned the "mere absence of significant affirmative evidence in support of a particular claim … does not translate into negative evidence 'against' it" (*Pearson II*).[31] The court continued in the vein of *Pearson I* holding a claim could not be "absolutely prohibited" if there was any "credible evidence" in support of the proposed claim. The court also noted the harm to Pearson outweighed any potential harm to the consumer, stating:

> At worst, any deception resulting from Plaintiffs' health claim will result simply in consumers spending money on a product that they might not otherwise have purchased, or perhaps spending more money on a product with a higher folic

---

[21] *Person v. Shalala*, 164 F.3d 650 (D.C. Circ. 1999).

[22] *Id.* at 653.

[23] *Id.* at 653–654.

[24] *Id.* at 654.

[25] *Id.* at 655.

[26] *Id.*

[27] *Id.* at 656–658.

[28] *Id.*

[29] *Id.* at 659.

[30] *Pearson v. Shalala*, 130 F. Supp. 2d 105, 108–109 (D.D.C. 2001).

[31] *Id.* at 115.

Fig. 5.1 *Pearson I* and *II* Analysis

acid content. This type of injury, while not insignificant, cannot compare to the harm resulting from the unlawful suppression of speech (*Pearson II*)[32]

Therefore under *Pearson I* and *II* the FDA cannot prohibit a claim where credible evidence exists to support it. A claim may only be banned where the qualitative evidence against the claim outweighs the qualitative evidence in support of it (see Fig. 5.1).

Latera, D.C. district court would suggest the FDA's ability to prohibit nutrient claims applied in rare cases. The plaintiffs in *Pearson* filed a second lawsuit following the FDA's decision not to authorize an antioxidant claim for a saw palmetto product. The FDA, aware of the *Pearson* framework, concluded the "weight of the scientific evidence against the relationship" between antioxidant's effect on cancer "was greater than the weight of the evidence in favor

of the relationship" (Whitaker 2002).[33] Such an analysis and conclusion the FDA hoped would fit squarely into the *Pearson I* framework. The district court, however, found the FDA failed to carry its burden under *Central Hudson Gas* by showing the "suppression of [the antioxidant vitamin claim was] the least restrictive means of protecting consumers against the potential of being misled…" (Whitaker 2002).[34] The court reasoned that the *Pearson I* framework only prohibited a claim in "very narrow circumstances" typically where there is "no qualitative evidence in support of the claim" or where the FDA can provide evidence that the public would still be misled even with a disclaimer (Whitaker 2002).[35]

Whitaker made clear the First Amendment only allows the FDA to prohibit claims in exceptional circumstances and prefers disclosure over suppression. The lesson is clear for manufacturers

---

[32] *Id.* at 119.

[33] 248 F.Supp. 2d 1, 7 (D.D.C. 2002).

[34] *Id.* at 8.

[35] *Id.* at 11.

seeking approval of claims or defending claims in enforcement actions: if there is no substantiation for the claim the FDA can act to prohibit the claim. The quality of evidence for claims remains a crucial and oft overlooked element of dietary supplement compliance.

Free Speech in the Revision of Disclaimers
In another set of twin cases the D.C. district court reviewed the FDA's efforts following *Pearson* to revise disclaimers rather than ban claims. The cases involved the Alliance for Natural Health, the same plaintiff's from *Pearson*, and the Coalition to End FDA and FTC Censorship. The plaintiffs' submitted a petition for qualified health claims "concerning the purported relationship between selenium and cancer" (*Alliance I*).[36] The agency denied seven of the ten claims, concluding under the *Pearson II* framework that no credible evidence exists to support the claims. Three of the claims were modified and the agency stated it would exercise enforcement discretion because NLEA only authorized the FDA to approve health claims meeting the SSA standard (*Alliance I*).[37]

The FDA entirely replaced one of the plaintiff's proposed claims. One claim proposed by the plaintiffs' read, "Selenium may reduce the risk of prostate cancer. Scientific evidence supporting this claim is convincing but not yet conclusive" (*Alliance I*).[38] The FDA rejected the claim, finding it "convincing but not yet conclusive," but rather than ban the claim it rewrote the claim to a level it stated allowed the agency to exercise enforcement discretion (*Alliance I*).[39] The FDA's claim read, "Two weak studies suggest that selenium intake may reduce the risk of prostate cancer. However, four stronger studies and three weak studies showed no reduction in risk. Based on these studies, FDA concludes that it is highly unlikely that selenium supplements

reduce the risk of prostate cancer" (*Alliance I*).[40] The FDA replaced a direct succinct claim with a confusing claim roughly double in length.

The court found the FDA's decision to entirely replace a claim with model language violated the commercial speech doctrine. It lashed out at the agency stating it "eviscerated [the] plaintiff's claim, with no explanation as to why a less restrictive approach would not be effective" (*Alliance I*).[41] It held the FDA's revision of the claim violated the reasonable requirement of *Central Hudson Gas* while, simultaneously ignoring the First Amendment's preference for disclosure over suppression. The FDA, it stated drafted a disclaimer that "contradicts the claim and defeats the purpose of making it in the first place" (*Alliance I*). The court directed the FDA to draft a new disclaimer that was "short, succinct, and accurate" (*Alliance I*).

In *Alliance II* the same plaintiffs argued an identical challenge to the FDA's efforts to reword and replace two of its qualified health claims. As with *Alliance I*, the court held that the FDA's action to replace and reword the claims failed for the same reasons as its prior attempt did in *Alliance I* (*Alliance II*).[42] Namely, the FDA's action could not require a disclaimer that "simply swallows the claim" (*Alliance II*).[43]

Alliance does not stand as a prohibition against the FDA's authority to revise or replace disclaimers. It holds the FDA cannot make an end-run around the decision in *Pearson* by requiring a disclaimer that effectively bans the claim. The FDA may revise disclaimers on a narrow basis that focuses on the strength of the evidence. If the revision "swallows," "contradicts," or "defeats the purpose of the claim" then it has not narrowly tailored its action to comply with the commercial speech doctrine (see Fig. 5.2).

Defending claims presents a daunting challenge. It requires a careful balance of scientific

---

[36] *Alliance for Natural Health US v. Sebelius*, 714 F. Supp. 2d 48, 57 (D.D.C. 2010).

[37] *Id.* at 57–58.

[38] *Id.* at Fn. 16.

[39] *Id.* at 71.

[40] *Id.*

[41] *Id.*

[42] *Alliance for Natural Health US v. Sebelius*, 786 F.Supp2d 1, 24 (D.D.C. 2011).

[43] *Id.*

Fig. 5.2 *Alliance I and II* Analysis

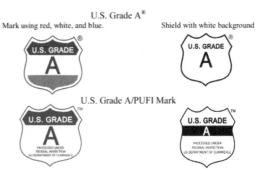

Fig. 5.3 NOAA Marks of Inspection

support and legal arguments. When scientific support is lacking the legal arguments become challenging. Although the focus was on health claims, this analysis is the same for other types of claims.

5.3 Other Areas of Specialized Regulations

Dietary supplements, albeit the largest and likely most complex, is not the only area of specialized enforcement. Other areas include seafood, juice, eggs, water and ice, milk, and FSMA. FSMA is often product-specific in its rules and is otherwise difficult to fit into the discussion of other chapters. To ensure the full FSMA landscape is covered an overview will be provided here.

5.3.1 Seafood and Juice HACCP

Seafood and juice are subject to additional regulations due to safety concerns. The FDA convened several committees and consulted with Congress about the risks posed by seafood and juice products. The FDA found that a history of repeated instances of serious public health issues related to juice and seafood warranted additional preventative controls (GAO 2008). In the case

of juice the FDA found a number of health risks from unpasteurized juice. In both cases the FDA promulgated new regulations to require hazard analysis and critical control point (HACCP) as a preventative control to minimize health hazards associated with juice and seafood. The FDA released the seafood HACCP regulation in 1995 and proposed a rule requiring HACCP for juice products in 1998, which was finalized in 2001.

Seafood and juice products are the only FDA product categories subjected to mandatory HACCP. For all other product categories the use of HACCP remains optional.

Seafood facilities may also elect to participate in a voluntary inspection program operated by the National Oceanic and Atmospheric Administration (NOAA). NOAA operates a seafood inspection program under the 1946 Agricultural Marketing Act. It inspects fish, shellfish, and fishery products (see, Fig. 5.3).

5.3.2 Eggs

Eggs are unique in that two different agencies control different aspects of the production and use. The USDA governs the laying facilities and grading. It also enforces production facilities of eggs products, liquid, frozen, and dehydrated, under the Egg Products Inspection Act (EPI). The FDA also plays a role. It is responsible for the laying hens' diet and for shelled eggs. When an outbreak linked to shelled eggs occurs it can cause quite a bit of confusion. Is the USDA responsible for hazards present in the laying facility, like manure, or is the FDA responsible since it regulates shelled eggs entering the market? Understanding the overlap and interplay between the agencies makes eggs an area of specialized regulation.

5.3.3   Water and Ice

Many are surprised to find that the FDA regulates bottled water. The FDA and EPA share responsibility for the safety of drinking water. The EPA regulates public drinking water and the FDA regulates bottled drinking water. Bottled water includes artesian, mineral, sparkling, and other bottled waters processed, purified, or distilled. The FDA regulates waters with added carbonation, soda water, tonic water, and seltzer as soft drinks. Flavored water needs to be carefully analyzed whether the added substance changes the classification, along with general considerations for nutrient labeling, ingredient listing, and the use of approved added substances.

Bottled water follows a unique set of GMPs. The GMPs are codified in 21 CFR Part 129. As with all FDA regulated facilities bottled water production plants are subject to inspection. The FDA is particularly concerned with water source and potential contamination.

Packaged ice is also regulated by the FDA. Unlike bottled water ice is subject to the GMP requirements of Part 110 conventional for human food rather than Part 129. There are also labeling requirements and exemptions for small producers.

5.3.4   Milk

Milk products sold across State lines are subject to FDA regulation. This would seem a rather straight forward category of regulation, but recent interests in raw milk raises serious health questions. The FDA requires pasteurization for all milk sold in interstate commerce. It also strictly prohibits the sale of raw or unpasteurized milk across State lines. Some States do allow raw milk sales, but only within the State. Raw milk is considered a public health hazard and linked to several recent outbreaks.

5.3.5   Food Safety Modernization Act (FSMA) Overview

The Food Safety Modernization Act is the first substantive change to food safety legislation since 1938. It is an amendment to the Food Drug and Cosmetic Act. Based on the discussion in Chapter 1 this already informs the savvy reader that the Amendment will not impact the USDA. The USDA continues to operate under its 1950s updates, while the FDA has been given a modern approach to regulating foods.

In addition to enhanced enforcement authority discussed in Chapter 3 FSMA required seven substantive rules. The seven rules represent the largest changes but are among a host of smaller changes, such as high-risk facility designation, tracing pilot program, grocery store recall notifications, and inspection frequency. A discussion of FSMA could truly fill every page of this textbook. This is what happens when changes are pentup for over 70 years! A brief overview of the seven rules is provided below. The rules often are over a 100 pages and drenched in detail. This overview should not be seen as an exhaustive of the rule's provisions.

- Produce Safety Rule: The FDA issued regulations as required under FSMA to regulate growing, harvesting, packing, and holding produce. This is the first time in U.S. history that these activities will be subject to regulation. The rule introduces new standards in key areas, including water, health, hygiene, equipment, facilities, and training. It is startling to consider that until the rule is in effect there is no requirement for restroom facilities or hand washing for personnel harvesting produce.
- Human Food Preventive Controls: This rule was discussed in Chapter 3. There it was described as requiring a HACCP like food safety plan and an update to food GMPs.
- Preventive Controls for Animal Feed: In a companion rule of sorts for the Human Food Preventative Control rule, animal

feed such as pet food is subject to regulation for the first time in history. This rule is similar to the human rule but lacks consideration for allergens, for example which are unique to humans. It also requires GMPs for facilities, which will be another first for the industry.

- Foreign Supplier Verification Program (FSVP): The foreign supplier verification rule will dramatically impact importers. Importers will need to assess risks associated with their products and take steps to verify the risks are controlled. This could include on-site inspections of the foreign manufacturer.
- Accreditation of Third-Party Auditors: This rule works in concert with the FSVP by allowing importers to rely on third-party audits. Those audits will only be recognized if from auditors or auditing agencies accredited under the new FDA rule.
- Intentional Adulteration of Food: This rule was also a discussion topic in Chapter 3. It focuses on food defense plans to mitigate the risks of intentional adulteration. A difficult rule to draft and enforce.
- Sanitary Transportation of Human and Animal Food: The FDA will begin inspecting and verifying shippers, carriers, receivers, and others transporting food comply with sanitation practices.

FSMA is a massive change girded by new detailed records requirements. All of the activities under FSMA, be it the seven substantive rules, or other provisions must be properly documented.

5.4 Comparative Law

The EU provides an analogous system compared to the US regulation of dietary supplements. The EFSA refers to dietary supplements as food supplements. The main regulation is found in a 2002 EU directive, Directive 2002/46/EC, which contains a list of permitted vitamin and mineral components for specific intended uses. The EU adopted regulations in 2006 for the

use of nutrition and health claims (see Regulation 1924/2006). Similar to the U.S. claims are subject to verification for substantiation and approval.

The EU system created a stir when it provided brief approval for a food supplement as a medical device. The approval albeit brief raised alarm at how the approval could be given by E.U. officials. It began when the manufacturer of a cranberry supplement gained Class IIb medical device status from an EU Notifying Body (Sterling 2014).[44] This allowed the manufacturer to make claims about treating urinary tract infections, a claim prohibited for functional foods or supplements. There are risks of a permissive system, including expansive approval of claims. It is also interesting to consider supplements serving device-like functions. Supplements are more commonly thought of as having drug-like functions.

Japan offers a modern example of heavy regulation of dietary supplements. It is often described as the most restrictive regulator of dietary supplements. Under the Japanese system all supplements are treated as foods. Supplements then are placed in one of four categories: Food for Special Dietary Uses, Food For Specified Health Uses, Health Foods, and Health Foods with Nutrient Function Claims (see, Fig. 5.4) (Japan Consumer Affairs Agency).[45]

The Japanese system is complex with few claims allowed. Structure/function claims are not permissible. Some health claims can be made but only for certain supplements. The health claims also must closely follow model claims to avoid regulation as drugs. With an aging population demanding more health care a call for change is growing. Some view supplements as a means both for preventative care and for low-cost functional ingredients. Yet, in doing so, it would seem a tacit admission that dietary supplements are quasi-drugs.

---

[44] Shane (April 2014).

[45] Consumer Affairs Agency, Food Labelling Division, Regulatory Systems of Health Claims in Japan (June 2011).

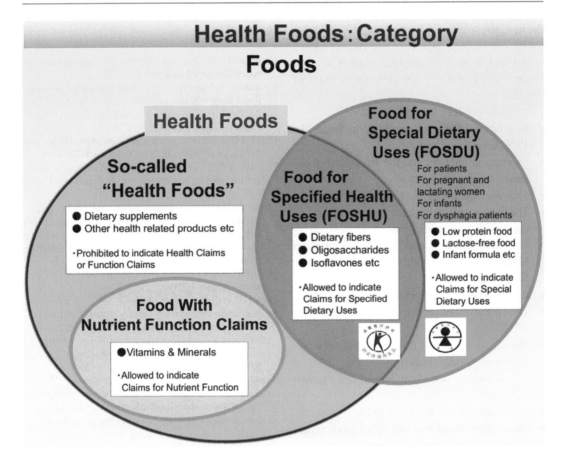

Fig. 5.4 Overview of Japanese supplement classification system (Consumer Affairs Agency/Food Labeling Division)

Dietary supplements are not an easy category to define or regulate. Some countries adopt a permissive attitude to allow a wide range of claims and benefits to be made. There, the onus is principally on consumers to evaluate products and determine which claims are accurate and appropriate for their use. Other countries adopt a restrictive view-finding supplements on the current market as open to abuse. In this model claims are restricted and identities of supplements standardized. In either system the concern is about safety and fraud. Fraud, perhaps more than any other issue, drives some countries to adopt a restrictive approach. Consumer fraud is a valid concern, especially as regulators struggle to define dietary supplements which often contain food based ingredients, but offer drug-like benefits.

5.5   Chapter Summary

This chapter focused heavily on dietary supplements. Through the legislative history a clear picture of the enforcement landscape that lead to DSHEA can be seen. It not only provides a sense of why dietary supplements are regulated, but an appreciation of the uniformity and consistency brought by DSHEA. Under the current system the definition of dietary supplement was explored, including examples of applications that fall outside the regulation of supplements. Labeling, claims, and new dietary ingredient notifications were also reviewed. The discussion of dietary supplements closed with a review of the constitutional issues raised by the FDA's regulation of claims and disclosures. In that discussion the need for substantiation and credible evidence for claims became apparent. The chapter closed with a review of other specialized areas of regulation, including the seven substantive rules of FSMA.

Overview of Key Points

- Regulation of dietary supplements under the 1906 Act
- Changes in industry and the struggle to regulate under the 1938 Act
- The role of the Nutrilite Consent Decree
- The FDA's aggressive enforcement policy and the corresponding litigation
- Early weak amendments, like the Vitamin and Mineral Amendment, that reacted to an outcry from industry
- Regulation of claims prior to NELA
- The benefits of making claims following NELA
- The *Pearson I* and *II* frameworks for reviewing constitutional challenges to claims
- The *Alliance I* and *II* review of the Pearson framework
- Seven substantive rules of FSMA
- Discussion of how and why other products are subject to specialized regulations.

5.6   Discussion Questions

1. The USDA/FSIS was omitted from the discussion of dietary supplements. Explain why FSIS should or should not be included in regulating supplements.
2. Identify a claim made on a dietary supplement product (vitamin, energy drink, etc.)? Begin by determining if the claim could be used to reclassify the product, and then look to determine how the claim is substantiated. Based on your analysis is the claim subject to FDA enforcement? Could it withstand scrutiny under *Central Hudson Gas, Pearson I and II, and Alliance I and II.*

References

Crawford CW (June 1951) Statement at his induction as commissioner of food and drugs

Consumer Affairs Agency Food Labelling Division, Regulatory Systems of Health Claims in Japan (June 2011) http://www.caa.go.jp/en/pdf/syokuhin338.pdf. Accessed 27 Aug 2014

Hutt PB (1984) Government regulation of the integrity of the food supply, 4 Ann. Rev. Nutrition 1–20

Hutt PB (1986) Government regulation of health claims in food labeling and advertising, 41 Food Drug Cosm. L.J., 3–73

FDA, Guidance for Industry, Distinguishing Liquid Dietary Supplements from Beverages (January 2014) http://www.fda.gov/food/guidanceregulation/guidancedocumentsregulatoryinformation/dietarysupplements/ucm381189.htm. Accessed 27 Aug 2014

FDA Regulatory Letter to ITT Continental Baking Company (October 1976)

FDA TC2-A (5 November 1945)

Federal Trade Commission Dietary (April 2001) supplements: an advertising guide for industry. http://www.business.ftc.gov/documents/bus09-dietary-supplements-advertising-guide-industry. Accessed 27 Aug 2014

Shane S (April 2014) Nutra ingredients, "French cranberry medical device status stripped". http://www.nutraingredients.com/Regulation/French-cranberry-medical-device-status-stripped

Food Additives

6

Abstract

This chapter grounds itself in a detailed discussion of the regulation of food additives. It explores the legislative history that animates and the current regulations as well as the practical aspects of GRAS and food additive petitions. It begins by defining food additives and the drive to ease the FDA's burden regulating added substances under the adulteration provisions of the Act. The chapter then explores key concepts of direct and indirect additives, GRAS, irradiation, interim food additives, and food additive petitions. It also includes a comparison of food additive regulation to color additive and new dietary ingredient regulation.

6.1 Introduction

Foods are a unique category of regulation in terms of pre market approval. In other areas of FDA regulation, such as medical devices and drugs, products cannot enter the market without some form of pre approval. Low risk medical devices still must submit a petition for FDA approval known as a 510(k). Higher risk devices typically must undergo lengthy clinical trials. The same is true for drugs, which in the majority of cases pass through a series of clinical trials to gain approval. Only in limited cases can a device or over-the-counter drug enter the market by simply following a regulation, such as a drug monograph. Food, at least as far as the term applies to FDA regulation, can be freely sold with little pre market oversight.

The Food Additives Amendment provided a response to the industrialization of food in the 1940s and 1950s. From 1906 to 1938, the food industry grew immensely in its transition from farm to factory. Following World War II, the manufacturers aimed to extend shelf-life, enhance taste, or improve natural products to gain market share. The 20 years between the 1938 Act and the Food Additives Amendment witnessed a dramatic increase in technology and use of intentional added substances. Examples of additives embroiled in controversy grew from nitrate-treated bacon to Alar on apples (alar was used to keep apples from falling off the trees before ripening). The FDA was ill equipped to carry out its mission of protecting the public from such harm. Its model of enforcing after-the-fact could not keep pace with the changes and created uncertainty in the industry.

Congress began an investigation into the food industry and what if any legislative changes could provide greater control. Estimates at the time found over 700 chemicals were intentionally used in food products, but only 428 were known to be safe. Many estimates placed the number of chemicals added to food much higher. The FDA

M. C. Sanchez, *Food Law and Regulation for Non-Lawyers,* Food Science Text Series,
DOI 10.1007/978-3-319-12472-8_6, © Springer International Publishing Switzerland 2015

could not provide an accurate estimate because it often only learned of added substances through widespread adverse events. This led industry to also push Congress for change. It sought change in order to avoid after-the-fact enforcement which invited negative press and potential liability in law suits.

Congress passed a series of amendments aimed at controlling the use of added substances including the Food Additives Amendment. The Amendment was originally codified as Section 409 but is currently found in Section 348. As will be discussed below the Amendment applies to a wide range of food substances, including food packaging materials. Classification and definitions again play a paramount role in assessing the procedures and approval criteria of the Amendment.

The Food Additives Amendment is rare because lacking a singular crisis the drafters required eight years to complete their work. Congress began hearings in 1950 and passed the Amendment in 1958. As it approaches its 60th anniversary many are beginning to debate whether the Food Additives Amendment continues to play a constructive role. As will be discussed in the sections below many of the exemptions raise questions about the FDA's ability to ensure safety.

The Food Additives Amendment works by requiring pre market approval for added substances. Congress eased the evidentiary burden for the FDA on proving whether added substances were adulterated (see Chapter 3). Instead the burden of proof—expert evidence, studies, and data that when interpreted showed the added substance is safe—now rests with the maker of the added substance. In most cases the demonstration must be made before entering the market.

Congress believed the Amendment would not only make food safer but encourage innovation. Under the Amendment, the FDA was given greater flexibility in approving food additives. It could authorize limited uses or levels of additives shown to have toxicity at higher levels. Thus, rather than ban any added substance deemed to be poisonous or deleterious, the Amendment allowed some additives below their known injurious levels.

6.1.1   Defining Food Additive

The Food Additives Amendment added a new definition to the Act. Prior to 1958 the Act did not need and did not define the term "food additive." When determining adulteration under the Act, the Agency needed to determine if the substance was added or non added and met the applicable evidentiary standard. The food additive definition is found in Section 201(s).

Food Additive Definition in Section 201(s)
(s) The term "food additive" means any substance the intended use of which results or may reasonably be expected to result, directly or indirectly, in its becoming a component or otherwise affecting the characteristics of any food (including any substance intended for use in producing, manufacturing, packing, processing, preparing, treating, packaging, transporting, or holding food; and including any source of radiation intended for any such use), if such substance is not generally recognized, among experts qualified by scientific training and experience to evaluate its safety, as having been adequately shown through scientific procedures (or, in the case of a substance used in food prior to January 1, 1958, through either scientific procedures or experience based on common use in food) to be safe under the conditions of its intended use; except that such term does not include—
1. a pesticide chemical residue in or on a raw agricultural commodity or processed food; or
2. a pesticide chemical; or
3. a color additive; or
4. any substance used in accordance with a sanction or approval granted prior to the enactment of this paragraph 4 pursuant to this Act … the Poultry Products Inspection Act … or the Meat Inspection Act…;

5. a new animal drug; or
6. an ingredient described in paragraph (ff) in, or intended for use in, a dietary supplement.

Components of Food

The Food Additives Amendment did not repeal any part of the 1938 Act raising questions about how to interpret some original definitions of the Act. This text introduced the definition of food under Section 201(f) in Chapter 1. There the definition provided read in part "articles used for food or drink... [including] components of any such article". This latter part, "components of food" could be read as an early definition of food additives. Although the pre market approval system of the Food Additives Amendment is lacking, Congress clearly intended for "components" of food to be regulated as food. Other definitions existing before the Amendment include the use of term "substance" in the adulteration prohibition of 402. The Amendment does not expressly revise any part of the 1938 Act but effectively displaces certain definitions. This concept will be explored further in Section 2 below.

Indirect Additives

Congress provided an inclusive definition of food additive. The definition covers the landscape of both intentional direct additives which are functional to intentional indirect additives. The definition in 201(s) expressly brings substances that may indirectly become a component or "otherwise affect" the characteristics of the food. Packaging is an example of an indirect or incidental additive. Other indirect additives require an analysis of how and to what extent they "otherwise affect" the food product.

Dietary Supplements

Dietary supplements complicate the food additive analysis. There are two ways dietary supplements require careful attention when assessing the Food Additives Amendment. The first is the classical question of whether the product is a food or a dietary supplement. This question features prominently in the food additive analysis. It does so in part because under DSHEA food additive regulation does not apply to dietary supplements. This emphasizes the importance of distinguishing between food and dietary supplements. It also raises the question when substances used in both foods and dietary supplements are approved. Under DSHEA new dietary ingredients use a strikingly similar analysis to the exemption under the food additives amendment. Not only it is important to apply the correct criteria, but also not to assume exemption status for one applies to the other. A more nuanced issue looks at when a food additive becomes a food. This is seen in particular with food additives sold as single-ingredient dietary supplements, but the case law carries wider applications. The question becomes, if the ingredient is approved for use in my double-fudge brownie?, is it also approved when isolated and encapsulated as a dietary supplement?

6.1.2 Generally Recognized as Safe

The food additive definition provides a seemingly small exemption. The definition states the term "food additive" does not include a substance generally recognized to be safe under the conditions of its intended use. It provides qualifying criteria on what "safe" means for the exemption. This is an import feature of the food additive regulatory framework. It is also an exemption that has experienced dramatic growth and popularity. If the exemption does not apply, the food additive approval criteria must be followed (see Fig. 6.1).

6.1.3 USDA and Food Additives

The FDA remains the primary agency for regulating the use of food additives even in meat and poultry. The FDA and FSIS share responsibility for the safety of food additives used in meat, poultry, and egg products. Under the Food Additives Amendment, all the proposed additives are first evaluated by the FDA. A secondary review may be conducted by the Risk, Innovations and Management division of FSIS.

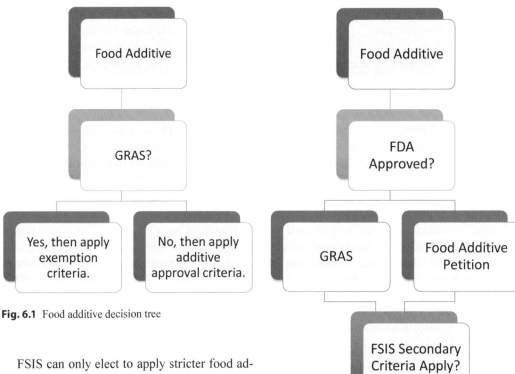

Fig. 6.1 Food additive decision tree

Fig. 6.2 Illustration of food additive decision tree for FSIS products

FSIS can only elect to apply stricter food additive standards than those adopted by the FDA. As an expert in meat, poultry, and egg products, FSIS believes it can make better determinations about the technical effects of additives proposed for FSIS-regulated products. Thus, it may elect to set higher standards than what the FDA sets for a food additive. When a higher standard is adopted, an additive must be approved first by the FDA then by FSIS. FSIS may never set a lower standard than the one set by the FDA (see Fig. 6.2). Chapter 3 provided an example of when FSIS asserted this authority.

lead to an enforcement action, including seizure of the product as adulterated. Wading through the wrong regulatory approval process, however, could squander value time and money. There is a balance to strike and an awareness to develop when utilizing any regulatory procedure.

6.2 Regulatory Framework

Understanding the regulatory framework of food additives provides an opportunity to understand when the Amendment requires pre market approval. As can be seen throughout this text food law involves as much time defining and outlining concepts and control mechanisms as it does describing how they work. In the case of pre market approval this is particularly important. Entering the market without the proper approval could

6.2.1 Components of Food

The Food Additives Amendment's pre market approval process applies to food additives. This preliminary question is as important as defining the term food in understanding the appropriate regulatory burden. As discussed in the introduction of this chapter, several definitions in the 1938 Act implied coverage of what the Amendment calls food additives. The definition of food used the term "components" of food and the adulteration prohibition applies to "substance[s]" added to food. The definitions are not repealed with applications outside of the Food Additives Amendment.

The Amendment effectively displaces the adulteration definition. When reading the prohibition against "added" substances carve outs are provided for approved food additives. This is reflected in the current codification of Section 402. The adulteration definition provides shelter to safe food additives, those approved under the Amendment. This is an added risk to unaware food additives, which are not in compliance.

Adulteration Prohibition in Section 342(a)

(a) Poisonous, insanitary, etc., ingredients
(1) If it bears or contains any poisonous or deleterious substance which may render it injurious to health; but in case the substance is not an added substance such food shall not be considered adulterated under this clause if the quantity of such substance in such food does not ordinarily render it injurious to health.1 (2)(A) if it bears or contains any added poisonous or added deleterious substance (**other than** a substance that is a pesticide chemical residue in or on a raw agricultural commodity or processed food, **a food additive**, a color additive, or a new animal drug) that is unsafe within the meaning of section 346 of this title; or (B) if it bears or contains a pesticide chemical residue that is unsafe within the meaning of section 346a(a) of this title; or (C) if it is or if it bears or contains (**i) any food additive that is unsafe within** the meaning of section 348 of this title; or

---

Defining food additive does not involve considering the substance's intended use. The definition provided in Section 201(s) signals a food additive is any "substance" that a manufacturer knows or should know, will become a component of the food, or "otherwise affect[t]" its characteristics. It is also clear from the 1950 Congressional hearing that the focus is centered on chemical additives. Consumers would consider food additives simply ingredients. Still the definition is broad enough to include commonly available food items, like tomatoes or salt in pasta sauce. A broad definition demands ample safe harbor and thus the use of GRAS.

Courts reviewing the definition find that it encompasses synthetic and natural substances added in small quantities no matter how incidental their effect is on the food. In *United States v. An Article Food, etc. Food Science Lab.* Food Science made various dietary supplements.[1] A line of supplements contained N, N-Dimethylglycine hydrochloride ("DMG"), which the FDA deemed adulterated (*Food Science Lab*).[2] Food Science argued DMG was a principal ingredient and outside what Congress intended to regulate with the Food Additives Amendment (*Food Science Lab*).[3] The court found Congress intended to draft a "very broad definition of food additive that could not be escaped by food purveyors claiming that particular substances were present in too small a quantity or had affected the food too indirectly" (*Food Science Lab*).[4] The court did not ban the use of DMG, but simply shifted the burden of proof to Food Science who would need to show the additive's safety or exemption from the pre market approval requirements of the Food Additives Amendment. The court, however, would not allow the products to simply escape classification as food additive.

Food additive is an inescapably expansive term. It encompasses a wide range of added substances. There is a temptation to quickly point to the exemptions and exclaim a product is not a food additive. Although there are exemptions, for example for foods in common use prior to 1958 (see Section 3 below), it is not accurate to describe those substances as non food additives. Taking the process step-by-step, the proper analysis first asks whether the substance is a food additive, then asks whether an exemption under

---

[1] 678 F.2d 735 (7th Cir. 1982).

[2] *Id.* at 735.

[3] *Id.* at 736.

[4] *Id.*

the Food Additives Amendment applies. It is a circuitous definition requiring a substance be classified as a food additive before being reclassified as outside the Amendment. Thus, to hastily skip-ahead to the exemption and claim the substance is not an additive begs the question, why look at the Food Additives Amendment at all?

6.2.2 Indirect Food Additives

Indirect additives prove more onerous in defining than direct additives. The food additive definition defines a food additive as a substance that indirectly affects the characteristic of a food. It provides examples of indirect additives, such as radiation, packaging, and transportation. The definition, however, leaves the more vital term "affects the characteristic" undefined. Intentionally broad, the definition leaves it to the discretion of the Agency and the wisdom of the courts to sort out when an additive indirectly affects the characteristics of food.

Early efforts by the FDA to regulate using this migration principle adopted a zero-tolerance policy. The FDA deemed any level of migration would render a substance a food additive. The statute did not set a threshold leaving the Agency broad discretion to set a high standard. As analytical methods improved, this policy became burdensome and difficult to enforce.

Early decisions focused on the likelihood of substances in packaging or other materials to migrate into the food product. For example, in *Natick Paperboard Corp. v. Weinberger,*[5] the First Circuit reviewed the PCB in paperboard packaging. It found the PCB as a food additive because evidence showed it would "migrate from paper packaging material to the contained food by a vapor phase phenomenon."[6] The court did not address at what levels the migrating additive must be found in the packaging court.

Migration requires more than a theory but evidence that it actually occurs. This was the hold-

ing in *Monsanto Co. v. Kennedy*[7] where the court reviewed the use of "acrylonitrile copolymer" to fabricate unbreakable beverage containers. The FDA conducted testing on the polymer in 1974. Following its testing, the FDA Commissioner held a formal hearing on the use of acrylonitrile copolymer for use in bottles. The FDA deemed the polymer a food additive and found it not to be safe for use (*Monsanto*).[8] Not long after issuing the regulations Monsanto challenged the decision in court. The court found the FDA relied solely on theory and it instead must be based on a "meaningful projection" from reliable data (*Monsanto*).[9]

Monsanto Co. v. Kennedy

This case arises on a petition for review of a Final Decision and Order of the Commissioner of Food and Drugs in which he ruled that a substance used to fabricate unbreakable beverage containers, acrylonitrile copolymer, is a "food additive" within the meaning of section 201(s) of the Federal Food, Drug, and Cosmetic Act (the Act). He further concluded that the data of record failed to provide the demonstration of safety established by section 409(c)(3) (A) of the Act as a precedent to FDA approval for use of any "food additive." The Commissioner's Final Order amended the pertinent FDA regulations to provide: "Acrylonitrile copolymers (of the type identified in the regulations) are not authorized to be used to fabricate beverage containers."

For the reasons set forth below, the decision of the Commissioner is affirmed in part, and in part is remanded to provide the opportunity for reconsideration.

I.

The FDA determination that acrylonitrile copolymers used in beverage contain-

---

[5] 525 F.2d 1103 (1st Cir. 1975).

[6] *Id.* at 1104.

[7] 613 F.2d 947 (D.C. Cir. 1979).

[8] *Id.* at 948.

[9] *Id.* 955.

ers are "food additives" within the statute is based on the finding that such containers invariably retain a residual level of acrylonitrile monomer that has failed to polymerize completely during the manufacturing process and that will migrate from the wall of the container into the beverage under the conditions of intended use. Although the administrative proceedings focused on beverage containers with a residual acrylonitrile monomer (RAN) level equal to or greater than 3.3 parts per million (ppm), the Commissioner made findings and conclusions applicable to all beverage containers manufactured with acrylonitrile, and the Final Order prohibited manufacture of such containers irrespective of their RAN levels.

FDA began to focus on acrylonitrile copolymer beverage containers in 1974, when the duPont Company submitted test results on a container fabricated from a somewhat different substance which alerted FDA to the possibility of significant migration from acrylonitrile containers. Subsequently, the Commissioner determined that, because of this putative migration, acrylonitrile copolymer was a "food additive" within the statute, and, on February 12, 1975, he published a regulation prescribing the conditions under which the chemical might be used safely in beverage containers: RAN levels in the wall of the container were limited to 80 parts per million (ppm), and acceptable migration of acrylonitrile monomer into the food was set at 300 ppb (parts per billion).

Two years later, FDA issued test results indicating that acrylonitrile caused adverse affects in laboratory animals. The Commissioner announced that he would lower the acceptable migration threshold for non-beverage containers to 50 ppb, and would withdraw approval entirely for acrylonitrile beverage containers, on the assumption that no such container could satisfy the 50 ppb migration limitation. Upon judicial review, this court held FDA's suspension of its food additive regulation without a hearing to be invalid. The court stayed the administrative action on March 18, 19778 and ordered that the required hearing be completed within 60 days... Subsequently, on a joint motion of the parties, the time limitation was extended by 120 days.

At the administrative hearing, petitioners introduced results from tests on a newly developed acrylonitrile beverage container having a RAN level of approximately 3.3 ppm. Tests on the container, employing a detection method sensitive to 10 ppb, detected no migration of acrylonitrile monomer. Nevertheless, the administrative law judge found that acrylonitrile copolymer was a "food additive," since migration had been detected from beverage containers composed of the same chemical compounds, though with higher RAN levels than those present in the "new" container. The Final Order prohibited manufacture of beverage containers containing acrylonitrile copolymer irrespective of their RAN levels.

II.

This case brings into court the second law of thermodynamics, which C. P. Snow used as a paradigm of technical information well understood by all scientists and practically no persons of the culture of humanism and letters. That law leads to a scientifically indisputable prediction that there will be Some migration of Any two substances which come in contact. The Commissioner's Final Decision, which upheld the ALJ's determination, is unclear on whether and to what extent reliance was placed on this "diffusion principle" rather than on a meaningful projection from reliable data. At one point in the Final Decision the Commissioner stated: "the migration of any amount of a substance is sufficient to make it a food additive" a passage evocative of the diffusion principle. Elsewhere, the Commissioner stated that he was able to make a finding of migration based on a projection from actual data on the assumption that a roughly linear relationship (as a function of time and temperature) existed between the RAN levels in a container and the concentration of acrylonitrile that would migrate into a test fluid. On this premise, though migration from the 3.3 ppm RAN container was itself below the threshold of detectability (10 ppb), it could be projected from the testing data obtained from containers with higher RAN levels.

This was a troublesome aspect of the case. As it was presented to us, the Commissioner had made a projection of migration from 3.3 ppm RAN containers without the support of any actual data showing that migration had occurred from such containers. One of petitioners' experts put it that the relationship might not be linear at very low RAN levels; but this was dismissed by the Commissioner as "speculative." One could not say that the expert's contention of no migration from very low RAN containers was improbable as a concept of physical chemistry, but it was put to us that the validity of this contention could neither be demonstrated nor refuted for 3.3 ppm RAN containers because, under the conditions of intended use, migration was projected to occur in amounts below the threshold of detectability.

Our own study showed the possibility of using experimental data to check the FDA's projection analysis. The FDA revealed that a projection of migration from low RAN containers had in fact been made for test conditions of prolonged duration and above-normal temperature. Under such conditions migration was projected in concentrations greater than 10 ppb, the threshold of detectability at the time of the Final Decision. Therefore, this court requested post-argument memoranda from the parties on whether tests had been per-

formed, or would be feasible, to confirm by actual data the hypothesis that migration occurs from containers with a RAN level of 3.3 ppm.

The responses to our inquiry have revealed the probable existence of data unavailable to counsel during the administrative proceedings that bear importantly upon the assumptions made by the Commissioner in reaching his findings and conclusions. This discovery buttressed our earlier conclusion that the Commissioner did not have sufficient support for his decision to apply the "food additive" definition in this case.

In light of the inadequacy of the agency's inquiry and in light of our view that the Commissioner has a greater measure of discretion in applying the statutory definitions of "food additive" than he appears to have thought, we remand this proceeding for further consideration.

III

The proceedings at hand are dramatic testimony to the rapid advance of scientific knowledge in our society. At the time of the administrative proceedings, the lowest concentration of acrylonitrile in a test fluid that could be detected with an acceptable degree of confidence was 10 ppb. There are now analytical techniques available that can detect acrylonitrile concentration of 0.1 ppb, an improvement of two orders of magnitude. Thus, on the issue of migration of acrylonitrile monomer it is now possible to generate "hard" data previously unobtainable.

In his post-argument testimony, Monsanto's expert claims, on the basis of such "hard" data, that the hypothesis which the Commissioner labeled as "speculative" may accurately describe the migration characteristics of containers with very low RAN levels, to wit, that in such containers the acrylonitrile monomer is so firmly affixed within the structure of the copolymer that no migration will occur under the conditions of intended use. If these assertions can be demonstrated to the satisfaction of the Commissioner, a modification of the current regulation is a likely corollary. The actual issuance of a regulation approving the production of a beverage container with an acceptable RAN level would presumably require both a container that had been developed and the appropriate petition. However, the Commissioner would have latitude to issue a statement of policy based upon the results of the proceeding or remand that would specify what in his review was an acceptable RAN level. This would serve a technology-forcing objective.

FDA opposes petitioners' post-argument motion for remand, asserting that the proffered new evidence will not affect the Commissioner's order insofar as that order precludes manufacture of beverage containers with RAN levels equal to or greater than 3.3 ppm the type of container already tested. FDA points out that the material submitted in response to this court's inquiry affirmatively supports the validity of the Commissioner's findings and conclusions. [20] FDA contends that a petition for modification of the regulation, or a similar procedure, would be the appropriate vehicle for presentation of any new evidence indicating that migration ceases when RAN levels fall below a certain threshold.

As a general rule, courts defer to administrative agency orders closing the record and terminating proceedings. The rule has applicability in cases involving scientific matters notwithstanding the possibility that advances and experiments will yield new material data. Indeed, the importance of finality as a matter of administrative necessity may be magnified by the possibility indeed probability of advance in at least some areas. Procedures for rehearing or modifying orders are generally available to

provide appropriate relief from any hardships or other harm.[21]

The general rule of finality applies in the usual case because the courts trust the administrator's ability to make a reasoned judgment that sufficient evidence has been submitted, that adequate time has been provided for rebuttal, and that the record should be closed. However, in this instance, the closing of the record did not reflect unfettered administrative judgment: FDA conducted these administrative proceedings under a time constraint dictated by an order of this court.

The Court is also concerned that the Commissioner may have reached his determination in the belief that he was constrained to apply the strictly literal terms of the statute irrespective of the public health and safety considerations. As we discuss below, there is latitude inherent in the statutory scheme to avoid literal application of the statutory definition of "food additive" in those De minimis situations that, in the informed judgment of the Commissioner, clearly present no public health or safety concerns.

In the usual case, the general doctrine of necessity and finality serves the public interest in immediate protection of the consuming public. But in this case production of acrylonitrile beverage containers was deferred voluntarily even when this court issued a stay of the FDA order, and in any event it is now prohibited pending further proceedings.

Finally, we are concerned that the record reflects a momentum toward a precipitate determination. Several factors bear on our judgment. One is the text of the decision, with its lack of precision as to basis. Another is the fact that the beverage container evaluated by the Commissioner was characterized by a migration level well below the agency's initial limit. It was offered by petitioners in the hearing as available as a result of ongoing technology, but the time constraint imposed by judicial mandate prevented the agency from scheduling the kind of administrative consideration that would ordinarily have been provided.

IV

Pretermitting various issues that should await conclusion of the remand proceedings, we turn to certain other important questions that are presented by the record, that have been fully briefed and argued, and that are ripe for resolution.

The statute requires a demonstration of safety precedent to FDA approval of any "food additive." The statutory definition of "food additive" which triggers that requirement contains a two part test. First, the Component element of the definition states that the intended use of the substance must be reasonably expected to result in its becoming a component of any food.[24] Second, the Safety element of the definition states that the substance must be not "Generally recognized (as) safe under the conditions of its intended use."

Petitioners are concerned that the Commissioner has determined, or will determine, that the component element of the definition may be satisfied solely by that application of the second law of thermodynamics called the diffusion principle: Any two substances that are in contact will tend to diffuse into each other at a rate that will be determined as a function of time, temperature, and the nature of the substances. Congress did not intend that the component requirement of a "food additive" would be satisfied by a mere recitation of the diffusion principle, a mere finding of any contact whatever with food. Petitioner's contention on this point is sound.

For the component element of the definition to be satisfied, Congress must have intended the Commissioner to determine with a fair degree of confidence that a substance migrates into food in more than insignificant amounts. We do not suggest that the substance must be toxicologically significant; that aspect is subsumed by the safety element of the definition. Nor is it necessary that the level of migration be significant with reference to the threshold of direct detectability, so long as its presence in food can be predicted on the basis of a meaningful projection from reliable data. Congress has granted to the Commissioner a limited but important area of discretion. Although as a matter of theory the statutory net might sweep within the term "food additive" a single molecule of any substance that finds its way into food, the Commissioner is not required to determine that the component element of the definition has been satisfied by such an exiguous showing. The Commissioner has latitude under particular circumstances to find migration "insignificant" even giving full weight to the public health and welfare concerns that must inform his discretion.

Thus, the Commissioner may determine based on the evidence before him that the level of migration into food of a particular chemical is so negligible as to present no public health or safety concerns, even to assure a wide margin of safety. This authority derives from the administrative discretion, inherent in the statutory scheme, to deal appropriately with De minimis situations.[26] However, if the Commissioner declines to define a substance as a "food additive," though it comes within the strictly literal terms of the statutory definition, he must state the reasons for exercising this limited exemption authority. In context, a decision to apply the literal terms of the statute, requires nothing more than a finding that the elements of the "food additive" definition have been satisfied.

In the case at hand, the Commissioner made specific rulings that the component element of the definition was satisfied with respect to acrylonitrile beverage containers having an RAN level of 3.3 ppm or more. These rulings were premised on a projection, based on an extrapolation from reliable data, of migration of acrylonitrile monomer in then-undetectable amounts. In light of the supplementary submission made in response to the post-argument inquiry of this court, we find that the determination can be made for the 3.3 ppm RAN containers with an appropriate degree of confidence, and with the support of the required quantum of evidence.

Turning to the safety element of the definition, the Commissioner determined that the scientific community had insufficient experience with acrylonitrile to form a judgment as to safety. Based on this lack of opinion, the Commissioner made a finding that acrylonitrile was not generally recognized as safe within the meaning of the statute. The Commissioner acted within his discretion in making such a finding, but we note that the underlying premise may be affected, perhaps weakened, perhaps strengthened, with time and greater experience with acrylonitrile.

This finding on the safety element will be open to reexamination on remand at the discretion of the Commissioner. He would have latitude to consider whether acrylonitrile is generally recognized as safe at concentrations below a certain threshold, even though he has determined for higher concentrations that in the view of the scientific community acrylonitrile is not generally recognized as safe.

V

Petitioners also made a claim of discriminatory treatment that the Commissioner is applying policies in the petitioners' case that have not been applied in other similar circumstances. However, there is no claim that the Commissioner was motivated by discriminatory intention to bring the petitioners before the agency and to focus on their product. Petitioners came before the agency in the ordinary course. Once the Commissioner undertook scrutiny, he shifted the lens of his microscope to a higher power but that is no ground for objection, so long as the final action remains within the legitimate scope of discretion.

The decision of the Commissioner is affirmed in part, and in part is remanded to provide the opportunity for reconsideration.

So ordered.

The court remanded the case back to the Commissioner to consider migration studies using improved methods. It also reminded the Commissioner of the Agency's inherent administrative discretion to ignore *de minmis* levels of migration where the evidence showed migration "is so negligible" as to present no public health or safety concerns (*Monsanto*).[10] Recall from earlier chapters, agencies are afforded enforcement discretion, which it elects when to exercise. The FDA conducted new studies and again deemed the acrylonitrile copolymer a food additive. In new regulations, the Agency deemed it safe under condition that eliminated the likelihood of migration.[11]

The FDA eventually adopted a regulatory definition for indirect food additives. In 21 CFR 170.4(e)(1), the FDA formally stated that if "no migration" occurs between the packaging component from the package to the food then the substance is not a food additive. The Agency also adopted new abbreviated exemption procedures that eased its zero-tolerance policy (21 CFR Part 170).[12] In the end it was not a court decision that changed the FDA, but an overwhelming workload. In the hearings in 1995, the FDA estimated

---

[10] *Id.* at 954–956.

[11] See, 21 CFR 177.1030.

[12] Food Additives; Thresholds of Regulation for Substances Used in Food-Contact Articles, 21 CFR 170.39.

Fig. 6.3 Food additive definition

75 % of food additive petitions were for indirect additives, which consumed a great deal of time for chemist and other personnel to analyze and test. Under the new abbreviated exemption procedure substance with a *de minimis* migration level, generally 0.5 parts per billion, are not required to submit a formal food additive petition. Documentation of the evidence supporting the exemption is still required.

Further regulation came in 1997 with the Food and Drug Administration Modernization Act (FDAMA). The FDAMA amended a number of sections of the FD&C, including Section 409. The amendment established a food contact substance (FCS) notification process (Food Contact Notification or FCN) that allows for faster review of a FCS qualifying as a food additive. It largely captures the indirect additives found in common packaging and preparation materials into a fast-track notification process. Not all FCSs, however, are subject to the abbreviated exemption process for *de minimis* migration levels. All of the considerations of Section 409, including the GRAS exemption, apply to new FCSs.

At this point a food additive can be described as any substance, natural or synthetic, intentionally added to food unless exempt from the amendment. Indirect additives are any substance that migrates from the indirect contact surface to the food based on reliable data (see Fig. 6.3).

6.2.3 Dietary Supplements

Dietary supplements are also subject to the Food Additives Amendment. As part of the food regulations the Amendment applies to all categories of supplements. This includes those products, such as dietary supplement beverages, that blur the line between conventional food and dietary supplement.

One important consideration relates to how a dietary supplement is defined. As discussed in Chapters 1 and 5, dietary supplements carry a unique definition from food in Section 201(f). In many ways the two products are the same, consumed orally with the nutrients absorbed in the stomach, but still are distinguished by the key factors. One of those factors are the substances allowed in dietary supplements, which can vary from what is allowed in conventional foods. A substance approved for use in one category is not necessarily the same for the other. Thus, if a dietary supplement cannot be distinguished from a conventional food it will be classified by the FDA. The FDA seeks the maximum authority possible and typically classifies the product as a food, which enables the Agency to find the product is adulterated.

A more perplexing issue is how to regulate food additives that become food. Recall that the Food Additives Amendment seeks to push the burden of proving safety off the FDA's shoulders and onto the lap of industry. This often creates a tension where the FDA attempts to broadly define food additives, often with dietary supplements, in order to place the onerous on the product manufacturer to demonstrate safety.

Two identical cases illustrate this conflict (*29 Cartons*; *Two Plastic Drums*).[13] In both the cases, the FDA seized shipments of black currant oil sold in capsule form on the basis the black currant

---

[13] See *United States v 29 Cartons of An Article of Food*, 987 F.2d 33 (1st Cir. 1993); *United States v. Two Plastic Drums*, 984 F.2d 814, (7th Cir. 1993).

oil was an unapproved food additive. The FDA reasoned the black current oil was a "component" of the capsule. Both courts found the FDA's interpretation of the Amendment "nonsensical" [14] and even called it an "Alice-in-Wonderland approach" to regulating additives (*29 Cartons*; *Two Plastic Drums*).[15] The courts reached identical conclusion finding that a "component" must exercise some effect on the final food to qualify as a food additive. Unintentionally the courts helped distinguish between "components of food" in Section 20(f) and "components" in 201(s). Thus, the black currant oil as the sole ingredient could not exercise an effect on the food since it was the food itself.

The court in *29 Cartons* provides one interpretation of the Amendment to define when a substance is an additive. The FDA's overreach was laid bare when it conceded black currant oil sold as liquid in bottles would not qualify as a food additive. The FDA attempted to use the gelatin capsule as the food since it was edible. The court ruled that such a distinction was not allowed under the Act (*29 Cartons*).[16] In interpreting the phrase "becoming a component or otherwise affecting the characteristics of any food," the FDA sought to make it a choice either the substance was a component or affected the characteristics. The makers of the black currant oil in *29 Cartons* sought a global definition where the substance affected the characteristics by becoming a component or "otherwise." The court adopted the manufacturer's definition observing the black currant oil did become a component, noting if it was removed there was no product, and it did not affect the characteristics of the gelatin capsule (*29 Cartons*).[17] The case cannot be too broadly interpreted. It stands on its own facts, but does suggest a test for measuring whether a substance is a food or a food additive. Namely, there must be more than one ingredient.

When multiple dietary ingredients are present in a capsule, the food additive definition becomes clearer. A federal district court, for example, deemed black currant oil an additive when encapsulated with fish oil, vitamins, and minerals (*Bulk Metal Drums*).[18] Other courts look at whether the capsules themselves are subject to the Food Additives Amendment. It is important to note the focus is not on the food but the individual added substance.

The passage of DSHEA largely ended the debate. It stated that if a substance can be classified as a dietary supplement or dietary ingredient, then it is excluded from the food additive definition (FD&C).[19] DSHEA mirrored the GRAS approval criteria for determining whether a dietary ingredient qualified as a "new dietary ingredient" (see Chapter 5). It is important not to assume GRAS also means exemption from the NDI notification requirements. Also, the judicial precedent prior to DSHEA remains relevant as novel foods enter the market. At this point the decision tree for navigating Food Additives Amendment is increasing in complexity (see Fig. 6.4).

6.3 Generally Recognized as Safe

The FDA only applies the Food Additive approval process to those added substances not generally recognized as safe (GRAS). These added substances can be thought of as exempt food additives, although widely considered non food additives. Exempt or non food additives will remain subject to the "may render injurious" standard discussed in Chapter 3 unless shown to be non added. Thus, exemption from the Food Additives Amendment does not represent approval, but rather a game of hot-potato where the burden of proof is quickly tossed back to the Agency. Aware of its evidentiary burden under the "may render injurious" standard, the Agency will be skeptical of GRAS claims. GRAS, like all statutory exemptions, must be carefully considered

---

[14] *29 Cartons Id.* at 39.

[15] See *Two Plastic Drums Id.* at 819.

[16] *29 Cartons Id.* at 35, 37. *Two Plastic Drums Id.* at 816–817, 819.

[17] 29 Cartons *Id. at 38.*

[18] See *United States v. 21 Apprx. 190 kg. Bulk Metal Drums*, 761 F. Supp. 180, 182 (D. Me. 1991).

[19] 21 USC 321(S)(6).

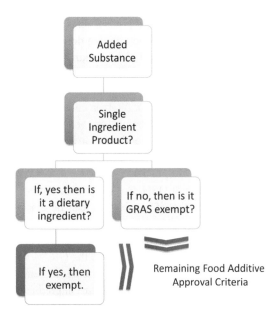

Fig. 6.4 Food additive petition decision tree with dietary supplements

and thoughtfully documented to ensure its survivability.

6.3.1 Defining GRAS

The starting place to define GRAS is the food additive definition. The definition in Section 201(s) states the food additives definition applies only to a substance that "is not generally recognized … to be safe under the conditions of its intended use." This standard is abbreviated as GRAS. The statute appears to suggest an added substance is not a food additive under the Act if it is generally recognized as safe for that purpose. Section 201(s) goes on to say a substance is only GRAS if "general" scientific agreement by experts about the safety either based on an appropriate analysis or common use in food prior to the Amendment that was enacted in 1958.

Following the passage of the Amendment, the FDA issued no clarifying regulations. This left the initial food additive landscape murky. For example, what level of scientific agreement qualifies for approval under the Amendment? The

FDA and the courts struggled with the GRAS exception. It grew from a loose informal concept to a rigid regulatory hurdle, then back again to relaxed regulation.

Nearly two decades after the passage of the Food Additives Amendment, the FDA issued the implementation of regulations. The regulations set a high-bar for scientific consensus. General recognition under the regulations requires "common knowledge" about the substance throughout the relevant scientific community. This immediately bars new or novel ingredients from the GRAS exemption.

Eligibility for classification as generally recognized as safe (GRAS).
21 CFR 170.30(a)
(a) General recognition of safety may be based only on the views of experts qualified by scientific training and experience to evaluate the safety of substances directly or indirectly added to food. The basis of such views may be either (1) scientific procedures or (2) in the case of a substance used in food prior to January 1, 1958, through experience based on common use in food. General recognition of safety requires common knowledge about the substance throughout the scientific community knowledgeable about the safety of substances directly or indirectly added to food.

The regulations also clarify the approval criteria for GRAS substances. GRAS approval depends on the substance and its intended use. The focus thus narrows to evidence demonstrating either a consensus among experts on safety or evidence of common use prior to 1958 that considers how the substance was used (cooked, raw, etc.) and in what amounts. Evidence for one use or at one level will not provide support for a different GRAS use or higher dietary intake. This evidence needs to more than simply testimony that the substance does not pose a risk to health.

Eligibility for classification as generally recognized as safe (GRAS).
21 CFR 170.30(b)-(c)(1)
(b) General recognition of safety based upon scientific procedures shall require the same quantity and quality of scientific evidence as is required to obtain approval of a food additive regulation for the ingredient. General recognition of safety through scientific procedures shall ordinarily be based upon published studies which may be corroborated by unpublished studies and other data and information.
(c)(1) General recognition of safety through experience based on common use in food prior to January 1, 1958, may be determined without the quantity or quality of scientific procedures required for approval of a food additive regulation. General recognition of safety through experience based on common use in food prior to January 1, 1958, shall be based solely on food use of the substance prior to January 1, 1958, and shall ordinarily be based upon generally available data and information. An ingredient not in common use in food prior to January 1, 1958, may achieve general recognition of safety only through scientific procedures.

The regulations also require significant evidence of "common use" prior to 1958. Initially the FDA required the "common use" to occur within the US. Given the growing number of imports, this is no longer practical. This policy was struck down on judicial challenge. The case involved an importer of "foods containing herbs traditional in China that have never been widely used in the United States" (*Famli Herb*).[20] The court held it was "illogical" to categorically exclude evidence outside the U.S. as "*never* provid[ing] probative evidence of safety" (*Famli Herb*).[21] The FDA's informal policy was

---

[20] *Fmali Herb, Inc. v Heckler*, 715 F.2d 1385, 1386(9th Cir. 1983).

[21] *Id.* at 1390 (emphasis original).

struck down and the Agency promulgated regulations outlining when a substance could be approved as GRAS based on foreign use.

Eligibility for classification as generally recognized as safe (GRAS).
21 CFR 170.30(c)(2)
(c)(2) A substance used in food prior to January 1, 1958, may be generally recognized as safe through experience based on its common use in food when that use occurred exclusively or primarily outside of the United States if the information about the experience establishes that the use of the substance is safe within the meaning of the act (see 170.3(i)). Common use in food prior to January 1, 1958, that occurred outside of the United States shall be documented by published or other information and shall be corroborated by information from a second, independent source that confirms the history and circumstances of use of the substance. The information used to document and to corroborate the history and circumstances of use of the substance must be generally available; that is, it must be widely available in the country in which the history of use has occurred and readily available to interested qualified experts in this country. Persons claiming GRAS status for a substance based on its common use in food outside of the United States should obtain FDA concurrence that the use of the substance is GRAS.

6.3.2   FDA's GRAS List

Prior to the passage of the Food Additives Amendment, the FDA established a partial list of GRAS substances. The list included familiar foods such as butter, coffee, cream, lard, and lemon juice. Outside of the technical confines of the regulations, consumers simply identify these items as foods or ingredients, not additives. Following the enactment of the Amendment, the FDA continued to list in its regulations additional GRAS ingredients. The list is now codified as Part 182 of the CFR. It has grown to such lengths that it is now organized by category, such an "anticaking agents" or "stabilizers." The FDA estimates that there are over 200 separate ingredients listed, which it believes is not exhaustive.

The list serves an important function beyond fast tracking the GRAS determination for a manufacturer. Under the Food Additives Amendment, the FDA is tasked with approving all food additives, both for the foods ordinarily under its jurisdiction and for FSIS-regulated meat, poultry, and egg products. It is an immense and diverse field to regulate. The GRAS list narrows the focus allowing the Agency to concentrate resources on chemical or novel substances. Common food items caught in the net of the Amendment are quickly released through the listing process for the benefit of the FDA more than for the use of industry. Regardless of this utilitarian aspect, listing provides a quick means to determine GRAS status for a host of common substances.

Those items deemed GRAS are still subject to FDA controls. In many cases, limitations are provided directly in the regulation. Deviating outside this limitation or for a substance with no limitations, a deviation outside the intended use voids the GRAS status provided in the list. GRAS is not a guarantee the substance will perpetually be exempt. The FDA cautions a GRAS status may require reconsideration and potential revision or repeal.

2013 Revocation of Partially Hydrogenated Oils—Trans Fats

Based on new scientific evidence and the findings of expert scientific panels, the FDA has tentatively determined that partially hydrogenated oils (PHOs), which are the primary dietary source of industrially-produced trans fatty acids, or trans fat, are not GRAS for any use in food based on current scientific evidence establishing the health risks associated with the consumption of trans fat, and therefore that PHOs are food additives. Although FDA has not listed the most commonly used PHOs, they have been used in food for many years based on self-determinations by industry that such use is GRAS. If finalized, this would mean that food manufacturers would no longer be permitted to sell PHOs, either directly or as ingredients in another food product, without prior FDA approval for use as a food additive. (FDA Notice 2013)[22]

This was the case in 1969 when the FDA began a systematic review of its hastily compiled GRAS list. As analytical methods improved and scientific studies on food additives gained new information questions were raised about items on the FDA's GRAS list. In particular safety concerns about cyclamate salts, which were considered GRAS, prompted President Nixon to direct the FDA to conduct a review of the GRAS list (FDA 2006).[23] Cyclamate would ultimately be banned under the Delaney Clause because evidence showed it caused cancer in the liver. The Agency completed a review using analytical standards at the time and any available safety information related to the substances. If the evaluation confirmed the substances GRAS status a new GRAS regulating affirming the finding was issued.

Revising or repealing a GRAS exemption is easier than revising or repealing a food additive regulation resulting from a food additive petition. The GRAS exemption only requires the FDA to follow the APA's notice-and-comment procedures, whereas food additive regulations require a formal evidentiary hearing.

---

[22] Notice by the FDA, Tentative Determination Regarding Partially Hydrogenated Oils; Request for Comments and for Scientific Data and Information, 78 FR 67169 (11/08/2013).

[23] FDA, How U.S. FDA's GRAS Notification Program Works (2006) available at: http://www.fda.gov/Food/IngredientsPackagingLabeling/GRAS/ucm083022.htm.

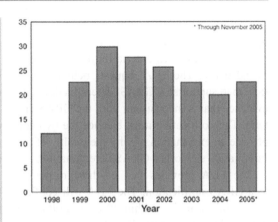

Fig. 6.5 FDA compiled number of GRAS notices filed by year, from 1998 to November 2005

Examples of Listed GRAS Substances Subject to Controls

Section 182.7255 Chondrus extract.

a. Product. Chondrus extract (carrageenin).

b. Conditions of use. This substance is generally recognized as safe when used in accordance with good manufacturing practice.

Section 182.1180 Caffeine.

a. Product. Caffeine.

b. Tolerance. 0.02%.

c. Limitations, restrictions, or explanation. This substance is generally recognized as safe when used in cola-type beverages in accordance with good manufacturing practice

6.3.3   FDA Approved GRAS

When a substance is not listed in Part 182, the FDA GRAS list, then a manufacturer may obtain GRAS status by applying to the FDA. Under current procedures a manufacturer can voluntarily notify the Agency of its GRAS determination in an attempt to receive feedback or affirmation of the substance's GRAS status. Prior to the current regulations the FDA considered all GRAS submissions by requiring a GRAS affirmation petition process. This was resource and time intensive.

The FDA shifted from requiring petitioning for approval to allowing notification of use. After nearly four decades of considering petitions in 1997 the FDA concluded it could no longer commit "substantial resources" to the GRAS petition system (FDA 2006).[24] It published a rule outlining a GRAS notification process to replace the petition process (See 21 CFR Part 170 and 186). The notification was voluntary and as will be discussed below allows for self-affirming GRAS determinations.

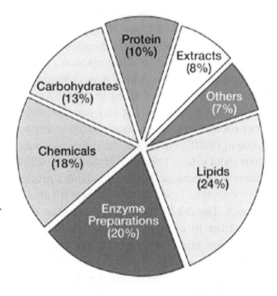

Fig. 6.6 FDA compiled break-out of types of substances for which GRAS notices have been submitted

The notification program simply provides the Agency a GRAS determination and allows comment if necessary. After nearly a decade of use the FDA only received 200 GRAS notices. Those notices are represented in volume and by substance-type in Figs. 6.5 and 6.6 above.

Submitting a GRAS notification to the FDA first involves preparing a GRAS affirmation. This provides the basis for claiming GRAS and its corresponding evidence. The notice will include information about the identity and properties of

---

[24] *Id.*

the substance, whether it GRAS based on common use or scientific analysis, and a discussion of meeting the GRAS criteria for the substance's intended use. The FDA requests both information that supports a GRAS determination and evidence that is inconsistent with the GRAS criteria. The notification must balance the two and provide an explanation of why the totality of the information supports a GRAS determination.

The FDA provides one of three responses to a GRAS notification. The FDA will confirm receipt of the notice within 30 days. Following the initial notification the FDA begins its evaluation of the notification to determine the adequacy of the GRAS determination. This may involve a consultation with FSIS for meat and poultry products. Following the review the FDA will provide one of three types of response. First it may respond that it does not question the basis for the notifer's GRAS determination. This is an approval response, which may contain limitations similar to those found in the GRAS list. Second the FDA may respond that the notice does not meet the GRAS requirements either because of insufficient data or the data raises questions about safety. This is a rejection letter which means the substance then must submit a petition for approval as food additive or alternatively not be used. The third letter follows a request from the notifier to cease evaluation, in which case the Agency responds to confirm the evaluation stopped. This can occur for a variety of reasons, including those such as bankruptcy that are unrelated to the sufficiency of the notification.

The Agency maintains a GRAS notices inventory available to the public. The page provides the public and industry information on types of notices received, FDA responses and copies of the FDA's correspondence. This helps bring transparency to the process.

6.3.4 Self-Affirming GRAS

GRAS determinations also may be made without notifying the FDA. All aspects of the GRAS process remain the same, but the GRAS determination is privately held. The FDA would only

learn about the self-affirming determination during an inspection or other post-market authority. Thus, it is important to complete a notification-style GRAS determination even if not submitting under the voluntary notification program. The Agency still expects consideration of the GRAS criteria, application to the substances intended use, and adequate evidence to support the determination.

The voluntary notification program and the freedom for self-affirmation raise concerns about the safety of current added substances. The process is beginning to look eerily familiar to 1958. The FDA no longer exercises strong control over added substances, which risks leaving it in the dark about what is currently used in the market. In light of the FDA's infrequent inspection rate, as Chapter 2 described occurring once every 5–7 years, the risk of injury from novel substances remains highly likely. The low barriers of entry for food importers and manufactures too often results in many facilities skimming the regulations, including the GRAS approval criteria. Nearly 60 years after the Food Additives Amendment many commentators wonder if it is time for a new system.

> **Is the GRAS System Broken**
>
> Questions are beginning to percolate about the effectiveness of GRAS. A system designed in 1958 may no longer be equipped to address the modern production of food. This is the issue the Pew Charitable Trust tackled in its new report, titled, "Fixing the Oversight of Chemicals Added to Our Food: Findings and Recommendations of Pew's Assessment of the US Food Additives Program."[25] The following except is from the Pew Assessment:
>
> "Our analysis focused on the overall regulatory system that is expected to ensure the safety of more than 10,000 chemical additives, rather than on concerns raised about specific substances. We evaluated

---

[25] (PEW Charitable Trust 2013).

FDA's ability to fulfill the mission of its food additive regulatory program to protect public health from chemicals intentionally added to food or food packaging. We did not evaluate whether specific chemicals or groups of substances, such as salt, trans fat, caffeine, bisphenol A (which is used to line the inside of cans), or artificial colors or flavorings, cause actual harm to the public. We also did not consider contaminants found in food from natural sources or because of pollution, because those are not intentionally added and are regulated under a different set of health and safety standards.

Our research found that the FDA regulatory system is plagued with systemic problems that prevent the agency from ensuring the use of food additives is safe. If one of these chemicals were causing health problems short of immediate serious injury it is unlikely that FDA would detect the problems unless the food industry alerted it. This is particularly true if the health consequences of ingesting the additive take years or decades to become manifest after the food is eaten. If the agency did identify a problem, it would still face challenges proving harm. Proof of harm was not the safety standard laid out by Congress in 1958. Under the law, a chemical may be used in food if competent scientists are reasonably certain that the use will cause no harm over a lifetime. In short, the question is whether it will cause no harm, rather than whether harm can be proven." (Internal citations omitted).

6.4 Approval Procedures

If a food additive is not exempt it must gain approval through a food additive petition. The petition process requires FDA involvement and can be resource intensive. It should be approached only after carefully considering the material

provided above. In addition to the GRAS exemption the Food Additives Amendment also excludes "prior sanctioned" substances.

6.4.1 Prior Sanctions Exceptions

Prior sanction provides a narrower and less common exemption from the Amendment. The statutory definition expressly excludes "any substance used in accordance with a sanction or approval granted prior to the enactment of this paragraph pursuant to" the FD&C, PPI, or FMI (FD&C).[26] The Food Additives Amendment was enacted on September 6, 1958 which is 9 months later than the GRAS "common use" cut-off date. The exemption, therefore, applies to any substance provided "prior sanction" prior to September 6, 1958. Prior sanction can be any evidence of FDA or FSIS, which is more likely the predecessors to the modern Agencies, that indicates official approval. This can include a standard of identity or simply correspondence indicating approval. Whatever the type of quantity of evidence, the burden of proving a prior sanction rests squarely on the party seeking the "prior sanction" exemption.

The prior sanctioned exemption provides narrow coverage. Like GRAS, where the focus is on the added substance, its intended use, and safe levels of consumption, prior sanction cannot be expanded to new intended uses or higher levels of use. Many courts supported this reading of the exception. For example, a court did not allow a prior sanction for potassium nitrate in meat to be widened to include beverages or another court holding the prior sanction of nitrates in other meats did not apply to use with bison meat (*Coco Rico, Inc.*; *Buffalo Jerky*).[27]

The FDA and USDA provide a list of known prior sanctions. The one example from the USDA is nitrates used both as a color fixative

---

[26] 21 U.S.C. § 201(s)(4).

[27] *United States v. An Article of Food...Coco, Rico, Inc.*, 752 F.2d 11, 16 (1st Cir. 1985); *United States v. Articles of Food...Buffalo Jerky*, 456 F. Supp. 207, 209–210 (D. Neb. 1978).

and preservative in cured meats (Nitrate Regulations.).[28] The FDA only prior sanctioned substances related to food packaging materials (FDA Prior Sanction List).[29] The FDA acknowledges the list of prior sanctions may not be a full accounting. This allows a party to request publication of its prior sanction. The FDA cautions a prior sanction will lose its status if not published and the Agency begins consideration of the substance as a food additive (FDA Lapse of Prior Sanction).[30] Nearly 60 years after the enactment of the Amendment, the likelihood of claiming such approval is all but extinguished.

A prior sanction remains subject to reversal and prohibition against adulteration. As with a substance deemed GRAS a prior sanctioned substance can lose its protected status. The FDA remains free to revoke the original approval using the Delaney Clause or other safety data. This was the case with the prior sanction for glycine. The FDA could also take action on the basis the added substance, which as a prior sanctioned substance is not a food additive, is adulterated under the "may render injurious" standard. Far from the passage of the Food Additives Amendment the prior sanctioned exemption's appeal is quickly fading. It is rapidly losing out to the GRAS exemption, which offers greater possibilities and protections.

6.4.2 Food Additive Petitions

A food additive petition requests the FDA issue a regulation authorizing a substance as a food additive for one or more intended uses. The food additive petition contains numerous elements that are beyond our limited introduction to food law. The statute outlines some of the required components of a food additive petition in Section 409(b) (2). In general the petition must identify the substance, its intended use including labeling, and

all relevant data describing the substances effect on food, residue detections and safety studies.

Statutory Elements of a Food Additive Petition

(2) Such petition shall, in addition to any explanatory or supporting data, contain—

a. the name and all pertinent information concerning such food additive, including, where available, its chemical identity and composition;

b. a statement of the conditions of the proposed use of such additive, including all directions, recommendations, and suggestions proposed for the use of such additive, and including specimens of its proposed labeling;

c. all relevant data bearing on the physical or other technical effect such additive is intended to produce, and the quantity of such additive required to produce such effect;

d. a description of practicable methods for determining the quantity of such additive in or on food, and any substance formed in or on food, because of its use; and

e. full reports of investigations made with respect to the safety for use of such additive, including full information as to the methods and controls used in conducting such investigations.

The FDA provides several Guidance Documents outlining the petition process. The Agency offers a total of five Guidance Documents that offer nonbinding guidance on how to document safe use in a petition. Those Guidance Documents include, "Toxicological Principles for the Safety Assessment of Direct Food Additives and Color Additives Used in Food" (referred to as the "Redbook II") issued in 1993 and Redbook 2000. The FDA also offers a general Guidance Document that answers frequent questions about the petitioning

---

[28] USDA, FSIS, Nitrate Regulations 21 CFR 181.33-.34.

[29] FDA Prior Sanction List 21 CFR 181.22-.32.

[30] FDA Lapse of Prior Sanction, 21 CFR 170.6.

process. This Guidance Document, titled "Guidance for Industry: Questions and Answers About the Petition Process," provides a second list of what the Agency expects in a petition.

> **FDA Guidance on the Petition Process**
> I. What are the essential elements of a good food and/or color additive petition?
> - The identity and composition of the additive
> - Proposed use
> - Use level
> - Data establishing the intended effect
> - Quantitative detection methods
> - Estimated exposure from the proposed use (in food, drugs, cosmetics, or devices, as appropriate)
> - Full reports of all safety studies
> - Proposed tolerances (if needed)
> - Environmental information (as required by the National Environmental Policy Act (NEPA), as revised (62 FR 40570; July 29, 1997)
> - Ensure that consistent information is presented throughout all sections of the petition, including those pertaining to:
>   - chemistry,
>   - toxicology,
>   - environmental science, and
>   - any other pertinent studies (e.g., microbiology)

The FDA provides public notice of a petition for a food additive regulation. In 21 CFR Part 171 the FDA provides binding regulations outlining the petition process. Under its regulations the FDA will provide a petitioner notice that its petition was either accepted or is incomplete within 15 days of receipt (FDA Part 171).[31] The FDA also must publish a notice of filing in the *Federal Register* containing the name of the petitioner and a brief description of the proposal. This must

Fig. 6.7 Food additive petition approval process

be done within 30 days of accepting the petition for review (FDA Part 171).[32] Although the petition is posted in the *Federal Register* as any new Agency rule would be, there is no statutory or regulatory basis for public hearing for a food additive petition. The FDA as a matter of policy will invite public comment in particular on safety data.(Fig. 6.7)

The Food Additives Amendment sets the timeline for the FDA to make a decision. Under Section 409(c) the FDA must issue a regulation approving the request for use of the food additive within 90 days of filing or provide the petitioner with the reasons for denial. If the Agency requires additional time, then it may notify the petitioner and extend the review by an additional 90 days. Given the time constraints and the need to review chemical, environmental, and toxicology, it is common for all three reviews to occur simultaneously. Parallel reviews allow the FDA to meet its statutory timeframe, but it is still rare for the Agency to complete a review even in 180-days.

> **Section 409(c)(2) Statutory Timeline for Approval**
> (2) The order required by paragraph (1) (A) or (B) of this subsection shall be issued within 90 days after the date of filing of the petition, except that the Secretary may

---

[31] 21 CFR 171.1(i)(1).

[32] 21 CFR 171.1(i)(2).

(prior to such 90th day), by written notice to the petitioner, extend such 90-day period to such time (not more than one hundred and 80 days after the date of filing of the petition) as the Secretary deems necessary to enable him to study and investigate the petition.

Once the FDA approves a petition it will invite public comment on the final regulation. The Food Additives Amendment in Section 409(f) requires the FDA open the regulation to public comment within 30-days of issuing a final rule. Subsection (f) allows "any person adversely affected" by the new regulation to file an objection. The receipt of an objection does not automatically require a hearing. The FDA typically only requires a hearing when material factual issues are raised. Hearings are fairly rare, but have occurred as was the case for the controversial approval of aspartame.

Section 409(f) Public Hearing on Final Rule
(f) Objections and public hearing; basis and contents of order; statement
1. Within 30 days after publication of an order made pursuant to subsection (c) or (d) of this section, any person adversely affected by such an order may file objections thereto with the Secretary, specifying with particularity the provisions of the order deemed objectionable, stating reasonable grounds therefor, and requesting a public hearing upon such objections. The Secretary shall, after due notice, as promptly as possible hold such public hearing for the purpose of receiving evidence relevant and material to the issues raised by such objections. As soon as practicable after completion of the hearing, the Secretary shall by order act upon such objections and make such order public.

2. Such order shall be based upon a fair evaluation of the entire record at such hearing, and shall include a statement setting forth in detail the findings and conclusions upon which the order is based.
3. The Secretary shall specify in the order the date on which it shall take effect, except that it shall not be made to take effect prior to the 90th day after its publication, unless the Secretary finds that emergency conditions exist necessitating an earlier effective date, in which event the Secretary shall specify in the order his findings as to such conditions.

6.4.3 Color Additive Petitions

Color additives are not food additives but it is worth comparing the two processes. Although color additives are added to food to become a "component" the food additive definition expressly excludes the substances from its definition. Instead color additives, under the Color Additives Amendment, follow their own approval procedures. There are many similarities between food additives and color additives that makes a comparison highly beneficial.

The Color Additive Amendment followed the Food Additives Amendment with passage in 1960. The Color Additive Amendment defined "color additive" and established a safety standard for all color additives. The Color Additive Amendment applied globally to the FDA's regulation of foods, drugs, cosmetics, and medical devices. It required the Agency to list all approved color additives and limited listing to only color additives that are "suitable and safe" (FD&C). Like the Food Additives Amendment color additives are subject a Delaney Clause prohibiting carcinogens.

Color Additive Definition 201(t)

(t)(1) The term "color additive" means a material which—

a. is a dye, pigment, or other substance made by a process of synthesis or similar artifice, or extracted, isolated, or otherwise derived, with or without intermediate or final change of identity, from a vegetable, animal, mineral, or other source, and

b. when added or applied to a food, drug, or cosmetic, or to the human body or any part thereof, is capable (alone or through reaction with other substance) of imparting color thereto; except that such term does not include any material which the Secretary, by regulation, determines is used (or intended to be used) solely for a purpose or purposes other than coloring.

(2) The term "color" includes black, white, and intermediate grays.

(3) Nothing in subparagraph (1) of this paragraph shall be construed to apply to any pesticide chemical, soil or plant nutrient, or other agricultural chemical solely because of its effect in aiding, retarding, or otherwise affecting, directly or indirectly, the growth or other natural physiological processes of produce of the soil and there-by affecting its color, whether before or after harvest.

Color additives must comply with the listing regulations. Any use of a color additive outside the intended use of the listed color additive or deviation from the purity and identity specifications of the regulation cause a product to be adulterated. There is no ratio of color to other ingredients balanced in the FDA's determination whether to take enforcement action. A small amount of non-compliant color can adulterate the entire product. Listed colors can be found in Parts 73, 74, and 82 of 21 CFR. The regulations describe in detail the chemical specifications, intended uses, restrictions on use, and labeling requirements.

The regulations describe two categories of color additives. Part 80 of 21 CFR describes color additives subject to FDA certification. The remaining color additives are exempt from certification. Certified color additives must gain approval for each batch manufactured. This group of color additives consists of synthetic dyes or lakes and is largely derived from petroleum. Certification exempt color additives primarily include dyes derived from plant or mineral sources. It can also include dyes derived from insects, such as with cochineal extract, also labeled "carmine."

Color additives approved for use in human food
Part 73, Subpart A: Color additives exempt from batch certification

21 CFR section	Straight color	EEC#	Year[2]approved	Uses and restrictions
§ 73.30	Annatto extract	E160b	1963	Foods generally
§ 73.40	Dehydrated beets (beet powder)	E162	1967	Foods generally
§ 73.75	Canthaxanthin[3]	E161g	1969	Foods generally, NTE[7] 30 mg/lb of solid or semisolid food or per pint of liquid food; May also be used in broiler chicken feed
§ 73.85	Caramel	E150a-d	1963	Foods generally
§ 73.90	β-Apo-8′-carotenal	E160e	1963	Foods generally, NTE[7]: 15 mg/lb solid, 15 mg/pt liquid

Color additives approved for use in human food
Part 73, Subpart A: Color additives exempt from batch certification

21 CFR section	Straight color	EEC#	Year[2] approved	Uses and restrictions
§ 73.95	β-Carotene	E160a	1964	Foods generally
§ 73.100	Cochineal extract	E120	1969	Foods generally
			2009	Food label must use common or usual name "cochineal extract"; effective January 5, 2011
	Carmine	E120	1967	Foods generally
			2009	Food label must use common or usual name "carmine"; effective January 5, 2011
§ 73.125	Sodium copper chlorophyllin[3]	E141	2002	Citrus-based dry beverage mixes NTE[7] 0.2% in dry mix; extracted from alfalfa
§ 73.140	Toasted partially defatted cooked cottonseed flour	–	1964	Foods generally
§ 73.160	Ferrous gluconate	–	1967	Ripe olives
§ 73.165	Ferrous lactate	–	1996	Ripe olives.
§ 73.169	Grape color extract[3]	E163?	1981	Nonbeverage food
§ 73.170	Grape skin extract (enocianina)	E163?	1966	Still & carbonated drinks & ades; beverage bases; alcoholic beverages (restrict. 27 CFR Parts 4 & 5)
§ 73.200	Synthetic iron oxide[3]	E172	1994	Sausage casings NTE[7] 0.1% (by wt)
§ 73.250	Fruit juice[3]	–	1966	Foods generally.
			1995	Dried color additive
§ 73.260	Vegetable juice[3]	–	1966	Foods generally.
			1995	Dried color additive, water infusion
§ 73.300	Carrot oil	–	1967	Foods generally
§ 73.340	Paprika	E160c	1966	Foods generally
§ 73.345	Paprika oleoresin	E160c	1966	Foods generally
§ 73.350	Mica-based pearlescent pigments[3]	–	2006	Cereals, confections and frostings, gelatin desserts, hard and soft candies(including lozenges), nutritional supplement tablets and gelatin capsules, and chewing gum
§ 73.450	Riboflavin	E101	1967	Foods generally
§ 73.500	Saffron	E164	1966	Foods generally
§ 73.575	Titanium dioxide	E171	1966	Foods generally; NTE[7] 1% (by wt)
§ 73.585	Tomato lycopene extract; tomato lycopene concentrate[3]	E160	2006	Foods generally
§ 73.600	Turmeric	E100	1966	Foods generally
§ 73.615	Turmeric oleoresin	E100	1966	Foods generally

Color additives approved for use in human food
Part 74, Subpart A: Color additives subject to batch certification

21 CFR Section	Straight color	EEC#	Year[2] approved	Uses and Restrictions
§ 74.101	FD&C Blue No. 1	E133	1969	Foods generally
			1993	Added Mn spec.
§ 74.102	FD&C Blue No. 2	E132	1987	Foods generally
§ 74.203	FD&C Green No. 3	–	1982	Foods generally
§ 74.250	Orange B[3]	–	1966	
§ 74.302	Citrus Red No. 2	–	1963	
§ 74.303	FD&C Red No. 3	E127	1969	Foods generally
§ 74.340	FD&C Red No. 40[3]	E129	1971	Foods generally
§ 74.705	FD&C Yellow No. 5	E102	1969	Foods generally
§ 74.706	FD&C Yellow No. 6	E110	1986	Foods generally

Unlike food additives, color additives lack a GRAS exemption under the Color Additive Amendment. As seen in the discussion above the GRAS exemption from the food additive definition provides coverage for the bulk of added substances otherwise covered by the Food Additives Amendment. The Color Additive Amendment lacks a similar exemption. A GRAS food additive, however, may be a color additive, but still must gain proper approval under the Color Additive Amendment. The FDA will also be sensitive to food additive petitions or GRAS determinations which suggest a substance's capability of imparting color to a food. If the intended use reaches into this capability, then the manufacturer must gain approval for use as a color additive. Understanding this boundary between the two categories is crucial for compliance.

Color certification provides the FDA control over synthetic color additives. The FDA certified batches representing a total of nearly 25 million pounds of color additives during fiscal year 2013 (Color Certification Reports)[33]. Certification requires the manufacturer of each batch of certifiable color additive submit a sample to the FDA Color Certification Branch. The FDA charges a fee for certification based on the batch weight. Until the sample is analyzed and approved the batch cannot be used.

The FDA conducts a battery of tests to approve a certifiable color additive. At least ten analyses are conducted to evaluate identity, purity, and strength. The process takes the FDA five days to complete, after which the results will be reviewed for compliance with the specifications in the listing regulation. If approved the FDA issues a certificate for the batch and assigns it a unique lot number.

Similar to the food additive petition new color additives must submit a color additive petition. The two petitions share a great deal in common, from stating identity and intended use to documenting various safety concerns. When evaluating a new color additive the FDA considers safety factors, such as cumulative effect in the diet and analytical methods for determining acceptable levels of impurities. The FDA will also decide whether to provide approval as a certified or exempt color additive. Generally, batch certification applies to synthetic colors, but can be assigned to any color composition that needs to be controlled to protect the public health. The petition process is regulated by 21 CFR Part 71.

A successful petition will result in the FDA promulgating a new listing regulation or amending an existing listing for a new use. As with food additives once the new regulation is announced there is a process for filing objections and holding public hearings. An approved color additive petition, like all color additives is continually monitored for safety, evidence of carcinogenic effects under the Delaney clause, and other data

---

[33] FDA, Color Certification Reports; Fiscal Year 2013, available at: http://www.fda.gov/ForIndustry/ColorAdditives/ColorCertification/ColorCertificationReports/default.htm.

that may indicate the color additive should be revoked.

Color additives require careful consideration and strict compliance. Deviations from listings will result in enforcement actions on the basis of adulteration. Even if a color additive derives from a natural plant or mineral source, and is likely to pose no significant risk, approval is required. This is one area where the FDA exercises sensitivity to following all applicable pre market requirements, in part because there is no ability for a facility to self-affirm a color as GRAS.

Comparison of food additive amendment and color additive amendment

	Food additive amendment	Color additive amendment
GRAS exemption	✓	
Prior sanctioned exemption	✓	
Petition for new additives	✓	✓
Batch certifications		✓
Adulterated if non compliant	✓	✓
Delaney clause	✓	✓

6.4.4 Irradiation

Irradiation is a unique food additive. Food irradiation involves the application of ionizing radiation to food in a process similar to pasteurizing milk irradiation extends shelf life and eliminates organisms that cause food borne illness. It enjoys a surprisingly long history in the US. The earliest accounts of using food irradiation date to 1905. The use was prevalent enough for Congress to include radiation in the 1958 Food Additives Amendment.

Congress expressly identified radiation when defining a food additive. The 1958 Food Additives Amendment stated that a food additive includes "any substance the intended use of which results or may reasonably be expected to result, directly or indirectly, in its becoming a component or otherwise affecting the characteristics of any food … including any source of radiation intended for any such use." In a report accompany the Amendment Congress went on to define sources of radiation as including radioactive isotopes, particle accelerators, and X-ray machines" (FDA; Irradiating Foods).[34] The report is informative when determining what "radiation" means under the Amendment.

Congress also added intentional irradiation as a form of adulteration. In section 402(a)(7) intentionally irradiation is added to the acts that constitute adulteration. It states that a food is deemed adulterated if "has been intentionally subjected to radiation, unless the use of the radiation was in conformity with a regulation or exemption in effect pursuant to" the Food Additives Amendment. Therefore, the statute informs industry that irradiation is a type of food additive that if not conforming to the 1958 Amendment or related regulations will be deemed adulterated under Section 402(a)(7).

What Type of Food Additive?
Before reading on, what type of food additive is irradiation—direct or indirect?
In your answer discuss how the type of food additive fits with the concept of adulteration. The adulteration definition suggests, like with other food additives, a food is adulterated because it contains an unapproved food additive. Does the irradiated food contain radiation? Is the adulteration definition expanding the concept to a process? Explain in your answer.

There are two ways for irradiation to gain approval. The FDA can issue a food additive regulation or a petitioner can use the Food Additive Petition process. The FDA does not consider irradiation for any intended use or at any level GRAS. All uses must be approved under existing food additive regulations or petitioning for new

---

[34] FDA, U. S. Regulatory Requirements for Irradiating Foods, May 1999, Presentation by George H. Pauli, available at: http://www.fda.gov/Food/GuidanceRegulation/GuidanceDocumentsRegulatoryInformation/IngredientsAdditivesGRASPackaging/ucm110730.htm.

regulations. The FDA rarely uses its discretion to promulgate a new irradiation regulation. All of the components discussed above in submitting a food additive petition apply. The petition must focus on intended use, narrowing the application of radiation to certain foods, and specifying what levels of radiation.

All irradiation additive regulations are promulgated in Part 179 of the CFR. The regulations cover radiation used to inspect food to ionizing irradiation used as a form of pasteurization. An example of the approved used of ionizing radiation is provided below. The FDA approved the use of irradiation at varying levels for beef, pork, poultry, molluscan shellfish (e.g., oysters, clams, mussels, and scallops), shell eggs, fresh fruits and vegetables, lettuce and spinach, spices and seasonings and seeds for sprouting (e.g., for alfalfa sprouts).

Use	Limitations
1. For control of *Trichinella spiralis* in pork carcasses or fresh, non heat-processed cuts of pork carcasses	Minimum dose 0.3 kiloGray (kGy) (30 kilorad (krad)); maximum dose not to exceed 1 kGy (100 krad)
2. For growth and maturation inhibition of fresh foods	Not to exceed 1 kGy (100 krad)
3. For disinfestation of arthropod pests in food	Do
4. For microbial disinfection of dry or dehydrated enzyme preparations (including immobilized enzymes)	Not to exceed 10 kGy (1 megarad (Mrad))
5. For microbial disinfection of the following dry or dehydrated aromatic vegetable substances when used as ingredients in small amounts solely for flavoring or aroma: culinary herbs, seeds, spices, vegetable seasonings that are used to impart flavor but that are not either represented as, or appear to be, a vegetable that is eaten for its own sake, and blends of these aromatic vegetable substances. Turmeric and paprika may also be irradiated when they are to be used as color additives. The blends may contain sodium chloride and minor amounts of dry food ingredients ordinarily used in such blends	Not to exceed 30 kGy (3 Mrad)
6. For control of food-borne pathogens in fresh (refrigerated or unrefrigerated) or frozen, uncooked poultry products that are: (1) Whole carcasses or disjointed portions (or other parts) of such carcasses that are "ready-to-cook poultry" within the meaning of 9 CFR 381.l(b) (with or without nonfluid seasoning; includes, e.g., ground poultry), or (2) mechanically separated poultry product (a finely comminuted ingredient produced by the mechanical deboning of poultry carcasses or parts of carcasses)	Not to exceed 4.5 kGy for non frozen products; not to exceed 7.0 kGy for frozen products
7. For the sterilization of frozen, packaged meats used solely in the National Aeronautics and Space Administration space flight programs	Minimum dose 44 kGy (4.4 Mrad). Packaging materials used need not comply with § 179.25(c) provided that their use is otherwise permitted by applicable regulations in parts 174 through 186 of this chapter
8. For control of foodborne pathogens in, and extension of the shelf-life of, refrigerated or frozen, uncooked products that are meat within the meaning of 9 CFR 301.2(rr), meat byproducts within the meaning of 9 CFR 301.2(tt), or meat food products within the meaning of 9 CFR 301.2(uu), with or without nonfluid seasoning, that are otherwise composed solely of intact or ground meat, meat byproducts, or both meat and meat byproducts	Not to exceed 4.5 kGy maximum for refrigerated products; not to exceed 7.0 kGy maximum for frozen products
9. For control of *Salmonella* in fresh shell eggs	Not to exceed 3.0 kGy
10. For control of microbial pathogens on seeds for sprouting	Not to exceed 8.0 kGy

Use	Limitations
11. For the control of Vibrio bacteria and other foodborne microorganisms in or on fresh or frozen molluscan shellfish	Not to exceed 5.5 kGy
12. For control of food-borne pathogens and extension of shelf-life in fresh iceberg lettuce and fresh spinach	Not to exceed 4.0 kGy
13. For control of foodborne pathogens, and extension of shelf-life, in unrefrigerated (as well as refrigerated) uncooked meat, meat byproducts, and certain meat food products	Not to exceed 4.5 kGy
14. For control of food-borne pathogens in, and extension of the shelf-life of, chilled or frozen raw, cooked, or partially cooked crustaceans or dried crustaceans (water activity less than 0.85), with or without spices, minerals, inorganic salts, citrates, citric acid, and/or calcium disodium EDTA	Not to exceed 6.0 kGy

All irradiated foods must follow certain labeling requirements. For a long period following the passage of the Food Additives Amendment there was no way for consumers to know if food was irradiated. The Amendment did not require any particular disclosure and the FDA did not issue labeling regulations specific to irradiation. The regulations in Part 179 now require the use of a symbol known as the "radura" (see Fig. 6.8). This is particularly the case when the process is not apparent, such as with whole fruits and vegetables. In addition the phrase "treated with radiation" or "treated by irradiation" is required. In some instances consumers still may not know if the food product or its ingredients were irradiated. For example, the radura is not required for irradiated ingredients co-mingled with non irradiated ingredients. FSIS provides the example of sausage with contains irradiated and non irradiated meat as an example where the radura is not required (FSIS Irradiation).[35]

Irradiation demonstrates even incredibly novel processes and additives can be managed by the current regulatory framework. Although the concepts may not intuitively fit, such as considering the irradiation processes as an added substance to food, the FDA provides a consistent framework to gain approval by demonstrating safety.

Fig. 6.8 Radura symbol used on to label irradiated foods

6.4.5   Interim Food Additives

Interim food additives is a category created for GRAS substances with new safety concerns. Created in 1972 the FDA issued new regulations creating the temporary category in light of improved analytical methods that often raised safety concerns (Interim Food Additive Regulations).[36] Under the regulations in Part 180 the FDA may deem an approved GRAS substance as an "interim" food additive.

21 CFR 180.1(a)
(a) Substances having a history of use in food for human consumption or in food contact surfaces may at any time have their safety or functionality brought into question by new information that in itself is not conclusive. An interim food additive regulation for the use of any such substance

---

[35] FSIS Irradiation; Irradiation and Food Safety Answers to Frequently Asked Questions available at: http://www.fsis.usda.gov/wps/portal/fsis/topics/food-safety-education/get-answers/food-safety-fact-sheets/production-and-inspection/irradiation-and-food-safety/irradiation-food-safety-faq.

[36] 21 CFR Part 180.

may be promulgated in this subpart when new information raises a substantial question about the safety or functionality of the substance but there is a reasonable certainty that the substance is not harmful and that no harm to the public health will result from the continued use of the substance for a limited period of time while the question raised is being resolved by further study.

Once deemed an interim food additive the FDA must follow the food additive petition process. The FDA identifies the safety issues raised and seeks at least one sponsor of the substance to undertake studies to resolve the safety concerns. If within 60 days a sponsor does not certify it is engaged in "adequate and appropriate" studies then the FDA will immediately revoke the interim food additive regulation (Interim Food Additive Regulations).[37] This action in effect pushes the burden of proving the substance in unsafe on to the FDA under the "added" substance standard for adulteration. When the studies are completed the FDA review the available data and either determines whether to issue a food additive regulation or ban the use of the substance.

21CFR 180.1(d)

(d) Promptly upon completion of the studies undertaken on the substance, the Commissioner will review all available data, will terminate the interim food additive regulation, and will either issue a food additive regulation or will require elimination of the substance from the food supply.

The interim food additive designation is highly criticized. Since 1972 the FDA regulated brominated vegetable oil (BVO) and other additives as "interim food additives" (Interim Food Additives Listing).[38] BVO is known for its use as a flame retardant chemical. Current regulations called for more safety data in particular toxicological studies. Nearly 42-years later BVO remains classified as an "interim food additive." The FDA continues to view it as safe within certain limitations, but does not deem its review and approval in a final regulation as a priority (Food Navigator 2013).[39]

6.4.6   Limitations from Delaney Clause

The Delaney Clause modifies the general safety standard for food and color additives. As discussed in Chapter 3 the Delaney Clause prohibits FDA approval for additives found to induce cancer. The carcinogenic effects must be addressed when considering safety during the pre market process. The Delaney Clause will also follow a substance through-out its existence, with any new data potentially used to ban the additive. The post-market enforcement under the Delaney Clause can be slow, often leading States to adopt its own standards.

As discussed in Chapter 3 the Delaney Clauses provides little guidance on when or how a substance should be deemed to induce cancer. The Delaney Clause stops short of stipulating the means, methods or thresholds to be used when deciding if an added substance induces cancer. It provides an open and flexible standard to allow scientific techniques to develop. This is both helpful and harmful. It affords the FDA ample flexibility in determining its obligations under the Delaney Clauses.

The discretion under the Delaney Clauses often leads to the FDA rejecting the need to ban carcinogens. There are numerous examples of the FDA exercising a balancing of benefits and risks to maintain approval of food and color additives. Chapter 3 provide early examples following the passage of the Color Additive Amendment with Orange No. 17 and Red No. 19. Other examples, include the controversy over 4-methylimidazole (4-MeI) used in Carmel coloring. Testing

---

[37] 21 CFR 180.1(c)(2).

[38] 21 CFR 180.22-.37.

[39] (FDA 2013).

continues to indicate a risk of cancer, but the FDA continues to evaluate and deems it safe (Consumer Reports 2014).[40]

6.5 Comparative Law

Food additive regulation varies in the degree of control from nation to nation, but remains a keystone element of food safety laws. Food additive approval provides regulators control and confidence in the safety of substances used on the market. In some cases the control is laxed and the confidence arguable lacking. While in other countries the regulations rigid and criticized as stifling innovation. This is often the balance with any regulation carefully controlled innovation.

The EFSA regulates all substance intentionally added to food products that perform "certain technological functions, for example to colour, to sweeten or to help preserve foods" (EFSA).[41] EFSA evaluates the safety of additives and assigns an "E Number" (EFSA). The E Number, for example Xantham Gum E 415, is required on the label in the ingredient list (EFSA). The definition of food additive used by Health Canada captures a similar intent to that of the Food Additives Amendment. Under Canadian regulations a food additive is a chemical substance added to food that "may reasonably be expected to result, in it or its by-products becoming a part of or affecting the characteristics of a food" (Canadian Regulations).[42] Health Canada also utilizes an informal term of "food processing aid" which functions similar to the concept of indirect additives (Canadian Policy 2008).[43] All food

additives are approved and listed by Health Canada (List of Permitted Additives).[44] Japan began using a concept of food additive when it passed its Food Sanitation Act in 1947 (MHLW Food Additives).[45] Under the FSA Japan regulates a wide range of substances as food additives, including, "both substances remaining in the final products, such as food colors and preservatives, and substances not remaining in the final products, such as microorganism control agents and filtration aids" (MHLW Food Additives).[46] All substances deemed additives, including vitamins and minerals, must be approved by the Ministry of Health, Labour and Welfare (MHLW) before use. The list of regulations could go on.

A common substance, such as BVO, provides a window to view how various systems regulate food additives. As mentioned above, the US continues to allow the use of BVO no longer as a GRAS substance but as a languishing interim food additive (see Section 6.4.5). The US is virtually alone in its approval of BVO. It was permitted in the United Kingdom at 80 ppm until being banned in 1970 following prohibitions in Germany and Netherlands (Bendig et al. 2012).[47] It is banned by the EFSA and Japan. It was toxicity studies conducted by the Canadian Food and Drug Director that lead to the US FDA revoking BVO's GRAS status (Bendig et al. 2012).[48] The presence of BVO on the US market begs the question, "are Americans unique immune to the toxicity of BVO or is the US food additive system broken?" Absent comparative law the opportunity for such question may be lost.

---

[40] Consumer Reports, Caramel Color: The Health Risk that May be in Your Soda (February 10, 2014) available at http://www.consumerreports.org/cro/news/2014/01/caramel-color-the-health-risk-that-may-be-in-your-soda/index.htm.

[41] EFSA, Food Additives (July 2014) available at: http://www.efsa.europa.eu/en/topics/topic/additives.htm.

[42] Canadian Food and Drug Regulations §B.01.001.

[43] Health Canada, Food and Nutrition, Policy for Differentiating Food Additives and Processing Aids (December

2008) available at: http://www.hc-sc.gc.ca/fn-an/pubs/policy_fa-pa-eng.php#fnb2-ref.

[44] Health Canada, Food and Nutrition, List of Permitted Food Additives, available at: http://www.hc-sc.gc.ca/fn-an/securit/addit/list/index-eng.php.

[45] Ministry of Health and Welfare, Food Additives available at: http://www.mhlw.go.jp/english/topics/foodsafety/foodadditives/.

[46] Id.

[47] (Bendig Paul et al. 2012).

[48] Id. at 678.

6.6   Chapter Summary

This chapter provides an overview of the US regulatory approach to food additives. It began by looking at the legislative history and the events that prompted the passage of the Food Additives and Color Additive Amendments. Once again it was crisis and consumer confidence leading the charge for change. The chapter then focused on key definitions, like food additive and GRAS, and central concepts like approval criteria and procedures. It finished its discussion by again raising the question of whether the GRAS system is broken and consumer safety open to unchecked risks.

Overview of Key Points
- Legislative history of the Food Additives and Color Additive Amendments
- Key definitions of food, dietary supplement, GRAS, and "components" of food
- Food additive concept including direct and indirect food additives
- Case law defining indirect food additives
- Regulatory framework for pre market approval of food additives
- GRAS concept and approval criteria
- Distinguishing dietary supplements from foods and "new dietary ingredients" from food additives
- Comparing color additives to food additives, including exemptions and approval criteria
- Limitation s from the Delaney Clause
- The unique regulation of irradiation as a food additive

6.7   Discussion Questions

1. Should foods be treated differently than other FDA regulated products? Explain your answer focusing on risk and consumer safety.
2. Why would Congress provided a GRAS exemption for food additives, but no exemptions for color additives?
3. Are there other substances like irradiation which could be considered "added" substances and possibly food additives despite not leaving a physical presence on the food?

4. Is the GRAS system broken? The PEW report and the example of the FDA banning trans fats seem to suggest gaps in the system. When answering the question identify examples that support or undermine your argument.

References

Bendig P et al (2012) Brominated vegetable oil in soft drinks—an underrated source of human organobromine intake. Food Chem 133:678–682 (679)

Consumer Reports (10 February 2014) Caramel color: the health risk that may be in your soda. http://www.consumerreports.org/cro/news/2014/01/caramel-color-the-health-risk-that-may-be-in-your-soda/index.htm. Accessed 28 Aug 2014

EFSA (23 October 2014) Food additives. http://www.efsa.europa.eu/en/topics/topic/additives.htm. Accessed 10 July 2014

FDA (2006) How U.S. FDA's GRAS notification program. works http://www.fda.gov/Food/IngredientsPackagingLabeling/GRAS/ucm083022.htm. Accessed 12 March 2014

FDA (29 January 2013) Brominated vegetable oil (BVO) is safe, so removing its interim status is 'not a priority', By Elaine WATSON, Food Navigator. http://www.foodnavigator-usa.com/Regulation/FDA-Brominated-vegetable-oil-BVO-is-safe-so-removing-its-interim-status-is-not-a-priority. Accessed 28 Aug 2014

PEW Charitable Trust (November 2013) Fixing the oversight of chemicals added to our food: findings and recommendations of Pew's assessment of the U.S. Food Additives Program. http://www.pewtrusts.org/~/media/legacy/uploadedfiles/phg/content_level_pages/reports/FoodAdditivesCapstoneReportpdf.pdf. Accessed 28 Aug 2014

Private Actions and Personal Liability

7

Abstract

The final chapter splits its focus between two related issues with different sources of authority. First, the concept of private lawsuits is explored. In particular, the concept of personal liability in the wake of serious illness or death following a foodborne illness is explored. Any discussion of private lawsuits would not be complete without describing the recent onslaught of litigation related to labeling claims. Second, the basis for criminal liability under the FD&C and FSIS enabling acts are reviewed. Both, private lawsuits and criminal liability represent the ultimate penalty for food law infractions. Private lawsuits, however, are rooted in State law and the reach of such actions are limited by preemption. There is no provision in any federal food law authorizing a lawsuit or passing the baton of enforcement. Criminal lawsuits remain tethered to the enabling acts discussed in previous chapters. Instead of warnings and seizures, the consequences are severe even for corporate officials acting in their official capacity.

7.1 Introduction

Private actions raise a question of whether the FDA abdicates its responsibility under the FD&C. This is particularly the case with labeling litigation. In the context of foodborne illnesses a valid critique surrounds the question of increased criminal prosecutions. Despite this it can be said that the FDA still acts by issuing recalls, seizures, and investigations. Yet when it comes to labeling and the enforcement of misleading or false claims the FDA acts rarely and often ineffectually. The Agency will issue "Dear Industry" letters calling for compliance or an *en masse* group of warning letters that are often seen as empty threats. The FDA commissioner and deputy commissioners in public statements all but admit defeat in enforcing misleading claims.

Courts reviewing FDA enforcement of labeling actions cannot be omitted from a discussion about the FDA's lack of meaningful actions. The commercial speech doctrine discussed in Chapters 4 and 5 demonstrated a formidable burden for the FDA in proving suppression of speech as the superior enforcement option. Even in its efforts to revise claims or disclaimers, the commercial speech doctrine poses a quagmire of issues for the Agency. The result is a significant investment of personnel and resources for any single case. Tackling the issue on an industry wide basis, where the scope of authority covers nearly every facet of food, beverage, and supplementation, is simply impractical if not wholly unimaginable.

Labeling litigation fills a policy gap where regulation fails to overcome constitutional rights. In some cases the litigation arises despite clear labeling regulations. In others it occurs where the FDA has yet to issue final regulations, clutching instead to informal policy. Where enforcement or regulation fails, for whatever reason, labeling litigation has begun to step-in. It attempts to enforce FDA labeling provisions using State law causes of action. In many regards, it can be said that labeling litigation may ultimately act as a greater deterrent than regulatory action by the FDA. Labeling litigation, as will be discussed in the sections further, and is notorious for settlements ranging from US$ 8 to 45 million. This is far more than any penalty discussed so far.

There are constraints on brining labeling lawsuits using State law. In general, states may only permit causes of action based on violations of State law that mirror the federal requirements. The concept of preemption, previously raised in Chapter 4, will be further expounded on in the context of labeling lawsuits. California, famous as a forum for labeling lawsuits, typically entertains labeling litigation under the Unfair Competition Law, False Advertising Law, Consumer Legal Remedies Act, or the Sherman Food Drug and Cosmetic Law (California's version of the FD&C). Each of these statutes must be carefully analyzed by courts hearing private labeling challenges to ensure the State statute used as the springboard for the suit is not preempted by the FD&C or other federal law.

Private lawsuits offer a complex but fascinating opportunity to learn about labeling and liability. It exposes policy weaknesses and policy choices. It is highly dependent on State law, but also requires a strong sense of FDA authority and regulations. Private lawsuits are also burdensome to bring as an individual, making large outbreaks and class actions the preferred vehicle for food poisoning and labeling lawsuits. FSMA may begin to chip away at the plaintiff's burden in food poisoning cases possibly making even minor bouts of illness open to litigation.

Criminal liability strikes fear into the hearts and minds of corporate officers. Many expect a corporation to provide a veil of protection to its officers and executives. Under two prominent cases the Supreme Court developed the Dotterweich Doctrine, which was later reaffirmed and renamed as the Park Doctrine. Under this doctrine corporate officers can be held personally criminally liable simply because of their position in the company. The doctrine is a blunt instrument and after a period of rapid use is rare to see in action. Most criminal actions are misdemeanors punishable by fines and probation. Other more severe violations are treated as felonies. The Park Doctrine though in rare use for criminal prosecutions, is finding new uses. Recently the FDA began citing the doctrine in warning letters for the principal within the doctrine of vicarious liability.

7.2 Private Actions

Private actions involve one or more individuals bringing a lawsuit. A myriad of questions are involved in bringing a private lawsuit ranging from selecting forum and venue to forming a class action lawsuit. Class action lawsuits are a defined term in federal civil procedure, which describe how a group of individuals can bring suit for the same injury against the same defendant. This text is aimed at the non-lawyer and also squarely on food law. Further discussion will explore what type of claims private individuals can bring and how those claims may be preempted by the FD&C. It will omit the nuts-and-bolts issues for another time.

7.2.1 Personal Liability for Foodborne Illnesses

There is no federal product liability law to provide a framework for foodborne illnesses cases to follow. The federal food laws provide regulation on various aspects of manufacturing and shipping food, but no cause of action or regulatory

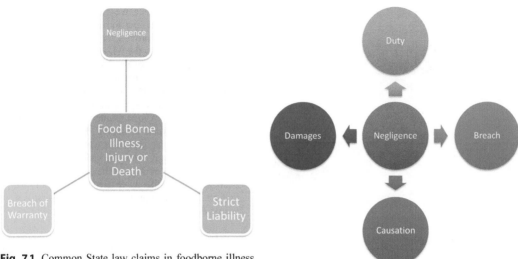

Fig. 7.1 Common State law claims in foodborne illness litigation

Fig. 7.2 Elements of a negligence claim

framework for ill consumers or their families to bring suit under. Product liability is the sole providence of State law and case precedent. Federal courts hearing foodborne illness cases will apply the State law of the jurisdiction where the court is seated. This approach results in inconsistent and unpredictable results.

All State law claims are some form of negligence (see Fig. 7.1). Negligence is a form of tort law, also known as personal injury law that requires a plaintiff to show that the defendant failed to act with the same care that a reasonable person in a similar situation would exercise. This is known as the "reasonable care" standard. In the typical negligence case plaintiffs must show four elements to establish the defendant's negligence. Those four elements are: duty of care, breach, causation, and damages (see Fig. 7.2). Causation can be a thorny issue. Other types of negligence involve strict liability and breach of warranty. Strict liability is a rare cause of action because it requires the defective product to be "unreasonably dangerous" which foods in most cases, even if adulterated, are not. Breach of warranty is not a negligence claim, but contractual. It states that the maker of adulterated food breached an implied or express warranty, such as marketability. It is usually added to complaints as a fail-safe cause of action.

Causation and Res Ipsa Loauitor

Causation provides a difficult hurdle for plaintiffs. Often plaintiffs will rely on FDA and CDC investigations into outbreaks, which typically will identify the responsible firm. Then experts are called to provide testimony aimed at attributing negligence of the firm as the cause of the outbreak. To prove negligence, plaintiffs must chow causation, which means the defendant's negligence is more likely than not the cause of their injury (Polin 1998).[1] How causation is demonstrated depends on State law. Typically a three prong approached is used under the theory of *res ipsa loquitor* (see Fig. 7.3). This doctrine looks at three points:

1. The incident was caused by "an agency or instrumentality under the exclusive control" of the defendant.
2. The incident must be the type that ordinarily does not happen unless someone is negligent.
3. The incident cannot be "due to any voluntary act or contributory fault of the plaintiff." (*Miller Meat Co.*)[2]

---

[1] Polin (1998).

[2] *Ford v. Miller Meat Co.*, 29 Cal.App.4th 1196, 1202-03 (Ca. 1994).

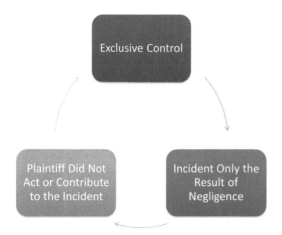

Fig. 7.3 *Res Ipsa Loquitur* elements

In *Ford* the court reviewed a negligence action against Miller Meat Company after the plaintiff broke a tooth on a bone fragment in a ground beef product produced by Miller Meat. The court using the three-prong test found that the plaintiff failed to show negligence. In particular the test noted the size of the bone fragment, which was around one-eighth of an inch, was so small that "no one could reasonably expect the vendor to remove [it] from ground beef" (*Miller Meat Co.*).[3] The decision in *Ford* is not only instructive on the elements of causation, but also signals to potential plaintiffs of how fact dependent negligence cases can be.

One concern with causation is whether the foodborne illness is attributable to the defendant's products and not another source. In some cases bacteria responsible for foodborne illnesses can remain dormant for two or more weeks before causing any symptoms. This was the case when a plaintiff sued Burger King after allegedly becoming ill eating a Whopper (*Hairston v. Burger King Corp.*).[4] The court dismissed the case largely on expert witness who testified that the plaintiff's illness could have been caused by something eaten an hour to a week before feeling ill (*Hairston v. Burger King Corp.*)[5]. The court reasoned:

The trial court did not rule that Hairston did not have food poisoning, but found that she failed to carry her burden of proving that her condition was caused by consuming the Whopper. Kamberov testified that the cause of the food poisoning could have been anything Hairston ate from one hour to one week before she started having problems. Additionally, there was no evidence that any of the other 800 people who consumed a Whopper that day developed food poisoning. Hairston's testimony merely established she was having severe gastric problems and that she had eaten a Whopper. Based on the entirety of the evidence, the trial court was not manifestly erroneous in determining that eating the Whopper did not cause Hairston's gastric distress.

Issues also arise when plaintiff's symptoms are chronic rather than acute. This was the case for a couple who sued a local restaurant, the Log Cabin a/k/a "Betty's," after becoming ill immediately following their dinner at the seafood buffet (*Burnett v. Essex Insurance Company*).[6] In addition to their acute reaction to the food one plaintiff continued to complain of chronic abdominal pain (*Burnett v. Essex Insurance Company*).[7] The court dismissed the case finding that their physician could not rule out other causes of the illness including treatment for similar symptoms prior to eating at Betty's. The court held:

Dr. Ghanta testified that if the Plaintiffs' history were correct, then he would relate their symptoms to the food consumed at the restaurant. However, Dr. Ghanta could not identify the source of the infection, and he could not rule out as other possible causes the local drinking water or from elsewhere in the community. Although a plaintiff need not scientifically identify the infection-producing organism to recover in a food poisoning case, he must still prove causation by a preponderance of the evidence. When the lack of more specific evidence of causation is considered in light of the Burnetts' medical histories, we find no error in the trial court's conclusion that they did not meet this burden. The evidence simply does not preponderate in their favor, given their propensity to gastric disorders and Dr. Ghanta's reliance solely on their accounts to formulate his opinion of causation.

---

[3] *Id.* at 1203.

[4] *Hairston v. Burger King Corp.*, 764 So.2d 176, 178 (La. App. Ct. 2000).

[5] *Id.*

[6] *Burnett v. Essex Insurance Company*, 773 So.2d 786 (La. 2000).

[7] *Id.* at 788.

There are numerous disadvantages to food-borne illness litigation. This often results in only the most severe instance of illness or death or egregious acts being brought to court. In other words this type of case is better suited for large outbreaks where class action status may be available. Litigation can be expensive and take several years to reach a resolution. The compensation, typically from the insurance companies, largely covers legal fees and court costs. Given these restraints the individual suits for foodborne illness are not brought to court.

FSMA and Negligence

The Food Safety Modernization Act may slightly ease a plaintiff's burden in negligence cases. The issue of causation will remain troublesome and difficult to prove. It may be aided by FSMA, which requires extensive record keeping that could prove to be a fertile ground for discovery. FSMA also requires new preventative measures and controls to be put in place. For example, the Preventative Controls rule discussed in Chapters 3 and 5 will require a food safety plan that outlines hazards and control of those hazards. This could be a roadmap for plaintiff's attorneys searching for the elements of a negligence claim. Other rules pose similar risks. Smaller provisions of the rule may also be magnified in negligence claims. The use of high risk facility designation or the use of mandatory recall authority could aid in both inferring causation and proving failure to follow preventative controls. FSMA by no means requires perfection, but the level of documentation will be both, a rich source for finding the elements of a negligence claim and also a basis to infer the lack of documentation indicating improper hazard control and thus likely causation of the incident. All of this remains to be tested in the court.

7.2.2   Labeling Litigation

The Origins of Labeling Litigation

Labeling lawsuits are barely a decade old. Unlike previous chapters that explored early case law, ranging from 1911 through the 1930s, 1940s, 1950s and on, labeling litigation is new to the food law scene. Labeling litigation's origins lie in 2005 following a precipitous drop in FDA enforcement of labeling and a corresponding increase in creative and non-compliant claims.

The market was ripe for the wave of new claims seen beginning in early 2000. The CDC began to issue reports on obesity in US adults, which the media began to report on and describe as an epidemic. The focus shifted to childhood obesity as an American epidemic. As the drumbeat of bad news carried on, a demand to surge for healthier foods began. Chapter 4 described the market share for "all natural" products, currently at US$ 41 billion, and "organic" currently at US$ 9 billion. The market demand is geared toward local, natural, and pure. It curiously demands that its processed foods should be as unprocessed as possible.

The FDA watched as the market flooded with new "eye-catching" claims but did little to enforce existing regulations. Amid this backdrop of market demand for more wholesome products, manufacturers began labeling products "all natural" "GMO free" "whole grains" or some other statement to imply the product was minimally processed or otherwise pure. The labeling created what one reporter described as a "halo effect" where the claims lead to a perception of healthfulness (NY Times 2008).[8] Consumers would accept the claims and not make any additional assessment about the wholesome or healthfulness of the product. Trusting the claims consumers ate the foods often exceeding the serving size because of a mistaken belief that the product was healthy. The FDA noted in a statement, "…with consumers' growing interest in eating healthy, we've seen the emergence of eye-catching claims and symbols on the front of food packages that may not provide the full picture on their products true nutritional value" (Hamburg 2010).[9] The FDA, however, could not effectively enforce the market.

---

[8] Tierney (2008).

[9] Margret Hamburg, M.D., Comm'r FDA, Remarks at the Atlantic Food Summit, Washington, D.C. (March 2010).

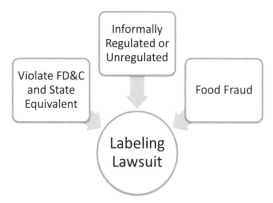

Fig. 7.4 Labeling litigation categories

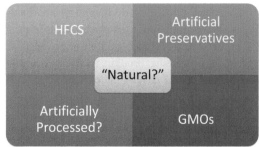

Fig. 7.5 Illustration of questions raised in "natural" suits

It is in this context that the first labeling lawsuit arose in 2005. In 2004 the non profit consumer advocacy group Center for Science in the Public Interest (CSPI) established a litigation department. The mission of the new litigation department would be "to fill the void left by the inactive government agencies" (CSPI Litigation). CSPI began pursuing a number of labeling claims and securing victories either in court or by settlement. It reached its first settlement in 2005 against Aunt Jemima's corporate parent Pinnacle Foods. It sought a misbranding and quasi-economic-adulteration case against Pinnacle for its blueberry waffles made suing "blueberry bits." In the settlement Pinnacle agreed to label the product as "artificially flavored" and stated that the blueberries were imitations. The suits and settlements by CSPI and others would continue to grow.

The cases multiplied rapidly. One estimate placed the number of labeling lawsuits at only 19 federal court cases in 2008 (U.S. Chamber 2013).[10] In the span of 4 years the number would grow to more than 102 federal court cases. The number continues to swell with many commentators now describing food labeling as replacing the era of litigation targeting tobacco companies with class actions.

The large number of cases can be difficult to sift through. It is helpful to separate the cases into three categories (see Fig. 7.4). The first category involves cases making claims that are legal but unregulated. In this category are claims like "all natural", "nutritious", or "healthful", which are permitted by the FDA under informal policy or no policy or regulation, but which are misleading. The second category centers on claims that violate State law equivalents to the FD&C and FDA regulations. Common claims for this category involve health and nutrient content claims, which are regulated by the FDA but also may violate State equivalent regulations or statutes. A third category involves food fraud. This unique category involves cases where a product claims to contain particular ingredients or amounts of ingredients, but upon analysis does not contain the ingredient, or not at the levels claimed.

"Natural" Suits

Natural lawsuits involve products labeled with claims like "all natural" on products containing some unnatural ingredient. This type of case fits in category one chiefly because the claim "all natural" is only subject to an informal policy. Products challenged using the "all natural" label typically contain artificial preservatives, high fructose corn syrup (HFCS), genetically modified organisms (GMOs), or other chemicals or unnatural ingredients (see Fig. 7.5).

The FDA refuses to issue a regulation defining the term "natural" under NLEA. The Agency considered issuing a rule in 1993, but cited resource limitations and other priorities as it abandoned the effort (Fed. Reg. 1993).[11] Instead, the

---

[10] U.S. Chamber Institute for Legal Reform (2013).

[11] 58 Fed. Reg. 2407 (Jan. 1993).

FDA adopted an informal policy which would assess the use of the term "natural" on a case-by-case basis. There would be no formal rule adopted. This decision has left courts to interpret the meaning of natural under the Act, often when considering labeling challenges.

Dozens of cases involve claims with some variant on "all natural." Given the frequent number of settlements there are only a few district or Circuit Court opinions to reference. Those Circuit Court options that are available frequently focus on the issue of preemption rather than a lengthy discussion of the substantive issue of whether the claims were misleading. Settlements for "all natural" claims reported in the media include most major brands. Kellogg's Kashi brand, for example, settled after a class action challenged the use of "all natural" and "nothing artificial" on products containing a long list of substances like pyridoxine hydrocholoride and tocopherols (American Bar Association).[12] Ben & Jerry's is another well-known brand sued for its claim that its ice cream was "all natural" (Ben & Jerry's Complaint).[13] Kellogg's also agreed to stop using the "all natural" claim. The complaint cited ingredients such as alkalized cocoa and other processed ingredients (Kashi Complaint).[14] The company settled for US$ 5 million. The lawsuits benefited from the FDA informal policy.

A number of cases specifically focused on whether the use of HFCS qualified as "natural." A case involving Snapple sought damages from the drink maker on the basis that it used HFCS and labeled its products "natural." HFCS, as the argument holds are not like natural sugars, but are unnatural or processed. Snapple is rare in that the plaintiffs lost, not because the court deemed HFCS "natural," but because they could not show that they suffered an economic harm by paying a premium for Snapple's "natural" drinks. Snapple did drop the claim from its label shortly after the suit was dismissed. Not the desired pay-day

sought by the plaintiff's, but it could be described as effective private regulation.

Recent "natural" suits focus on the use of GMOs. Here again the FDA does not have a regulation or definition for GMOs. It regulates GMOs the same as any other food. The lawsuits, however, see a distinction calling GMOs unnatural foods. ConAgra Foods, offers an example of a GMO "natural" suit. ConAgra sells its Wesson line of cooking oils with the claim that it's "100 % natural" (Food Safety News).[15] The complaint asked the court to find the GMO ingredients in the product artificial or non natural, thus making the claim misleading. Other GMO suits involve General Mills and Frito Lay. There is yet to be a definitive answer on the subject from the courts.

Unregulated Misleading Claims

A number of labeling lawsuits allege false or misleading statements that are not regulated by the FDA such as "wholesome" or "balanced breakfast." This type of suit also fits into category one. Critics of this type of lawsuit argue that the suits are frivolous with no true deception because the ingredients listing and nutrition fact panel provide the consumer all the necessary information to understand the context of the claim. Courts often reject this view finding compliance with ingredient listing and fact panels that cannot cure misleading claims without making the misbranding provisions of the Act superfluous. Unlike the "natural" claims cases where the question is whether the product truly is natural, the question in this type of case is whether the product actually is "wholesome" or "healthy."

Nutella provides a much talked about example of this type lawsuit. Nutella maker, Ferrero USA Inc. settled for US$ 3 million following a class action law suit alleging that it mislead consumers with deceptive claims (NPR).[16] The claims in question states that the Nutella spread was part of "balanced breakfast" and described the product as "healthy" and "nutritious." Not mentioned in its labeling claims or marketing was its high sugar

---

[12] American Bar Association (ABA)(2010).

[13] *Astiana v. Ben & Jerry's Homemade, Inc.*, Case No. 10-cv-4387 (N.D. Ca. filed Sept. 29, 2010).

[14] *E.g., Bates, et. al. v. Kashi Co., et. al.*, No. 3:11-cv-01967 (S.D. CAL. Filed Aug. 24, 2011).

[15] Food Safety News (2011).

[16] NPR (2012).

and fat content. This lead the lead plaintiff to decry the product that was the "next best thing to a candy bar" and far from the healthy, balanced snack depicted in marketing (NPR).[17] Despite compliance with ingredient listing and nutrition facts Ferrero saw risk in defending the suit and settled.

Many other lawsuits follow a similar pattern as Nutella. Tropicana was sued for its claim on orange juice products that stated it was "100% pure" and "natural" (Tropicana Complaint).[18] Plaintiffs argued this was misleading because the juice was pasteurized, deaerated, colored, and flavored. Quaker Oats faced a lawsuit alleging that its claims on oatmeal and oatmeal bars were misleading (Quaker Oats Complaint).[19] A wide variety of claims were cited by plaintiff's including "wholesome" and "All the Nutrition of a Bowl of Instant Oatmeal." The plaintiffs pointed to partially hydrogenated oils, a form of trans fats as an evidence that the claims mislead the public about the nutrition and healthfulness of its products. As can be seen, its not incredibly novel claims that are causing trouble. It is an odd combination of modern processed food ingredients with claims typically used for unprocessed products.

A surge of suits focused on the use of evaporated cane juice (ECJ), on labels instead of sugar or dried cane syrup. Plaintiffs brining suits challenging the use of ECJ stated it sounded like a healthier product, but they actually hide the sugar content. They also argued it violated the FDA's standard of identity regulations. Suits were brought against Chobani and Trader Joe's to name a few. Prior to the suits, the FDA issued a Draft Guidance, which is the lowest rung of authority the Agency can use, stating that the term juice does not encompass ECJ (FDA 2014).[20] It never finalized the Guidance, but did issue warning letters stating that the use of ECJ was false and misleading because it did not adequately inform the consumer that the product was sugars or syrups and not juices.

Unsubstantiated Health and Nutrition Claims

In the second category of suits are two related types of lawsuits. The claims in question are health or nutrient content claims, but the wrongdoing varies between unsubstantiated and unauthorized. If the claim is an unauthorized claim it is one that failed to follow the procedure set by the FDA under NLEA. If it is an unsubstantiated claim it simply is not supported by current studies or science.

CSPI sued several major U.S. companies for unsubstantiated health and nutrition claims. In 2007, it sued Coca-Cola and Nestlé for claims made on a green tea drink called Enviga (CSPI Press Release).[21] Enviga was labeled "the calorie burner" and in marketing materials states it aided in weight loss. The claims went so far as to claim the product contained "negative calories." In its suit CSPI alleged the claims were unsupported and lacked any evidence that consumers would experience any calorie burning benefit. The suit led to a settlement and investigations by 28 States Attorneys General.

In another example from CSPI it threatened suit against Smart Balance, Inc. for labeling on its Smart Balance Blended Butter Sticks. The butter sticks were labeled prominently with the claim "Plant Sterols to Help Block Cholesterol in the Butter" (CSPI Press Release 2014).[22] This type of claim dances on the line between health claim and disease or new drug claim. In characterizing the relationship between an ingredient and an effect on a disease it was also a health claim. CSPI alleged that the health claim did not follow the FDA regulations and was misleading because the products contained a disqualifying amount of saturated fat to be a health claim.

The proper use of health claims can provide a potential shield from labeling litigation. Yet,

---

[17] *Id.*

[18] *Lynch v. Tropicana Prods., Inc.*, No. 2:11-cv-07382 (DMC)(JAD) (D.N.J. June 12, 2013).

[19] *Chacanaca v. The Quaker Oats Company*, Docket No. 5:10-cv-00502 (N.D. Cal. Feb. 3, 2010).

[20] FDA, Draft Guidance for Industry on Ingredients Declared as Evaporated Cane Juice; Reopening of Comment Period; Request for Comments, Data, and Information (March 5, 2014).

---

[21] CSPI (2007).

[22] CSPI (2014).

the misuse of such claims provides fodder for consumer advocacy groups and private plaintiffs to bring suit. Typically the fear of noncompliance would be a regulatory correction from the FDA. In the current environment, concern is secondary to potential exposure to litigation. Occasionally the FDA will follow-up on labeling litigation or settlements with warning letters. More commonly litigation quickly follows a Warning Letter.

Unauthorized Nutrient Content Claims

Another class of nutrient content claims rather that not following the requirements of a regulation are completely unauthorized. As discussed in Chapter 4 only the nutrient content claims provided in the regulations may be used on labeling. If a desired claim, which qualifies as a nutrient content claim, is not in the regulations then a process must be followed to gain approval. In some cases the unauthorized health claims border on disease claims. Typically, one would expect the FDA to enforce the use of unapproved nutrient content claims. The Agency does issue warning letters, but continues to struggle keeping pace with the market.

An example of a warning letter leading to a labeling lawsuit is seen in the case of Cheerios. The cereal from General Mills received a warning letter in 2009 for the use of unauthorized health claims. The cereal maker boldly stated, "you can Lower Your Cholesterol 4% in 6 weeks." The FDA not only found the claim to be an unauthorized health claim, but also demonstrating an intended use to "prevent, mitigate, or treat" hypercholesterolemia and coronary heart disease. A swarm of class action labeling lawsuits shortly followed the FDA publication of the letter (Brookings).[23]

A 2012 lawsuit against Bumble Bee provides another lesson stating the use of unauthorized health claims. In 2012 a lawsuit alleged Bumble Bee made unapproved nutrient content claims about the properties of Omega 3 (Bumble Bee Complaint).[24] It labeled products as "Rich in Natural Omega-3" or "Excellent Source of Omega-3" which were not previously authorized by the FDA. The use of "rich" and "excellent" are similar to the use of "good source" under NELA. These types of claims need a daily value to compare what level is expected in the diet. The FDA has not established a daily value for Omega-3. There was no defense for Bumble Bee using the claim since it lacked authorization and could not point to persuasive authority as there was no daily value recommended for the ingredient.

Food Fraud

Food fraud can take many forms. The term fraud carries a negative connotation that may suggest an intentional deception or other nefarious act. In most cases food fraud involves economic adulteration. The pure push to reduce costs and increase profits leads to a conflicting aim to claim wholeness while using the cheapest ingredients possible. Other claims, likewise seeking to gain market share, stretch product's attributes to the limit. When these claims fall flat consumers sue for fraud. In most cases the issue is not safety, but solely deception. For example, products falsely claiming to contain superior or healthy ingredients, or offering benefits not seen by a consumer or supported by any scientific analysis.

There are numerous examples of food fraud suits alleging some form of economic adulteration. The introduction to this topic provided the example of Aunt Jemima's blue berry waffles only containing "blue berry bits." The suit against an importer of grape seed oil sought damages because the product contained less than 25% grape seed oil (Marquez Complaint).[25] A similar lawsuit against Kangadis Food Inc. alleges it passed off pomace oil as "100% Pure Olive Oil." This type of case is rooted solely in State law (Class Action Notice).[26] There is typically no need to cite FDA regulations or provisions of the Act.

---

[23] Brookings (2014).

[24] *Ogden v. Bumble Bee Foods*, No. CV12-01828 (N.D. Cal. April 12, 2012).

[25] *Marques v. Overseas Food Distributors*, No. BC 535015 (Cal. Super., Los Angeles County Jan. 31, 2014).

[26] Class Action Notice, If You Purchased Capatriti 100% Pure Olive Oil, A Class Action Lawsuit May Affect Your Rights available at: http://kangadislawsuit.com/.

General Mills, Inc. 5/5/09

Department of Health and Human Services

Public Health Service
Food and Drug Administration

Minneapolis District Office
Central Region
250 Marquette Avenue, Suite 600
Minneapolis, MN 55401
Telephone: (612) 758-7114
FAX: (612) 3344142

May 5,2009

WARNING LETTER

CERTIFIED MAIL
RETURN RECEIPT REQUESTED

Refer to MIN 09 -18

Ken Powell
Chairman of the Board and CEO
General Mills
One General Mills Boulevard
Minneapolis, Minnesota 55426

Dear Mr. Powell:

The Food and Drug Administration (FDA) has reviewed the label and labeling of your Cheerios®
Toasted Whole Grain Oat Cereal. FDA's review found serious violations of the Federal Food,
Drug, and Cosmetic Act (the Act) and the applicable regulations in Title 21, Code of Federal
Regulations (21 CFR). You can find copies of the Act and these regulations through links in
FDA's home page at http://www.fda.gov.

Unapproved New Drug

Based on claims made on your product's label, we have determined that your Cheerios®
Toasted Whole Grain Oat Cereal is promoted for conditions that cause it to be a drug because
the product is intended for use in the prevention, mitigation, and treatment of disease.
Specifically, your Cheerios® product bears the following claims ort its label:

• "you can Lower Your Cholesterol 4% in 6 weeks" "
• "Did you know that in just 6 weeks Cheerios can reduce bad cholesterol by an average of 4

percent? Cheerios is ... clinically proven to lower cholesterol. A clinical study showed that eating two 1 1/2 cup servings daily of Cheerios cereal reduced bad cholesterol when eaten as part of a diet low in saturated fat and cholesterol."

These claims indicate that Cheerios® is intended for use in lowering cholesterol, and therefore in preventing, mitigating, and treating the disease hypercholesterolemia. Additionally, the claims indicate that Cheerios® is intended for use in the treatment, mitigation, and prevention of coronary heart disease through, lowering total and "bad" (LDL) cholesterol. Elevated levels of total and LDL cholesterol are a risk factor for coronary heart disease and can be a sign of coronary heart disease. Because of these intended uses, the product is a drug within the meaning of section 201(g)(1)(B) of the Act [21 U.S.C. § 321 (g)P)(B)]. The product is also a new drug under section 201(p) of the Act [21 U.S.C. § 321(p)] because it is not generally recognized as safe and effective for use in preventing or treating hypercholesterolemia or coronary heart disease. Therefore, under section 505(a) of the Act [21 U.S.C. § 355(a)], it may not be legally marketed with the above claims in the United States without an approved new drug application.

FDA has issued a regulation authorizing a health claim associating soluble fiber from whole grain oats with a reduced risk of coronary heart disease (21 CFR 101.81). Like FDA's other regulations authorizing health Claims about a food substance and reduced risk of coronary heart disease, this regulation provides for the claim to include an optional statement, *as part of* the health claim, that the substance reduces the risk of coronary heart disease through the intermediate link of lowering blood total and LDL cholesterol. See 21 CFR 101.81(d)(2),-(3). Although the lower left corner of the Cheerios® front label contains a soluble fiber/coronary heart disease health claim authorized under 21 CFR 101.81, the two claims about lowering cholesterol are not made as part of that claim but rather are presented as separate, stand-alone claims through their location on the package and other label design features. The cholesterol claim that mentions the clinical study is on the back of the Cheerios® box, completely separate from the health claim on the front label. Although the other cholesterol claim is on the same panel as the authorized health claim, its prominent placement on a banner in the center of the front label, together with its much larger font size, different background, and other text effects, clearly distinguish it from the health claim in the lower left corner.

Additionally, even if the cholesterol-lowering claims were part of an otherwise permissible claim, under 21 CFR 101.81, the resulting claim language still would not qualify for the use of the soluble fiber health claim. To use the soluble fiber health claim, a product must comply with the claim specific requirements in 21 CFR 101.81, including the requirement that the claim not attribute any degree of risk reduction for coronary heart disease to diets that include foods eligible to bear the claim. See 21 CFR 101.81(c)(2)(E). However, the label of your Cheerios® cereal claims a degree of risk reduction for coronary heart disease by stating that Cheerios® can lower cholesterol by four percent in six weeks. High blood total and LDL cholesterol levels are a surrogate endpoint for coronary heart disease; therefore, the cholesterol-lowering claims on the Cheerios® label attribute a degree of risk reduction for coronary heart disease because if total and LDL cholesterol levels decline, the risk of coronary heart disease declines as well.

Misbranded Food:

Your Cheerios ® product is misbranded within the meaning of section 403(r)(1)(B) of the Act [21

U.S.C. § 343(r)(1)(B)] because it bears unauthorized health claims in its labeling. We have determined that your website www.wholegrainnation.com is labeling for your Cheerios® product under section 201(m) of the Act [21 U.S.C. § 321 (m)] because the website address appears on the product label. This website bears the following unauthorized health claims:

- "Heart-healthy diets rich in whole grain foods, can reduce the risk of heart disease."

This health claims misbrands your product because it has not been authorized either by regulation [see section 343(r)(3)(A)-(B) of the Act [21 U.S.C. § 343(r)(3)(A)(B)]] or under authority of the health claim notificati6n provision of the Act [see section'343(r)(3)(C) of the Act [21 U.S.C. § 343(r)(3)(G)]]. Although FDA has issued a regulation authorizing a health claim associating fiber-containing grain products with a reduced risk of coronary heart disease (21 CFR 101.77), the claim on your website does not meet the requirements for this claim. For example, under section 101.77(c)(2), the claim must state that diets low in saturated fat and cholesterol and high in fiber-containing fruit, vegetable, and grain products may reduce the risk of heart disease. The claim on your website leaves out any reference to fruits and vegetables, to fiber content, and to keeping the levels of saturated fat and cholesterol in the diet low. Therefore, your claim does not convey that all these factors together help to reduce the risk of heart disease and does not enable the public to understand the significance of the claim in the context of the total daily diet (see section 343(r)(3)(B)(iii) of the Act [21 U.S.C.§ 343(r)(3)(B)(iiiII].

In addition to the health claim authorized by regulation in 21 CFR 101.77, other health claims linking the consumption of whole grain foods to a reduced risk of heart disease have been authorized through the notification procedure in section 403(r)(3)(C) of the Act. Of those authorized claims, the one closest to the claim on your website states: "Diets rich in whole grain foods and other plant foods, and low in saturated fat and cholesterol, may help reduce the risk of heart disease.1" Although the claim on your website also concerns whole grains and reduced risk of heart disease, it is different from the authorized claim in significant ways. To meet the requirements of the authorized claim, the claim must state that diets that are (1) rich in Whole grains and other plant foods, and (2) low in saturated fat and cholesterol will help reduce the risk of heart disease) Instead, the claim on your website only states that diets rich in whole grains can reduce the risk of heart disease, with no mention of other plant foods or of low saturated fat and cholesterol.

- "Including whole grain as part of a healthy diet may ... [h]elp reduce the risk of certain types of cancers. Regular consumption of whole grains as part of, a low-fat diet reduces the risk for some cancers, especially cancers of the stomach and colon."

This health claim misbrands your product because it has not been authorized either by regulation [see section 343(r)(3)(A)-(B) of the Act [21 U.S.C. § 343(r)(3)(A)(B)]] or under authority of the health claim notification provision of the Act [see section 343(r)(3)(C) of the Act [21 U.S.C. § 343(r)(3)(C)]]. Although FDA has issued a regulation authorizing a health claim associating fiber-containing grain products with a reduced risk of cancer (21 CFR 101.76), the claim on your website does not meet the requirements for the authorized claim. For example, under section 101.76(c)(2) the claim must state that diets high in fiber-containing grain products, fruits, and vegetables may reduce the risk of some cancers. The claim on your website leaves out any reference to fruits, vegetables, and fiber content. Therefore, your claim does not convey that all these factors together help to reduce the risk of heart disease and does not enable the public to understand the significance of the claim in the context of the total daily diet [see section 343(r)(3)(B)(iii) of the Act [21 U.S.C. § 343(r)(3)(B)(iii)]].

In addition to the health claim authorized by regulation in 21 CFR 101.76, a health claim linking the consumption of whole grain foods to a reduced risk of certain cancers has been authorized through the notification procedure in section 403(r)(3)(C) of the Act. The authorized claim is: "Diets rich in whole grain foods and other plant foods ... may help reduce the risk of... certain cancers."2 Although the claim on your website also concerns whole grains and reduced risk of some cancers, it is different from the authorized claim in significant ways. For example, the authorized claim states that diets rich in whole grain foods and "other plant foods" may help reduce the risk for certain cancers. However, the claim on your website does not mention "other plant foods." Also, by using the language "especially cancers of the stomach and colon" the claim on your website emphasizes the relationship between whole grain foods and stomach and colon cancers as compared to other cancers, suggesting a greater degree of risk reduction or stronger evidence for the relationship between whole grain foods and risk of those two cancers. The claim authorized through the notification procedure does not emphasize the relationship between whole grain foods and stomach and colon cancer as compared to other cancers.

This letter is not intended to be an all-inclusive review of your products and their labeling. It is your responsibility to ensure that all of your products are in compliance with the Act and its implementing regulations.

Failure to promptly correct the violations specified above may result in enforcement action without further notice. Enforcement action may include seizure of violative products and/or injunction against the manufacturers and distributors of violative products.

Please advise this office in writing 15 days from your receipt of this letter of the specific steps you have taken to correct the violations noted above and to ensure that similar violations do not occur. Your response should include any documentation necessary to show that correction has been achieved. If you cannot complete all corrections before you respond, state the reason for the delay and the date by which you will complete the corrections.

Please send your reply to the attention of Tyra S. Wisecup, Compliance Officer, at the address in the letterhead. If you have any questions regarding this letter, please contact Ms. Wisecup at (612) 758-7114.

Sincerely,

/s/

W. Charles Becoat
Director
Minneapolis District

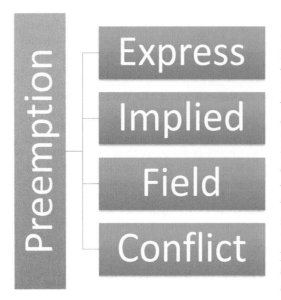

Fig. 7.6 Types of preemption

The case simply asks whether consumers were defrauded by the products.

One of the largest settlements involved food fraud. In 2010 Dannon settled for US$ 45 million when confronted with a suit alleging false and misleading claims for its Activia probiotic yogurt (ABC News).[27] The Activia and DanActive brands made claims that their yogurt was "clinically" and "scientifically" proven to improve digestion and boost immunity. Dannon was unable to prove that the claims were anything other than empty marketing ploys.

Preemption

One tool to defend against a labeling lawsuit is preemption. It is perhaps the only instrument to shield a manufacturer from a full trial or more likely a settlement in lieu of trial. Preemption, as described in Chapter 4, is a doctrine unique to the U.S. It holds private actions, which rely on State law, which may not proceed if preempted by federal law. A handful of labeling lawsuits raised the challenge to stave off a settlement or trial.

Preemption is a complex area of law involving several grounds to bar State causes of actions. Preemption can be expressed, where Congress in the federal statute states an intent to preempt State law. State causes of action may not be barred if the parallel claim exception applies. The parallel claim exception allows a State law claim that is identical, or parallel, to the requirements of the federal law. Implied preemption evaluates the Congressional intent of passing legislation to determine if preemption was implied. Field preemption occurs where Federal regulation is so pervasive as to dominate the field leaving no room for State law, parallel or otherwise. Finally, conflict preemption describes the scenario where Federal and State law conflict make compliance between an either–or, but not both. This type of preemption can also be expressed or implied. (see Fig. 7.6)

Conflict preemption provides insight into the FDA's role in "all natural" lawsuits. The Snapple example provided above was challenged in court (*Snapple*).[28] Snapple argued that the claims were preempted since it could not comply with both, State law and the FDA's informal policy and enforcement history. The court found that the FDA's informal policy was insufficient, lacking "preemptive weight." (*Snapple*).[29] The court noted that the FDA did not appear settled and the definition could change. This varies from the express preemption in NELA which remains settled since its enactment. Snapple waived its right to express preemption, but did attempt to state that HFCS was an artificial color and preempted the State law claim. The court did not decide the issue but noted that the FDA treats HFCS as a sweetener not a flavoring (*Snapple*).[30] In the absence of clear definitive policy from the FDA, defining the term "all natural" preemption will be a defense subject to the case precedent of the district and circuit courts. This provides little predictability for firms making such claims or facing these suits.

---

[27] ABC News (2010).

[28] *Holk v. Snapple Beverage Corp.*

[29] *Id.* at 340-342.

[30] *Id.* at 336 n.3.

Holk v. Snapple Beverage Corp., 575 F.3d 329
SMITH, Circuit Judge.

This appeal presents three issues related to the federal preemption of state causes of action. Plaintiff-appellant Stacy Holk brought several State law claims against defendant-appellee the Snapple Beverage Corporation in the Superior Court of New Jersey. After removing Holk's lawsuit to the United States District Court for the District of New Jersey, Snapple sought to dismiss Holk's complaint on, inter alia, the grounds of express preemption, implied field preemption, and implied conflict preemption. The District Court granted Snapple's motion on the basis of implied preemption. For the reasons discussed below, we will reverse.

I.

...

B.

Snapple Beverage Corporation ("Snapple") manufactures a variety of beverages, including a number of juice and tea-based drinks. In its marketing and advertising materials, Snapple represents that these beverages are "All Natural." As the FDA has acknowledged, "[t]he word `natural' is often used to convey that a food is composed only of substances that are not manmade and is, therefore, somehow more wholesome." Food Labeling: Nutrient Content Claims, General Principles, Petitions, Definition of Terms, 56 Fed.Reg. 60,421, 60,466 (Nov. 27, 1991). Snapple products, however, contained high fructose corn syrup ("HFCS"), an ingredient manufactured from processed cornstarch.

Stacy Holk bought two bottles of Snapple on May 4, 2007. She paid $ 1.09 for each bottle. She had purchased other Snapple products over the preceding 6 years. Holk contends that the labels on these products are deceptive. She argues that consumers "have been, and continue to be, easy prey for Snapple's unlawful activities because

of their willingness to pay a premium price for foods and beverages, including Snapple beverages, that are represented to be 'All Natural.'"

C.

Holk filed a class action lawsuit against Snapple in the Superior Court of New Jersey, asserting claims on the basis of: (I) the New Jersey Consumer Fraud Act; (II) unjust enrichment and common law restitution; (III) breach of express warranty; and (IV) breach of the implied warranty of merchantability. Holk's claims were predicated on her belief that a number of statements on Snapple's labels were misleading. She argued that (1) Snapple products were not "All Natural" because they contained HFCS; (2) Snapple products were not "Made from the Best Stuff on Earth," as indicated on the label; and (3) Snapple falsely labeled some beverages, for example, calling one drink "Acai 333*333 Blackberry Juice," despite the fact that the drink contained neither acai berry juice nor blackberry juice.

Snapple removed the case to the United States District Court for the District of New Jersey pursuant to the Class Action Fairness Act, 28 U.S.C. 1453(b). It then filed a motion to dismiss. The parties subsequently agreed that Holk could amend her complaint, rather than respond to Snapple's motion. In October 2007, Holk filed an Amended Complaint, which reasserted that Snapple's labels were misleading because they claimed the products were "All Natural" and because Snapple advertised some products as containing juice that was not in the beverages. The Amended Complaint did not allege any claims based on Snapple's use of the phrase "Made From the Best Stuff on Earth." Snapple filed a second motion to dismiss, arguing that Holk's claims were preempted, that the claims should be dismissed under the doctrine of primary jurisdiction, and that the allegations failed to state a claim. Holk responded by dropping the argument related to the juice components of Snapple beverages, leaving only the claim that Snapple products containing HFCS were deceptively labeled "All Natural."

The District Court heard oral argument on Snapple's motion to dismiss in June 2008. On June 12, 2008, the District Court dismissed Holk's complaint. It held that Snapple's claims were preempted. In its opinion, the District Court correctly identified and discussed the three types of preemption. It also noted that Snapple argued that all three types of preemption were present in this case, as Snapple contended that (1) NLEA expressly preempted state labeling requirements that are not identical to federal requirements; (2) the comprehensive nature of the FDCA and its implementing regulations demonstrate that Congress intended the federal government to occupy the field; and (3) that State law stands as an obstacle to the purposes underlying the FDCA. Next, the District Court rejected Snapple's express preemption argument, stating that there was not "specific preemptive language" in the FDCA that covered the claims. Nonetheless, the Court ruled that "Plaintiff's claims in this case are impliedly preempted by the detailed and extensive regulatory scheme established by the [FDCA] and the FDA's implementing regulations."

The District Court stated that the FDA has used the broad authority granted to it under the FDCA to issue comprehensive regulations governing the labeling and naming of juice drinks. The Court declared that the comprehensive nature of these regulations demonstrate that "the FDA has carefully balanced beverage industry and consumer interests and created a complex regulatory framework to govern beverage labeling." Though it acknowledged that the FDA has not defined "natural," it found

that the "FDA has in fact contemplated the appropriate use of the term," as indicated by the FDA's definition of "natural flavor" and its informal policy regarding use of the term "natural." The Court also noted that the FDA has the authority to enforce the FDCA and regulations issued pursuant to it. In the Court's view, these factors counseled in favor of its conclusion "that the [FDCA] and FDA regulations so thoroughly occupy the field of beverage labeling at issue in this case that it would be unreasonable to infer that Congress intended states to supplement this area."

Finally, the District Court deferred to the agency's expertise in the regulation of food and beverages. It asserted that it would be inappropriate for the Court to set rules, which the FDA "with all of its scientific expertise" has not yet done. Thus, the District Court concluded that the claims were "impliedly preempted" because "permitting states through statutes or common law causes of action to impose additional limitations and requirements on beverage labels such as described here would create obstacles to the accomplishment of Congress's objectives...."

Holk filed this timely appeal...

III.

Snapple argues that the District Court's dismissal must be upheld "whether analyzed under the doctrine of express preemption, implied 'field' preemption, or implied 'obstacle' preemption." The preemption doctrine is rooted in Article VI of the United States Constitution, which states that the laws of the United States "shall be the supreme Law of the Land; ... any Thing in the Constitution or Laws of any State to the Contrary notwithstanding." U.S. Const. art. VI, cl. 2. Under the Supremacy Clause, federal law may be held to preempt State law where any of the three forms of preemption doctrine may be properly applied: express preemption, field preemption, and implied conflict preemption … We are guided in our preemption analysis "by the rule that `[t]he purpose of Congress is the ultimate touchstone in every preemption case.'"

…Additionally, we must begin our analysis by applying a presumption against preemption…"In areas of traditional state regulation, we assume that a federal statute has not supplanted State law unless Congress has made such an intention clear and manifest." …This requires that, if confronted with two plausible interpretations of a statute, we "have a duty to accept the reading that disfavors pre-emption."

Health and safety issues have traditionally fallen within the province of state regulation. This is true of the regulation of food and beverage labeling and branding… The federal government did not begin to regulate the labeling of food products until 1906, when Congress passed the Wiley Act. Nonetheless, Snapple argues that the presumption against preemption should not be applied "because of the century-long tradition of federal regulation over food and beverage misbranding, and the expansive scheme of juice-beverage labeling regulation in particular." The Supreme Court, however, rejected a similar argument in Levine and applied the presumption… Accordingly, all of Snapple's preemption arguments must overcome the presumption against preemption, as food labeling has been an area historically governed by State law.

A.

Snapple argues that Holk's State law claims are expressly preempted by NLEA, specifically 21 U.S.C. § 343-1(a)(3). As a threshold matter, however, we must consider whether this issue is properly before us. As stated above, Holk initially argued that Snapple's labels were misleading on several grounds, namely because Snapple claimed the products were "All Natural"

despite containing HFCS and because Snapple advertised some products as containing juice that was not in the beverages. Holk subsequently dropped the argument related to the juice contents of certain Snapple beverages. This prompted Snapple to concede, during oral argument before the District Court, that it was no longer arguing express preemption: "[T]here's only one preemption argument left because of the dropping of the juice claims…. There was expressed [sic] preemption, there was implied field preemption, and now there's implied obstacle preemption. And it's implied obstacle preemption that applies to the high fructose corn syrup natural claims." Yet, on appeal, Snapple again raises express preemption.

Holk argues that because Snapple did not raise express preemption before the District Court in relation to her HFCS argument, Snapple has waived its express preemption argument. Snapple counters that "[w]here a new ground would support affirmance, this Court may invoke it so long as it is supported by the record."

First, we note that the District Court did not rule in Snapple's favor on express preemption. The Court stated that it "agrees with Plaintiff that Congress has not explicitly preempted Plaintiff's claims by inserting any specific preemptive language into the [FDCA]…." It also noted that "Snapple's express preemption arguments were directed at Plaintiff's claims concerning the fruit juices contained in Snapple beverages, which Plaintiff has withdrawn." Because the District Court did not rule in Snapple's favor on its express preemption argument, we do not have an express preemption claim to affirm.

Second, Snapple is correct that this Court has held that "we may affirm a correct decision of the district court on grounds other than those relied upon by the district court." … However, this rule does not apply to cases in which the party has waived the issue in the district court. This Court has stated: "We may affirm the lower court's ruling on different grounds, provided the issue which forms the basis of our decision was before the lower court."

…We conclude that Snapple has waived its express preemption argument with regard to Holk's HFCS claims. Though Snapple contended in its two motions to dismiss that Holk's juice content claims were expressly preempted by 21 U.S.C. § 343-1(a)(3), it did not raise this provision with regard to Holk's HFCS claim. In fact, it did not raise any express preemption argument in response to the HFCS claim and explicitly disclaimed the applicability of express preemption to this claim. This clearly demonstrates that the issue was not before the District Court. For this reason, we conclude that the issue is waived.

B.

Field preemption occurs when State law occupies a "field reserved for federal regulation," leaving no room for state regulation… It may also be inferred when "an Act of Congress `touch[es] a field in which the federal interest is so dominant that the federal system will be assumed to preclude enforcement of State laws on the same subject.'" …Nonetheless, for field preemption to be applicable, "congressional intent to supersede State laws must be `clear and manifest.'" … Snapple asserts that Holk's claims are preempted because federal law occupies both the field of beverage regulation and the field of juice drinks regulation.

First, we note that NLEA declares that courts may not find implied preemption based on any provision of NLEA. It states that the Act "shall not be construed to preempt any provision of State law, unless such provision is expressly preempted under [21 U.S.C. § 343-1] of the Federal Food, Drug, and Cosmetic Act." … Accordingly, if we are to find that Holk's claims are impliedly

preempted, we must do so based on provisions of federal law other than NLEA.

Given this limitation, Snapple argues that the FDCA, preNLEA, broadly addressed labeling and the misbranding of food and beverage products.[5] Snapple has also argued, both in its brief and during oral argument, that the FDA has promulgated, pursuant to its authority under the FDCA, "exhaustive" regulations regarding juice products in particular. Finally, Snapple asserts that the FDA has addressed HFCS and declared it to be "natural." Snapple submits that "[f]ederal law thus comprehensively regulates misbranding of food in general, juice beverages in particular, the distinction between natural and artificial, and even the specific question of whether HFCS can be `natural.'" For this reason, Snapple maintains that the District Court's analysis was correct.

Holk argues that the field in this case is not juice regulation, but rather food and beverage labeling. She contends that NLEA forecloses the implied preemption of State law in the food and beverage field. She reasons that the limited nature of the express preemption provision in NLEA, which applies only to those federal laws specifically enumerated, "would serve no purpose and would simply be surplus if Congress had intended to occupy the entire field of food and beverage labeling." She also cites NLEA's legislative history to demonstrate that Congress intended to preserve state authority in the food and beverage labeling field.

As discussed briefly above, field preemption requires a demonstration that "Congress… left no room for state regulation of these matters." … It does not appear that Congress has regulated so comprehensively in either the food and beverage or juice fields that there is no role for the states. First, there was no express preemption provision in the FDCA prior to enactment of the NLEA… Thus, we are lacking a "clear and manifest" expression of Congressional intent to occupy either field.

Second, as Holk argues, NLEA's express preemption provision demonstrated that Congress recognized the existence of State laws relating to beverages generally and juice products specifically. See, e.g., 21 U.S.C. § 343-1(a)(2) (preempting State laws that conflict, inter alia, with federal law requiring foods to indicate: (1) the name and location of the manufacturer, as well as the weight or quantity of food contained in a package; and (2) the percentage of fruit or vegetable juice contained in a beverage). NLEA plainly states that the Act "shall not be construed to preempt any provision of State law, unless such provision is expressly preempted under [21 U.S.C. § 343-1] of the Federal Food, Drug, and Cosmetic Act." … Furthermore, NLEA declares that its express preemption provision "shall not be construed to apply to any requirement respecting a statement in the labeling of food that provides for a warning concerning the safety of the food or component of the food," thereby preserving state warning laws…These provisions demonstrate that Congress was cognizant of the operation of State law and state regulation in the food and beverage field, and it therefore enacted limited exceptions in NLEA. As the Supreme Court instructed in Levine, "'[t]he case for federal pre-emption is particularly weak where Congress has indicated its awareness of the operation of State law in a field of federal interest, and has nonetheless decided to stand by both concepts and to tolerate whatever tension there [is] between them.'"

Furthermore, we note that the FDA has stated that it does not intend to occupy the field of food and beverage labeling, even with regard to regulations affecting juice products…In a final rule published in 1986 concerning sulfiting agents, a substance present in some juice drinks, the FDA responded to a comment that it should adopt a policy that would result in the preemption of State law with regard to the labeling of food products containing sulfites. Food Labeling; Declaration of Sulfiting Agents, 51 Fed.Reg. 25,012, 25,016 (July 9, 1986). There, the FDA stated: "The agency does not use its authority to preempt State requirements unless there is a genuine need to stop the proliferation of inconsistent requirements between the FDA and the States. FDA is not persuaded that such a need now exists with regard to sulfite labeling." Id. Similarly, in two proposed rules regarding nutrition labeling on food and beverage products, the FDA acknowledged the receipt of numerous comments that urged the FDA to explicitly preempt contrary state labeling regulations. Food Labeling; Mandatory Status of Nutrition Labeling and Nutrient Content Revision, 55 Fed.Reg. 29,487, 29,509 (July 19, 1990) (seeking to amend the nutrition label as it pertains to the listing of nutrients); Food Labeling; Serving Sizes, 55 Fed.Reg. 29,517, 29,528 (July 19, 1990) (seeking to amend the nutrition label as it pertains to serving size). In both cases, the FDA responded:

The preemption issue is complex and divisive: whether a uniform, national label is necessary for consumers and manufacturers to function in the marketplace versus whether States should be 339*339 permitted to require additional information for their residents. The input of States, as well as consumers, businesses, and other concerned parties is essential in evaluating this matter. FDA therefore requests comment on the issue of whether preemption is appropriate.

Food Labeling; Mandatory Status of Nutrition Labeling and Nutrient Content Revision, 55 Fed.Reg. at 29,509; Food Labeling; Serving Sizes, 55 Fed.Reg. at 29,528.

Finally, we are reluctant to find field preemption predicated solely on the comprehensiveness of federal regulations. The Supreme Court has repeatedly stated that "the mere existence of a federal regulatory scheme," even a particularly detailed one, "does not by itself imply pre-emption of state remedies."…To conclude otherwise would be "virtually tantamount to saying that whenever a federal agency decides to step into a field, its regulations will be exclusive."

In the instant case, not only do we lack a "clear and manifest" statement from Congress of its intent to preempt, but we also note that the claims in this case are governed by the presumption against preemption. These factors, along with the Supreme Court's direction that we should not infer field preemption from the comprehensiveness of a regulatory scheme alone, lead us to conclude that neither Congress nor the FDA intended to occupy the fields of food and beverage labeling and juice products.

C.

Implied conflict preemption is present when it is "impossible for a private party to comply with both state and federal requirements." … Alternatively, conflict preemption results when State law "stands as an obstacle to the accomplishment and execution of the full purposes and objectives of Congress." … With regard to the latter, "'[i]f the purpose of the act cannot otherwise be accomplished—if its operation within its chosen field else must be frustrated and its provisions be refused their natural effect—the State law must yield to the regulation of Congress within the sphere of its delegated power.'" … Both federal statutes and regulations have the force of law and can preempt contrary State law…

Snapple submits that Holk's claims are preempted because they stand as an obstacle to federal law. It contends that the FDA has adopted a policy regarding the use of the term "natural" and that this policy would be undermined by Holk's suit. Specifically, it alleges that liability in Holk's suit would result in the imposition of "additional conditions not contemplated by the federal regime." Additionally, Snapple argues that State law must yield if it undermines federal efforts to create uniform standards.

Holk counters that her state causes of action do not serve as an obstacle to federal objectives because there "are no federal 340*340 requirements in place regarding the term 'natural.'" She also asserts that her claims do not conflict with federal law because, even if she obtained a favorable verdict, Snapple would not be required to undertake a specific corrective action.

To determine whether Holk's claims present an obstacle to federal law, we must as an initial matter consider whether the FDA has regulations or has otherwise taken actions that are capable of having preemptive effect. In *Fellner v. Tri-Union Seafoods, L.L.C.,* we declared "that it is federal law which preempts contrary State law; nothing short of federal law can have that effect." … We recognized that "there is no doubt that federal regulations as well as statutes can establish federal law having preemptive force."…Beyond this, however, we noted that "in appropriate circumstances, federal agency action taken pursuant to statutorily granted authority short of formal, notice and comment rulemaking may also have preemptive effect over State law." …For example, agency adjudications could have the force of law because agencies have the choice to address issues via rulemaking or adjudication. That said, we declared that not every agency action or statement would have preemptive effect.

In determining whether an agency action is entitled to deference, we will be guided by the Supreme Court's pronouncement that "'[i]t is fair to assume generally that Congress contemplates administrative action with the effect of law when it provides for a relatively formal administrative procedure tending to foster the fairness and deliberation that should underlie a pronouncement of such force.'" ... Accordingly, we declined in Fellner to "afford preemptive effect to less formal measures lacking the 'fairness and deliberation' which would suggest that Congress intended the agency's action to be a binding and exclusive application of federal law." Id. Finally, with respect to agency letters, we noted that "we have found no case in which a letter that was not the product of some form of agency proceeding and did not purport to impose new legal obligations on anyone was held to create federal law capable of preemption." Id.

In this case, we must determine whether the FDA's policy statement on the use of the word "natural" has preemptive effect. In 1991, the FDA announced that it was considering defining the term "natural" for the purpose of future rulemaking. Food Labeling: Nutrient Content Claims, General Principles, Petitions, Definition of Terms, 56 Fed.Reg. 60,421, 60,466 (Nov. 27, 1991). At that time, the FDA recounted its existing "informal policy" on the use of the term:

[T]he agency has considered "natural" to mean that nothing artificial or synthetic (including colors regardless of source) is included in, or has been added to, the product that would not normally be expected to be there. For example, the addition of beet juice to lemonade to make it pink would preclude the product being called "natural." Id. (emphasis added).

We conclude that the FDA's policy statement regarding use of the term "natural" is not entitled to preemptive effect. First, the FDA declined to adopt a formal definition of the term "natural." After soliciting comments on the use of the term "natural," the

FDA recognized that the use of the term "is of considerable interest to consumers and industry." Food Labeling: Nutrient Content Claims, General Principles, Petitions, Definition of Terms; Definitions of Nutrient Content Claims for the Fat, Fatty Acid, and Cholesterol Content of Food, 58 Fed. Reg. 2,302, 2,397 (Jan. 6, 1993).

It also stated that it believed "that if the term 'natural' is adequately defined, the ambiguity surrounding use of this term that results in misleading claims could be abated." Id. Nevertheless, the FDA declined to do so: "Because of resource limitations and other agency priorities, FDA is not undertaking rulemaking to establish a definition for 'natural' at this time." Id. This hardly supports preemption...

Though the FDA declined to adopt a formal definition of "natural," it declared that it would continue to adhere to the informal policy previously announced. Food Labeling: Nutrient Content Claims, General Principles, Petitions, Definition of Terms; Definitions of Nutrient Content Claims for the Fat, Fatty Acid, and Cholesterol Content of Food, 58 Fed.Reg. at 2397. This too, however, lacks preemptive weight. The FDA's request for comments on use of the term "natural" makes clear that the FDA's informal policy predated the request for notice and comment. Food Labeling: Nutrient Content Claims, General Principles, Petitions, Definition of Terms, 56 Fed.Reg. at 60,466. Because a search of the Federal Register results in neither earlier references to this policy nor other requests for comments on the use of the term "natural," the record demonstrates that the FDA arrived at its policy without the benefit of public input. Additionally, after requesting comments on the use of the term "natural," the FDA did not appear to consider all the comments received. For instance, the FDA noted that one comment questioned whether restrictions on the use of "natural" could raise First Amendment concerns. Food Labeling: Nutrient Content Claims, General Principles, Petitions, Definition of Terms; Definitions of Nutrient Content Claims for the Fat, Fatty Acid, and Cholesterol Content of Food, 58 Fed. Reg. at 2397. The FDA did not respond to this comment, as it declared it moot in light of its decision not to proceed with a definition. Id. In fact, despite numerous public comments, the FDA announced that it would adhere to its preexisting policy on the use of the term "natural" and make no changes. Id. at 2407. Finally, the FDA stated that it was declining to define "natural," in part, because there were still "many facets of the issue that the agency will have to carefully consider if it undertakes a rulemaking to define the term 'natural.'" Id. This statement alone demonstrates a lack of the kind of "fairness and deliberation" contemplated by Fellner.

Despite these shortcomings, Snapple argues that the FDA's policy is entitled to preemptive effect because the FDA has enforced the informal policy. In its briefs to this Court, Snapple directed our attention to several letters in which the FDA told a food or beverage manufacturer to remove the term "natural" from one of its labels for violating the FDA policy on the use of the term "natural." We do not think these letters are sufficient to accord the policy the weight of federal law. In Fellner, we recognized that Congress likely intended to give administrative action the effect of law when the agency adhered to "a relatively formal administrative procedure tending to foster the fairness and deliberation that should underlie a pronouncement of such force.'" …Thus, we were predominately focused on the process by which the agency arrived at its decision, rather than on what happened after that decision was made. In this case, the deficiencies inherent in the process by which the FDA arrived at

its policy on the use of the term "natural" are simply too substantial to be overcome by isolated instances of enforcement.

We believe that neither the FDA policy statement regarding the use of the term "natural" nor the FDA's letter indicating that some forms of HFCS may be classified as "natural" have the force of law required to preempt conflicting State law. Both lack the formal, deliberative process contemplated in Fellner. As a result, there is no conflict in this case because there is no FDA policy with which State law could conflict.

IV.

For the reasons discussed above, we conclude that Holk's claims are not preempted. We will reverse the judgment of the District Court, and remand to the District Court for further proceedings consistent with the foregoing opinion.

7.3 Criminal Actions

The discussion thus far focused on corporate responsibility to individuals brining a lawsuit. This section markedly shifts the discussion to corporate officer criminal liability in an action brought by the FDA or USDA. It can be unsettling to consider a CEO, President or Vice-President of a company bearing criminal sanctions for actions in their official corporate capacity. This action represents the pinnacle of federal enforcement and is currently in exceedingly rare use.

Criminal actions vary on severity. Many involve misdemeanor offenses, but some cases escalate to felony offenses. Within the FDA the Office of Criminal Investigations (OCI) is responsible for reviewing all matters that the civil division of the FDA recommends for criminal investigation. The OCI handles all criminal matters for the FDA regardless of the complexity or level of offense. The Regulatory Procedures Manual (RPM) in Subchapter 6-5 outlines the FDA's policy on referring civil cases for criminal enforcement. Within FSIS, the Office of Investigation, Enforcement and Audit (OIEA), conducts all criminal investigations. The OIEA develops and follows various FSIS directives to conduct criminal investigations and initiate prosecution.

Courts hearing criminal enforcements follow two doctrines, both articulated by the Supreme Court. The second emerged following a Fourth Circuit opinion that attempted to reign in the earlier doctrine only to be struck down by the Supreme Court. This alternative interpretation of personal criminal liability offers a tantalizing idea of what could have been and perhaps still what could be.

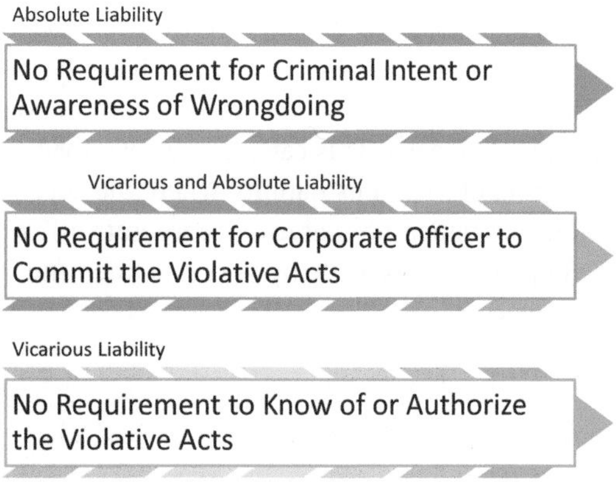

Absolute Liability

No Requirement for Criminal Intent or Awareness of Wrongdoing

Vicarious and Absolute Liability

No Requirement for Corporate Officer to Commit the Violative Acts

Vicarious Liability

No Requirement to Know of or Authorize the Violative Acts

Fig. 7.7 The Dotterweich Doctrine

7.3.1 Dotterweich Doctrine

The first doctrine follows only five years after the passage of the 1938 Act. It involved a criminal prosecution under Section 303(a) of the Act against the president and general manager. The charges alleged two counts of shipping adulterated drugs and one count of shipping misbranded drugs. The criminal complaint alleged Buffalo Pharmacal Co., Inc. of purchasing drugs from a wholesale manufacturer and repackaging them to fill orders (*Dotterweich*).[31] The president of the company Mr. Dotterweich did not personally ship the orders identified in the criminal complaint, but as the company president he managed the actions of the company. At the criminal trial the jury acquitted the corporation, but found Mr. Dotterweich guilty on all three counts. He was sentenced a fine of US$ 500 for each count, with payment suspended on the second and third counts, and 60-day probation for each count to run concurrently (Dotterweich Circuit Opinion).[32] Mr. Dotterweich appealed to the Second Circuit and secured a reversal of his conviction. The Second Circuit held that he was not a "person" within the meaning of the Act (*Dot-*

terweich Circuit Opinion).[33] The case was again appealed to the Supreme Court.

In a split-decision, 5-4, the Supreme Court reinstated the conviction and provided the Dotterweich Doctrine. The Dotterweich Doctrine holds a corporate officer criminally liable under the FD&C even though they are not conscious of wrongdoing, knew of or authorized the violative acts, and did not commit the acts. The court reasoned the Act that placed the burden of compliance on the corporate officers and the violation was enough to invoke the criminal sanctions of the Act (*Dotterweich*).[34] The court further held that the president was a person subject to responsibility under the Act because under Section 301 a corporation and "all persons who aid and abet its commission" are subject to criminal penalties if found guilty by a jury (*Dotterweich*).[35]

The Dotterweich Doctrine, like all interpretations of the Act, lays down broad markers for finding criminal liability (see Fig. 7.7). Under the doctrine, the distribution of adulterated or misbranded products regulated by the Act is a crime not requiring criminal intent or conscious awareness of wrongdoing. This is a doctrine of absolute liability where the prohibited act is sufficient to find criminal culpability. The Doctrine further holds a corporate official does not need to

---

[31] *United States v. Buffalo Co.*, 131 F.2d 500, 501 (2d Cir. 1942).

[32] *Id.* at 501.

[33] *Id.* at 502.

[34] *Dotterwich*, at 281.

[35] *Id.* at 284.

personally cause the violation or even be aware of it. Under this concept of absolute liability the official need only "aid and abet" the corporation in the violative act. This can also be described as a doctrine of vicarious liability where the acts of employees under the charge of a corporate officer are imputed to acts made by the corporate officer themselves.

United States v. Dotterweich, 320 U.S. 277 (1943)

Mr. Justice FRANKFURTER delivered the opinion of the Court.

This was a prosecution begun by two informations, consolidated for trial, charging Buffalo Pharmacal Company, Inc., and Dotterweich, its president and general manager, with violations of the Act of Congress of June 25, 1938... known as the Federal Food, Drug, and Cosmetic Act.

The Company, a jobber in drugs, purchased them from their manufacturers and shipped them, repacked under its own label, in interstate commerce. (No question is raised in this case regarding the implications that may properly arise when, although the manufacturer gives the jobber a guaranty, the latter through his own label makes representations.) The informations were based on 301 of that Act, 21 U.S.C. 331, 21 U.S.C.A. 331, paragraph (a) of which prohibits 'The introduction or delivery for introduction into interstate commerce of any... drug... that is adulterated or misbranded'. 'Any person' violating this provision is, by paragraph (a) of 303, 21 U.S.C. 333, 21 U.S.C.A. 333, made 'guilty of a misdemeanor'. Three counts went to the jury-two, for shipping misbranded drugs in interstate commerce, and a third, for so shipping an adulterated drug. The jury disagreed as to the corporation and found Dotterweich guilty on all three counts. We start with the finding of

the Circuit Court of Appeals that the evidence was adequate to support the verdict of adulteration and misbranding...

Two other questions which the Circuit Court of Appeals decided against Dotterweich call only for summary disposition to clear the path for the main question before us. He invoked 305 of the Act requiring the Administrator, before reporting a violation for prosecution by a United States attorney, to give the suspect an 'opportunity to present his views'. We agree with the Circuit Court of Appeals that the giving of such an opportunity, which was not accorded to Dotterweich, is not a prerequisite to prosecution. This Court so held in United States v. Morgan... in construing the Food and Drugs Act of 1906, 34 Stat. 768, 21 U.S.C.A. 1 et sEq. and the legislative history to which the court below called attention abundantly proves that Congress, in the changed phraseology of 1938, did not intend to introduce a change of substance... Equally baseless is the claim of Dotterweich that, having failed to find the corporation guilty, the jury could not find him guilty. Whether the jury's verdict was the result of carelessness or compromise or a belief that the responsible individual should suffer the penalty instead of merely increasing, as it were, the cost of running the business of the corporation, is immaterial. Juries may indulge in precisely such motives or vagaries...

And so we are brought to our real problem. The Circuit Court of Appeals, one judge dissenting, reversed the conviction on the ground that only the corporation was the 'person' subject to prosecution unless, perchance, Buffalo Pharmacal was a counterfeit corporation serving as a screen for Dotterweich. On that issue, after rehearing, it remanded the cause for a new trial. We then brought the case here, on the Government's petition for certiorari... because this construction raised questions of importance in the enforcement of the Federal Food, Drug, and Cosmetic Act.

The court below drew its conclusion not from the provisions defining the offenses on which this prosecution was based (301(a) and 303(a), but from the terms of 303(c). That section affords immunity from prosecution if certain conditions are satisfied. The condition relevant to this case is a guaranty from the seller of the innocence of his product. So far as here relevant, the provision for an immunizing guaranty is as follows:

'No person shall be subject to the penalties of subsection (a) of this section... (2) for having violated section 301(a) or (d), if he establishes a guaranty or undertaking signed by, and containing the name and address of, the person residing in the United States from whom he received in good faith the article, to the effect, in case of an alleged violation of section 301(a), that such article is not adulterated or misbranded, within the meaning of this Act, designating this Act....'

This Circuit Court of Appeals found it 'difficult to believe that Congress expected anyone except the principal to get such a guaranty, or to make the guilt of an agent depend upon whether his employer had gotten one.'... And so it cut down the scope of the penalizing provisions of the Act to the restrictive view, as a matter of language and policy, it took of the relieving effect of a guaranty.

The guaranty clause cannot be read in isolation. The Food and Drugs Act of 1906 was an exertion by Congress of its power to keep impure and adulterated food and drugs out of the channels of commerce. By the Act of 1938, Congress extended the range of its control over illicit and noxious articles and stiffened the penalties for disobedience. The purposes of this legislation thus touch phases of the lives and health of people which, in the circumstances of modern industrialism, are largely beyond self- protection. Regard for these purposes should infuse construction of the legislation if it is to be treated as a working instrument of government and not merely as a collection of English words... The prosecution to which Dotterweich was subjected is based on a now familiar type of legislation whereby penalties serve as effective means of regulation. Such legislation dispenses with the conventional requirement for criminal conduct-awareness of some wrongdoing. In the interest of the larger good it puts the burden of acting at hazard upon a person otherwise innocent but standing in responsible relation to a public danger. And so it is clear that shipments like those now in issue are 'punished by the statute if the article is misbranded (or adulterated), and that the article may be misbranded (or adulterated) without any conscious fraud at all. It was natural enough to throw this risk on shippers with regard to the identity of their wares....'

The statute (303) makes 'any person' who violates 301(a) guilty of a 'misdemeanor'. It specifically defines 'person' to include 'corporation'. 201(e). But the only way in which a corporation can act is through the individuals who act on its behalf...And the historic conception of a 'misdemeanor' makes all those responsible for it equally guilty...a doctrine given general application in 332 of the Penal Code, 18 U.S.C. 550, 18 U.S.C.A. 550. If, then, Dotterweich is not subject to the Act, it must be solely on the ground that individuals are immune when the 'person' who violates 301(a) is a corporation, although from the point of view of action the individuals are the corporation. As a matter of legal development, it has taken time to establish criminal liability also for a corporation and not merely for its agents. See *New York Central & H. R.R. Co. v. United*

States, supra. The history of federal food and drug legislation is a good illustration of the elaborate phrasing that was in earlier days deemed necessary to fasten criminal liability on corporations. Section 12 of the Food and Drugs Act of 1906, 21 U.S.C.A. 4, provided that, 'the act, omission, or failure of any officer, agent, or other person acting for or employed by any corporation, company, society, or association, within the scope of his employment or office, shall in every case be also deemed to be the act, omission, or failure of such corporation, company, society, or association as well as that of the person.' By 1938, legal understanding and practice had rendered such statement of the obvious superfluous. Deletion of words-in the interest of brevity and good draftsmanship1-superfluous for holding a corporation criminally liable can hardly be found ground for relieving from such liability the individual agents of the corporation. To hold that the Act of 1938 freed all individuals, except when proprietors, from the culpability under which the earlier legislation had placed them is to defeat the very object of the new Act. Nothing is clearer than that the later legislation was designed to enlarge and stiffen the penal net and not to narrow and loosen it. This purpose was unequivocally avowed by the two committees which reported the bills to the Congress. The House Committee reported that the Act 'seeks to set up effective provisions against abuses of consumer welfare growing out of inadequacies in the Food and Drugs Act of June 30, 1906'... And the Senate Committee explicitly pointed out that the new legislation 'must not weaken the existing laws', but on the contrary 'it must strengthen and extend that law's protection of the consumer.'...If the 1938 Act were construed as it was below, the penalties of the law could be imposed only in the rare case where the corporation is merely an individual's alter ego. Corpo-

rations carrying on an illicit trade would be subject only to what the House Committee described as a 'license fee for the conduct of an illegitimate business.' A corporate officer, who even with 'intent to defraud or mislead' (303(b), introduced adulterated or misbranded drugs into interstate commerce could not be held culpable for conduct which was indubitably outlawed by the 1906 Act. See, e.g., *United States v. Mayfield, D.C.*, 177 F. 765. This argument proves too much. It is not credible that Congress should by implication have exonerated what is probably a preponderant number of persons involved in acts of disobedience-for the number of non-corporate proprietors is relatively small. Congress, of course, could reverse the process and hold only the corporation and allow its agents to escape. In very exceptional circumstances it may have required this result... But the history of the present Act, its purposes, its terms, and extended practical construction lead away from such a result once 'we free our minds from the notion that criminal statutes must be construed by some artificial and conventional rule'.

...The Act is concerned not with the proprietary relation to a misbranded or an adulterated drug but with its distribution. In the case of a corporation such distribution must be accomplished, and may be furthered, by persons standing in various relations to the incorporeal proprietor. If a guaranty immunizes shipments of course it immunizes all involved in the shipment. But simply because if there had been a guaranty it would have been received by the proprietor, whether corporate or individual, as a safeguard for the enterprise, the want of a guaranty does not cut down the scope of responsibility of all who are concerned with transactions forbidden by 301. To be sure, that casts the risk that there is no guaranty upon all who according to settled doctrines of criminal law are responsible for the commission of a misdemeanor. To read the guaranty section, as did the court below, so as to restrict liability for penalties to the only person who normally would receive a guaranty-the proprietor-disregards the admonition that 'the meaning of a sentence is to be felt rather than to be proved'... It also reads an exception to an important provision safeguarding the public welfare with a liberality which more appropriately belongs to enforcement of the central purpose of the Act.

The Circuit Court of Appeals was evidently tempted to make such a devitalizing use of the guaranty provision through fear that an enforcement of 301(a) as written might operate too harshly by sweeping within its condemnation any person however remotely entangled in the proscribed shipment. But that is not the way to read legislation. Literalism and evisceration are equally to be avoided. To speak with technical accuracy, under 301 a corporation may commit an offense and all persons who aid and abet its commission are equally guilty. Whether an accused shares responsibility in the business process resulting in unlawful distribution depends on the evidence produced at the trial and its submission-assuming the evidence warrants it-to the jury under appropriate guidance. The offense is committed, unless the enterprise which they are serving enjoys the immunity of a guaranty, by all who do have such a responsible share in the furtherance of the transaction which the statute outlaws, namely, to put into the stream of interstate commerce adulterated or misbranded drugs. Hardship there doubtless may be under a statute which thus penalizes the transaction though consciousness of wrongdoing be totally wanting. Balancing relative hardships, Congress has preferred to place it upon those who have at least the opportunity of informing themselves of the existence of conditions imposed for the protection of consumers before sharing in illicit commerce, rather than to throw the hazard on the innocent public who are wholly helpless.

It would be too treacherous to define or even to indicate by way of illustration the class of employees which stands in such a responsible relation. To attempt a formula embracing the variety of conduct whereby persons may responsibly contribute in furthering a transaction forbidden by an Act of Congress, to wit, to send illicit goods across state lines, would be mischievous futility. In such matters the good sense of prosecutors, the wise guidance of trial judges, and the ultimate judgment of juries must be trusted. Our system of criminal justice necessarily depends on 'conscience and circumspection in prosecuting officers,' ... even when the consequences are far more drastic than they are under the provision of law before us .... For present purpose it suffices to say that in what the defense characterized as 'a very fair charge' the District Court properly left the question of the responsibility of Dotterweich for the shipment to the jury, and there was sufficient evidence to support its verdict.

Judgment reversed.

Court's decisions following *Dotterweich* continued to paint corporate criminal liability with broad strokes under the newly formed doctrine. In 1947 the cosmetic company faced criminal charges because one product, hair lacquer pads, contained a deleterious substance (*Parfait Powder Puff Co.*)[36] The defendant corporation argued on appeal that it was not its actions, but a contract manufacturer who added the "deleterious" substance. The court described his arrangement

---

[36] *United States v. Parfait Powder Puff Co.*, 163 F.2d 1008 (7th Cir. 1947), *cert. denied*, 332 U.S. 851 (1948).

with the contract manufacturer Helfrich Laboratories:

> Defendant, engaged in the manufacture and sale of cosmetic products, in 1943, entered into a contract with Helfrich Laboratories whereby the latter agreed to manufacture, place in packages and distribute to defendant's customers hair lacquer pads. Defendant supplied Helfrich with jars, caps, labels, display cards, flannel pads and shipping containers. Helfrich impregnated the pads with a shellac lacquer, placed them in labeled jars bearing defendant's name, shipped the packages, in accord with shipping directions furnished by defendant, consigned by defendant to its purchasers as consignees, and rendered bills to defendant for the commodity. (*Parfait Powder Puff Co.*)[37]

The court held the defendant corporation vicariously liable despite the contractual relationship, lack of knowledge, and utter absence of any participation. The court reasoned:

> It argues that Helfrich was not its agent, but an independent contractor, for whose acts it is not responsible. But we are not concerned with any distinction between independent contractors and agents in the ordinary sense of those words. It is clear that defendant was engaged in procuring the manufacture and distribution of the article in interstate commerce. It saw fit to create out of Helfrich's activities in its behalf an instrumentality and to avail itself of the acts of that instrumentality, which effected an introduction into commerce of an adulterated article violative of the standards fixed by the Act. This we think it could not do without incurring the criminal penalty imposed by the statute. The liability was not incurred because defendant consciously participated in the wrongful act, but because the instrumentality which it employed, acting within the powers which the parties had mutually agreed should be lodged in it, violated the law. (*Parfait Powder Puff Co.*)[38]

The decision in *Kordel v. United States* takes the Dotterweich Doctrine to an extreme of finding even scientific disagreement criminally punishable (Kordel).[39] *Kordel* provided the focal point in other parts of this text, but never a discussion of its implications on criminal law. As previously

discussed, *Kordel* involved the use of pamphlets that were separately shipped from his supplements, which were later classified as drugs based on the marketing claims. The dissent and many reviewers of the *Kordel* opinion, questionned the fairness of the majority opinion. The basis of the case rests on whether the pamphlets were criminal violations. Some may disagree with the statements and others may point to dangers of advising consumers to take a supplement rather than seek professional treatment, but the Dotterweich Doctrine does not provide a balancing of the risks and benefits. It is an absolute liability doctrine that holds any violation of the Act, which could be a basis for criminal penalty.

The Dotterweich Doctrine only requires a corporate officer holding the position with the corporate entity to be liable. The vicarious liability element of the doctrine does not require knowledge or involvement in the violative act. In one case the Ninth Circuit Court concluded that this did not require the president and general manager of the plant to be present at the facility during the time the violation occurred (Golden Grain Macaroni Co.).[40] The court dispensed of the issue stating that "the criminal responsibility of a corporate officer having broad authority such as that possessed by the defendant does not depend upon his physical presence" (*Golden Grain Macaroni Co.*).[41] This reasoning along with additional findings supported the conviction, which imposed a fine on the president of US$ 5000 (*Golden Grain Macaroni Co.*).[42]

Subsequent cases affirmed that defendants did not need to know or participate in the violation or develop any criminal intent. In *United States v. H. Wool & Sons, Inc.*, a corporation and corporate officer, Mr. Herbert Wool, was held criminally liable for a misbranding violation involving underweight repacked butter (*H. Wool & Sons, Inc.*).[43] The defendants claimed no knowledge

---

[37] *Id.* at 1009.

[38] *Id.*

[39] 335 U.S. 345 (1948).

[40] *Golden Grain Macaroni Co. v. United States*, 209 F.2d 166 (9th Cir. 1953).

[41] *Id.* note 34 at 168.

[42] *Id.*, 168.

[43] 215 F. 2d 95 (2d Cir. 1954).

that the butter was underweight but the court held the testimony, which showed that Mr. Wool was the "dominating factor" in the company and "intimately concerned in its affairs" (*H. Wool & Sons, Inc.*).[44] The lack of knowledge was immaterial. Mr. Wool was a corporate officer in control of operations when the violation occurred. In *United States v. Diamond State Poultry Co.* two corporate officers were convicted for selling decomposed and diseased chickens (*Diamond State Poultry Co.*).[45] The officers sought refuge in the fact that they did not participate in the violative acts, but the court cited *Dotterweich* to hold:

> Under the Food, Drug, And Cosmetic Act, proof of personal participation of an individual defendant is not required to establish guilt if the individual is the responsible person for the operation of the business out of which the violation grows. (*Diamond State Poultry Co.*)[46]

It was the defendant's job description, not the actual participation that mattered. Finally in *United States v. Wiesenfeld Warehouse Co.*, the Supreme Court found a corporation criminally liable with no evidence of criminal intent (*Wiesenfeld Warehouse Co.*).[47] The Supreme Court stated food and drug regulation "dispenses with the conventional requirement for criminal conduct—awareness of come wrongdoing" (*Wiesenfeld Warehouse Co.*).[48] Even in the criminal context the Act would be broadly construed.

As can be seen, *Dotterweich* imposed a draconian doctrine whose only real limit was the discretion of the FDA or USDA to seek criminal prosecution. It would not be until 1974 that a court attempted to impose any limitation on the reach of the doctrine.

7.3.2 Park Doctrine

The Park Doctrine began as an FDA inspection in late 1971. The FDA inspected the Baltimore warehouse of Acme Markets, Inc. whose president was John R. Park. It made a second follow-up inspection in early 1972. On both occasions the FDA inspectors found evidence of rodent infestation of food, which violated Section 301(k) of the Act. The FDA sought criminal penalties against both Acme Markets and Mr. Park. In 1973 the FDA and DOJ filed a five count complaint charging both the corporate entity and its president. Acme Markets plead guilty and Mr. Park pled innocent. He was found guilty on all five counts following a jury trial and fined US$ 250. Park appealed the Fourth Circuit (*Park 1974*).[49]

The Fourth Circuit trimmed the Dotterweich Doctrine down. It reasoned that the doctrine could not base criminal conviction "solely upon a showing that the defendant, Park, was the President of the offending corporation" (*Park 1974*).[50] The court chided the Government for conflating the element of "awareness of wrongdoing" and the element of "wrongful action" (*Park 1974*).[51] The court held *Dotterweich* and eliminated the "awareness of wrongdoing" element, but kept intact the element of "wrongful action" (*Park 1974*).[52] In an accompanying footnote the court defined wrongful action as, "acts of the accused which cause the adulteration of such food" (*Park 1974*).[53] The lack of awareness of wrongdoing prong remained, but the court required some "wrongful action." The court remanded the case back to the trial court with the following instruction:

> Upon a subsequent trial the jury should be instructed that a finding of guilt must be predicated upon some wrongful action by Park. That action may be gross negligence and inattention in discharging his corporate duties and obligations or any of a host of other acts of commission or omission which would 'cause' the contamination of the food. (*Park 1974*).[54]

---

[44] *Id.* at 99.

[45] 125 F. Supp. 617 (D. Delaware 1954).

[46] *Id.* at 620.

[47] 376 U.S. 86 (1964).

[48] *Id.* at 91.

[49] *United States v. Park*, 499 F. 2d 839 (4th Cir. 1974).

[50] *Id.* at 841.

[51] *Id.*

[52] *Id.*

[53] *Id. at* fn. 4.

[54] *Id. at* 842.

The court not only required the instruction asking the jury to find a wrongful action, but that the defendant bore responsibility for the action. The court stated:

> As a general proposition, some act of commission or omission is an essential element of every crime. For an accused individual to be convicted it must be proved that he was in some way personally responsible for the act constituting the crime. The Supreme Court recognized this in Dotterweich… The criminal acts which were the subject of Acme's conviction cannot be charged to Park without any proof that he participated directly or constructively therein or that the acts were done for some further criminal conspiracy in which he took part…It is the defendant's relation to the criminal acts, not merely his relation to the corporation, which the jury must consider; 21 U.S.C. 331 is concerned with criminal conduct and not proprietary relationships… (*Park 1974*).[55]

The Park Doctrine articulated by the Fourth Circuit provides a framework similar to what plaintiffs face in bringing foodborne illness cases. It places the Government on the same level as private plaintiffs. Unfortunately the Fourth Circuit decision would not become law.

On appeal to the Supreme Court the Fourth Circuit was reversed. The Dotterweich Doctrine, if renamed and reaffirmed, remains a good law. Its use has become rather rare despite the low threshold it imposes on the FDA. Still there are modern examples of its use and the new ways in which it is being applied.

United States v. Park, **499 F. 2d 839 (4th Cir. 1974)**
BOREMAN, Senior Circuit Judge:
   John R. Park, President of Acme Markets, Inc. (hereafter Acme), was tried and convicted by a jury of violating 21 U.S.C. 331(k)—causing the adulteration of food which had traveled in interstate commerce and which was held for sale. The evidence is undisputed that an inspector of the Food and Drug Administration (F.D.A.) made an

inspection of Acme's Baltimore warehouse in November and December of 1971 and again in March of 1972. On both of these occasions the inspector found evidence of rodent infestation of food stored in the warehouse. As a result of these inspections an informal hearing was held in June 1972 at the F.D.A.'s Baltimore office. Although Park was not present, he was invited to attend the hearing and was represented by Robert W. McCahan, Baltimore Divisional Vice President of Acme.

   In March of 1973, a five-count information was filed charging Acme and Park with the offenses cited above; four counts stemmed from the 1971 inspection and the fifth count from the 1972 inspection. Prior to trial Acme pleaded guilty to all counts. Park was tried on the theory that he 'was a corporate officer who, under law, bore a relationship to the receipt and storage of food which would subject him to criminal liability under *United States v. Dotterweich*, 320 U.S. 277 (64 S.Ct. 134, 88 L.Ed. 48) (1943).' The jury found Park guilty on all counts and he was fined a total of $ 250.

   Park appeals his conviction alleging (1) the court erred in its instructions to the jury and (2) prejudicial evidence of warnings of alleged prior violations of the Act was improperly admitted.

   The court charged the jury that the sole question was 'whether the Defendant held a position of authority and responsibility in the business of Acme Markets'; that Park could be found guilty 'even if he did not consciously do wrong' nd even though he had not 'personally participated in the situation' if it were proved beyond a reasonable doubt that Park 'had a responsible relation to the situation.' We conclude that this charge does not correctly state the law of the case, that the conviction of Park on all counts must be reversed and a new trial awarded.

---

[55] *Id.* t841.

The Government asserts that this case is controlled by the decision of the Supreme Court in *United States v. Dotterweich*, 320 U.S. 277, 64 S.Ct. 134, 88 L.Ed. 48 (1943), and contends that Dotterweich specifically held that the Federal Food, Drug, and Cosmetic Act (Act), 21 U.S.C. 301–392, 'dispenses with the conventional requirement for criminal conduct—awareness of some wrongdoing.' 320 U.S. at 281, 64 S.Ct. at 136. From this the Government argues that the conviction may be predicated solely upon a showing that the defendant, Park, was the President of the offending corporation. The error here is that the Government has confused the element of 'awareness of wrongdoing' with the element of 'wrongful action'; Dotterweich dispenses with the need to prove the first of those elements but not the second.

As a general proposition, some act of commission or omission is an essential element of every crime. For an accused individual to be convicted it must be proved that he was in some way personally responsible for the act constituting the crime. The Supreme Court recognized this in Dotterweich: 'The offense is committed... by all who do have such a responsible share in the furtherance of the transaction which the statute outlaws.....' 320 U.S. at 284. The criminal acts which were the subject of Acme's conviction cannot be charged to Park without proof that he participated directly or constructively therein or that the acts were done to further some criminal conspiracy in which he took part. To use the language of Dotterweich, 'under 301 (21 U.S.C. 331) a corporation may commit an offense and all persons who aid and abet its commission are equally guilty.' 320 U.S. 284, 64 S.Ct. at 138. It is the defendant's relation to the criminal acts, not merely his relation to the corporation, which the jury must consider; 21 U.S.C. 331 is concerned with criminal conduct and not proprietary relationships.

In sum, the court told the jury that Park would be guilty if it were shown that he 'had a position of authority and responsibility in the situation out of which these charges arose.' This instruction, taken in combination with the other parts of the charge related above, might well have left the jury with the erroneous impression that Park could be found guilty in the absence of 'wrongful action' on his part.

Upon a subsequent trial the jury should be instructed that a finding of guilt must be predicated upon some wrongful action by Park. That action may be gross negligence and inattention in discharging his corporate duties and obligations or any of a host of other acts of commission or omission which would 'cause'6 the contamination of the food.7 'Whether an accused shares responsibility in the business process resulting in unlawful distribution depends on the evidence produced at the trial and its submission—assuming the evidence warrants it—to the jury under appropriate guidance.' Dotterweich, supra, 320 U.S. at 284, 64 S.Ct. at 138.

It is argued by the prosecution that the requirement of such proof will make enforcement more difficult. Nevertheless, the requirements of due process are intended to favor fairness and justice over ease of enforcement. We perceive nothing harsh about requiring proof of personal wrongdoing before sanctioning the imposition of criminal penalties.

We find another ground for reversal of Park's conviction. He contends that the admission in evidence of a warning by F.D.A. as to conditions alleged to have existed in 1970 in Acme's Philadelphia warehouse was prejudicial error requiring reversal. There was no evidence of prosecution of either Acme or Park following

F.D.A.'s warning. In the recent case of *United States v. Woods*, 484 F.2d 127 (4 Cir. 1973), this court examined in detail the admissibility of evidence relating to prior alleged offenses not the subject of conviction and we will not attempt to elaborate here.

The Woods majority adopted a modern and liberal approach concerning the admissibility of such evidence. This court refused to decide the issue by 'pigeonholing' the evidence under one of a number of recognized exceptions to the general rule that such evidence is inadmissible. Opting for a more flexible balancing test the court cited McCormick on Evidence 190, p. 453 (Cleary Ed.1972).

'The problem is not merely one of pigeonholing, but one of balancing, on the one side, the actual need for the other crimes evidence in the light of the issues and the other evidence available to the prosecution, the convincingness of the evidence that the other crimes were committed and that the accused was the actor, and the strength or weakness of the other crimes evidence in supporting the issue, and on the other, the degree to which the jury will probably be roused by the evidence to overmastering hostility.'

In determining the admissibility of 'prior crimes' evidence the Woods majority balanced the relevancy, persuasiveness and need for such evidence against the prejudice resulting to the defendant because of its admission. In a dissenting opinion in Woods Judge Widener favored the application of a traditional and less liberal balancing test.9 Regardless of the balancing test applied, we are convinced that the evidence concerning the Philadelphia incident was inadmissible under the theory on which the case was tried.

Initially we conclude that there was no actual need for the Philadelphia evidence.

In his jury charge the district judge stated that:

'You need not concern yourselves with the first two elements of the case. The main issue for your determination is only with the third element, whether the Defendant held a position of authority and responsibility in the business of Acme Markets.'

Thus, as this case was submitted to the jury and in light of the sole issue presented, need for the Philadelphia evidence is not apparent. Absent a showing of such need we are of the opinion that, even under the liberal balancing test of Woods, the prejudicial effect of this evidence outweighed any possible relevancy or persuasiveness it might have had.10

We note in passing, without deciding, that in light of our comments above evidence of 'prior crimes' might become sufficiently necessary and relevant to warrant its admission on retrial. Our conclusion that the prosecution must show some wrongful act by the accused may affect the result of the balancing test as prescribed in Woods by increasing the need for 'prior crimes' evidence. Whether the need for and the persuasiveness and relevance of such evidence may outweigh its prejudicial effect will, in large part, depend on the prosecution's new approach to the presentation of its case. Thus, on retrial, it will be incumbent upon the district court to determine the admissibility of this 'prior crime' evidence in light of developments.

Reversed and remanded for a new trial.

7.3.3   Modern Applications of the Park Doctrine

Odwalla, Sara Lee, and Jensen

The FDA has become reluctant to use criminal prosecutions. Following the hay-day of the 1940s through the early 1970s criminal prosecutions cooled a bit. In particular in the area of food and dietary supplement regulation the FDA rarely uses its broad authority under *Park*. More common are criminal cases in the drug area of regulation. Some commentators point to a lack of resources, for which a criminal investigation and prosecution require a great deal to adequately complete. Others suggest the change is attributable to a change in attitude about when a food violation merits criminal prosecution. In either case it has led to criticism when the publicly available facts scream criminal sanctions. The *Salmonella* outbreak tied to Peanut Corporation of America in 2008 and 2009, for example, did not see criminal charges filed until 2013. This provides some sense of how resource intensive and slow the cases can be.

Other recent examples involve Odwalla, Sara Lee, and Jensen Farms. Odwalla plead guilty to 16 misdemeanor charges following an outbreak of E.coli in 1996 (Chicago Tribune 1998).[56] The outbreak resulted in the death of one child. Sara Lee pled guilty to one misdemeanor charge in 2001 for an outbreak linked to contaminated hotdogs and deli meats (Chicago Tribune 2001).[57] The outbreak resulted in 15 deaths. In 2013 Jensen Farms owners Eric and Ryan Jensen plead guilty to misdemeanor charges following an outbreak linked to cantaloupes. (Harvest Public Media).[58] The outbreak was described as the deadliest in the past 20 years with 33 people killed.

Warning Letters and the Risk of Contract Manufacturing

Beginning in 2013, the FDA began issuing warning letters citing to *Dotterweich* and Park not for holdings on criminal liability, but to emphasize vicarious liability. The letters, to dietary supplement companies relying on contract manufacturers or labelers, largely point out the facility's attention to its potential liability for GMP violations made by its contracted party. Recall that this was the case for Parfait Powder Puff Co. in

1947. The GMP requirement for dietary supplements is new with regulations only fully in effect since 2009. The warning letters appear to foreshadow prosecution as a last resort to add teeth to the GMP regulations and guidance. It could also be the FDA's attempt to use a low resource option for maximum deterrent effect. The doctrine certainly provides the FDA the pathway to prosecution, but a great deal in American society has changed since the Parfait Powder Puff conviction nearly 70 years ago. It is open to a debate whether a jury today would find a company vicarious liable under *Dotterweich* and *Park* for the violated acts of its contract manufacturer or labeler.

Excerpt from Bhelliom Enterprises Corp Warning Letter (5/29/14)

As a distributor that contracts with other manufacturers to manufacture, package, or label dietary supplements that your firm releases for distribution under your firm's name, your firm has an obligation to know what and how these activities are performed so that you can make decisions related to whether your dietary supplement products conform to established specifications and whether to approve and release the products for distribution [72 Fed. Reg. 34752, 34790 (Jun. 25, 2007)]. Your firm introduces or delivers, or causes the introduction or delivery, of dietary supplements into interstate commerce in their final form for distribution to consumers. As such, your firm has an overarching and ultimate responsibility to ensure that all phases of the production of those products are in compliance with dietary supplement CGMP requirements.

Although your firm may contract out certain dietary supplement manufacturing operations, it cannot, by the same token, contract out its ultimate responsibility to ensure that the dietary supplement it places into commerce (or causes to be placed into commerce) is not adulterated for failure to

---

[56]   Chicago (1998).

[57]   Chicago (2001).

[58]   Harvest Public Media (2013).

comply with dietary supplement CGMP requirements (see *United States v. Dotterweich*, 320 U.S. 277, 284 (1943) (explaining that an offense can be committed under the Act by anyone who has "a responsible share in the furtherance of the transaction which the statute outlaws"); *United States v. Park*, 421 U.S. 658, 672 (1975) (holding that criminal liability under the FD&C Act does not turn on awareness of wrongdoing, and that "agents vested with the responsibility, and power commensurate with that responsibility, to devise whatever measures are necessary to ensure compliance with the Act" can be held accountable for violations of the FD&C Act)). In particular, the Act prohibits a person from introducing or delivering for introduction, or causing the delivery or introduction, into interstate commerce a dietary supplement that is adulterated under section 402(g) for failure to comply with dietary supplement CGMP requirements (see 21 U.S.C. §§ 342(g) and 331(a)). Thus, a firm that enters into a contract with other firms to conduct certain dietary supplement manufacturing, packaging, and labeling operations for it is responsible for ensuring that the product is not adulterated for failure to comply with dietary supplement CGMP requirements, regardless of who actually performs the dietary supplement CGMP operations.

7.4   Comparative Law

By this point in the text it has become clear that different cultures produce different laws. The same certainly holds true in the context of litigation and criminal prosecutions. Every culture approaches both topics in ways utterly unique to their history, culture, and system of government. The USA is perhaps known as the most litigious, but even then its citizens are not alone in the pursuit of accountability through private law suits. The more permissive a regulatory system is in

allowing claims or ingredients, the greater the risk of litigation. The same can be said about the degree of control over inspections and sampling.

Canada and Germany offer comparative examples of labeling litigation. In Germany a group of citizens called the German Federation of Consumer Organisations (Verbraucherzentrale Bundesverband) claimed that Nutella's labeling was misleading (Carreño 2012).[59] The Frankfurt Court of Appeals agreed and entered a judgment on October 2011 requiring a revision to the Nutella brand's nutrition labeling (Carreño 2012).[60] The Frankfurt Court reasoned that the label was misleading despite complying with the German nutrition labeling regulation, because it led the consumer to believe that it was low in fat and sugar and high in vitamins and minerals (Carreño 2012).[61] A different Western legal system, but still the same accusations emerge about how the label was misleading. Canada offers another example of labeling litigation. A Montreal woman sued Danone Inc. over its Activia and DanActive brands claiming that the probiotic yogurt was not effective (CBC News).[62] The suit settled in 2012 for US$ 1.7 million with Danone agreeing to revise its labeling (CBC News). The Canadian and German examples demonstrate instances where foreign litigation parallels the US lawsuits. As brands launch global products with similar claims the risk of litigation rises dramatically.

Criminal law is highly variable depending both on the legal system and the primary food law. Many nations do empower regulators to pursue both civil and criminal penalties for violations. Japan for example allows its regulators to pursue criminal investigations and prosecutions. This was the case in 2000 following an outbreak linked to dairy products suspected in 14,500 people becoming ill (The Japan Times).[63] Police pursued an professional investigation; this was despite the clear link to the Food Sanitation Law

---

[59] Carreño (2012).

[60] *Id.*

[61] *Id.*

[62] CBC News (2012).

[63] The Japan Times (2000).

(The Japan Times).[64] In the wake of the horse meat scandal in Europe, the UK considered a "food crime unit" as part of the Food Standards Agency (BBC News).[65] Despite similarities to the OCI in the US, the report recommending the crime unit suggested Denmark, Holland, and Northern Ireland as models for the police force (BBC News).[66] Criminal law may vary in its approach, but where a robust food regulatory system exists criminal penalties are rarely omitted.

7.5 Chapter Summary

This chapter expounded on the ideas of corporate liability. It began by looking at liability under State law for foodborne illnesses using negligence claims then turned to the tsunami of labeling litigation. The chapter explored the full landscape of State law causes of action, including the limitation of State law under the concept of preemption. The topic then shifted away from civil tort liability to criminal liability under the FD&C, FMI, and PPI. Criminal liability exposes individual corporate officers to personal liability. The two doctrines at the core of this concept ensnare nearly every violation in the name of absolute and vicarious liability. This chapter concludes the full arc that began in Chapter 1 introducing the basic framework of food law by ending with the ultimate consequences for violating the terms of that regulatory system.

Overview of Key Points:
- The concept of State tort and personal liability law
- Elements of a negligence cause of action;
- Challenges to bringing food cases involving foodborne illnesses
- Potential changes to outbreak litigation under FSMA
- The background and spark for labeling litigation

- Types of labeling lawsuits and examples of each type
- Types of preemption and application to labeling litigation
- Introduction to the Dotterweich and Park Doctrine
- Examination of an alternative doctrine developed by the Fourth Circuit
- Historical and modern examples of the Doctrine to demonstrate its ubiquitous application
- The use of the Doctrine in warning letters for the principal of vicarious liability

7.6 Discussion Question

1. Do you agree nutrition fact panels and ingredient listings cannot cure a misleading claim? Explain your answer.
2. Has the FDA abandoned its role as a regulator of food law claims to private plaintiffs? If so what can be done to remedy the problem? If not, what can be done to protect companies from private litigation?
3. The Supreme Court decision in *Park* was split 5-4, if you were a Justice on the Court with the deciding vote, would you agree with the majority or with the minority (who agreed with the Fourth Circuit's opinion)? Explain your answer.

References

ABC News (26 February 2010) Dannon to pay $ 45 M to settle Yogurt Lawsuit. http://abcnews.go.com/Business/dannon-settles-lawsuit/story?id=9950269. Accessed 29 Aug 2014

American Bar Association (ABA) (30 April 2012) Section of litigation: class actions and derivative lawsuits, Dawn Goulet, Confusion in Court over "All Natural" claims. http://apps.americanbar.org/litigation/committees/classactions/articles/spring2012-0412-all-natural-labels-mean-marketing.html. Accessed 29 Aug 2014

BBC News (Dec 2013) Horsemeat scandal: review urges UK Food Crime Unit. http://www.bbc.com/news/business-25347633. Accessed 29 Aug 2014

---

[64] *Id.*

[65] BBC News (2013.

[66] *Id.*

Brookings Governance Studies Negowetti Nicole E (June 2014) Food Labeling Litigation: Exposing Gaps in the FDA's Resources and Regulatory Authority. http://www.brookings.edu/~/media/research/files/papers/2014/06/26%20food%20labeling%20litigation/negowetti_food%20labeling%20litigation.pdf. Accessed 12 Aug 2014

Carreño I (2012) German court orders change to nutrition labeling on nutella due to its misleading nature. Eur J Risk Regul 3:91

CBC News (September 2012) Danone agrees to pay $ 1.7 M in Yougurt health claims case. http://www.cbc.ca/news/canada/montreal/danone-agrees-to-pay-1-7m-in-yogurt-health-claims-case-1.1206623. Accessed 29 Aug 2014

Chicago Tribune (24 July 1998) Odwalla pleads guilty In fatal '96 juice contamination $ 1.5 Million criminal fine is largest in food injury case, U.S, officials say. http://articles.baltimoresun.com/1998-07-24/news/1998205017_1_fresh-apple-juice-odwalla-juice-food. Accessed 29 Aug 2014

Chicago Tribune, Young Alison (22 June 2001) Sara Lee to plead to criminal charge in meat recall. http://articles.chicagotribune.com/2001-06-22/business/0106220187_1_bil-mar-plant-sara-lee-hot-dogs-and-deli. Accessed 29 Aug 2014

CSPI (1 February 2007) Watchdog group sues coke, Nestlé for bogus "Enviga" claims. http://www.cspinet.org/new/200702011.html. Accessed 29 Aug 2014

CSPI (8 January 2014) Lawsuits loom for three major food companies. http://www.cspinet.org/new/201401081.html. Accessed 29 Aug 2014

Food Safety News (24 August 2011) Michele Simon ConAgra sued over GMO '100% Natural' cooking oils. http://www.foodsafetynews.com/2011/08/conagra-sued-over-gmo-100-natural-cooking-oils/#.U_XZZfldV8E. Accessed 29 Aug 2014.

Harvest Public Media Luke Runyon (22 October 2013) Cantaloupe farmers plead guilty to criminal charges. http://harvestpublicmedia.org/content/cantaloupe-farmers-plead-guilty-criminal-charges#.U_XWsv-ldV8E. Accessed 29 Aug 2014

NPR (26 April 2012) The salt, Ted Burnham, Nutella maker may settle deceptive ad lawsuit for $ 3 million. http://www.npr.org/blogs/thesalt/2012/04/26/151454929/nutella-maker-may-settle-deceptive-ad-lawsuit-for-3-million. Accessed 29 Aug 2014

Polin D (1998) Proof of liability for food poisoning, 47 Am.Jur. Proof of Facts 3d Sec. 18

The Japan Times (August 2000) Negligence suspected: snow's offices searched in criminal liability probe. http://www.japantimes.co.jp/news/2000/08/31/national/snows-offices-searched-in-criminal-liability-probe/#.VAB79_ldV8E. Accessed 29 Aug 2014

Tierney J (8 December 2008) Health halo can hide the calories. N.Y. Times

U.S. Chamber Institute for Legal Reform (October 2013) The New Lawsuit Ecosystem: Trends, Targets and Players. http://www.instituteforlegalreform.com/uploads/sites/1/The_New_Lawsuit_Ecosystem_pages_web.pdf. Accessed 10 Aug 2014

Index

CPSIA information can be obtained at www.ICGtesting.com
Printed in the USA
LVOW01*1439030515

437052LV00002B/117/P

9 783319 124711